Industrial Crops
Bioresources to Biotechnology

Aliyu Ahmad Warra

Centre for Entrepreneurial Development
Federal University Gusau, Nigeria

Majeti Narasimha Vara Prasad

Professor (Retd), Department of Plant Sciences
University of Hyderabad, Hyderabad, India

CRC Press
Taylor & Francis Group
Boca Raton London New York

CRC Press is an imprint of the
Taylor & Francis Group, an **informa** business

First edition published 2024
by CRC Press
2385 NW Executive Center Drive, Suite 320, Boca Raton FL 33431

and by CRC Press
4 Park Square, Milton Park, Abingdon, Oxon, OX14 4RN

© 2024 Taylor & Francis Group, LLC

CRC Press is an imprint of Taylor & Francis Group, LLC

Library of Congress Cataloging-in-Publication Data (applied for)

ISBN: 978-0-367-36046-7 (hbk)
ISBN: 978-1-032-72969-5 (pbk)
ISBN: 978-0-429-34627-9 (ebk)

DOI: 10.1201/9780429346279

Typeset in Times New Roman
by Innovative Processors

Preface

Many crop plant raw materials and bio-products are resourceful for various sectors of biotechnology and industry and provide diversified income for entrepreneurs and enterprises worldwide. Plentiful crop plant resources relative to energy demand especially bioenergy crops are nowadays exploited. Bioenergy crops provide carbon neutrality and an alternative to fossil fuels. The crops have been recognized since antiquity in many parts of the world for their valuable resource with multiple applications ranging from industry to environment. This has persuaded the zeal to explore the prospects of emerging bioenergy technologies for sustainable development. Due to the organic richness of most of these industrial crops research has been in increased. An attempt was made to have insight into how products of value addition can be derived from the development of the cultivation of the crops, genetic improvement, and better processing of their products that could substantially be useful for industrial applications: agricultural, pharmaceuticals, biofuels, etc. Plant Genetic Resources (PGRs) have demonstrated a significant function in areas of agriculture, environmental restoration, and protective measures. Recombinant DNA Technology has made it possible to modify plants at genomic and molecular considerations so as to make improvements in plant traits which before was through conventional breeding methods. Genetic transformation techniques are adopted to achieve the desired results. Transgenic industrial crops are very useful in the production of products such as biofuel which are included biodiesel, bioethanol, bio-engine oil for jets, bioplastics, fibres, starches, chemicals agricultural feedstock, etc., through bioprocessing which contributes to innovation, research and development and productivity gains. Crops generally can be utilized for biomass combustion or gasification, for heat, electricity, or in the case of biomass for fermentation processes with combined heat and power. Some literature focuses on plant species with desired traits, their propagation, and breeding to obtain the needed feedstock for utilization in producing different finished products mainly bioenergy and bioproducts. The greater amount of energy feedstock is obtainable through small- and large-scale industries. However, considerations of checks and balances are to put across the crisis of using some industrial crops, most importantly valued food crops for energy purposes due to cultural reasons and avoidance of competition with the food chain. Plant Tissue culture has pushed biotechnology knowledge forward, especially in the areas of farming, cultivation, plant rearing, ranger service, physical cell hybridization, phytopathology, and creation of plant metabolites by mechanical means, etc. Recently, in the middle of two decades plant cell, tissue, and organ culture quickly expanded and turned into a successful biotechnological tool in agribusiness, cultivation, ranger service, and industry. In this maiden edition, various uses of crop plant products in energy production, food, pharma, industry, and environment were explored. Emphasis was on the provision of an overview of the potential of plant-derived products in greener biotechnologies and industries.

Aliyu Ahmad Warra
Majeti Narasimha Vara Prasad

Introduction

Industrial crops are so-called, primarily because of their industrial uses and they require industrial processing before they can be utilized. However, they may also have secondary domestic applications such as in gardening, horticulture and environmental protection. Though, non-food uses are commonly descriptive of industrial crops. Industrial crops wealth ranging from fruits, nuts, seeds, tubers, fibres, plant oils, de-oiled cakes, etc. have contributed significantly in raising the standard of human life for many years and have maintained their valuable composition in foods, pharmaceuticals and personal care products, textile and dyes, polymers and composites and by large biotechnological applications (Warra 2012, Warra 2014, Warra 2017). Recombinant DNA Technology is opening a window for man prospects for the improvement of products from the agricultural sector environmental protection via the reduction of agro-chemicals usage like fertilizers, pesticides, etc. In the attainment of environmental sustainability by using environment-friendly crops such as insect-resistant, herbicide-tolerant species and crops that can fix nitrogen leading to the purification of the environment biotechnology has played an important role (Soetan 2011). With the recent introduction of molecular markers, it is now feasible to create straight inferences about genetic divergence and inter-relationships amongst organisms at the DNA level, devoid of misleading environmental influences or imperfect pedigree accounts (Gantait et al. 2014). Many crops are used to create non-food products: oils, resins, fibres, clothing, energy, cosmetics and plastics to list but a few. Even though attempts to diversify the energy portfolios of developed countries with green technologies have brought competition between food and fuel for crop production resources to the forefront of public policy debates biofuel is still one of the global alternative fuels (Williams et al. 2013). Among the alternative energy sources, bioenergy is emerging as the most promising as compared to atomic, solar, and wind. In as much as required strategies and technologies are formulated and validated for commercialization in cost-effective ways bioenergy including bioethanol and biodiesel can be produced from cellular biomass, starch, sugar, and oil derived from several plants and plant products in huge amounts (Goldman and Cole 2014). Various areas of industrial applications of Genetic manipulation were reported in the literature (Devi et al. 2019).

Contents

Energy and Multipurpose Crops

Bioenergy crops are sources of some varieties of fiber, biofuel, biopharmaceuticals, etc. (Fig. 1.1). There has been a rapid transformation in the utilization of transgenic energy crops for the development of therapeutic compounds from being laboratory and field experiments to a potential and enterprise-viable process prepared for the delivery of products vital for animal and human therapies—areas such as novel bioactive peptides preparations therapy, and edible oral vaccines are being explored (Joshi et al. 2014). These energy feedstock crops require many genetic engineering protocols for improvement. Improving biomass yield and quality by genetic breeding methods adds to their economic value.

1.1 Castor crop (*Ricinus communis* L.)

The castor (*Ricinus communis* L.) crop belonging to the family Euphorbiaceae is now globally regarded as a remedy for agricultural problems in tropical and subtropical regions, providing solutions and viable returns from commercial crops with low input costs. Castor is a crop that easily survives in varied climatic conditions because planting it is a simple process, it is drought resistant, can tolerate different soil, and has a high yield. It is a multipurpose crop with high prospects in the pharmaceutical and agricultural industries (Salihu et al. 2014). The *Ricinus communis* L. plant has great potential for phytoremediation, reclamation of lands contaminated with mine wastes, and as a resource for remediation of constrained lands and production value addition products (Warra and Prasad 2018, Prasad and Kiran 2017, Warra and Prasad 2016). Castor is a perennial crop, typically less than 2 meters in height. It possesses large leaves which are glossy and often red or bronze. Flowers appear in clusters of white and yellow colors. The fruit contains seeds in a spiny pod that is oval and light brown (Fig. 1.2). There are two varieties of seeds for the castor bean and one variety for the wild castor (Bolaji et al. 2014).

The castor plant is fast growing. However, the seeds require a long frost-free season to mature. Autumnal sowings are likely favorable for castor grown as a semi-perennial crop to allow saving irrigation water (Anastasia et al. 2015). Castor oil is obtained from the seed by means of extraction and has numerous uses. The oil content depends on the genotype, environmental conditions, indigenous practices, and harvesting time (Omohu and Omale 2017). The oil is usually more than 80% ricinoleic acid, making it a resourceful feedstock for various industrial applications ranging from soaps, lubricants, coatings, synthesis of biodegradable polymers, cosmetics, biofuels, pharmaceuticals, and biorefineries, etc. (Warra et al. 2020, Mensah et al. 2018, Patel et al. 2016, Kanti et al. 2015). About 900 million pounds of Castor oil and its derivatives are used yearly in producing multipurpose soaps, lubricants for engines, fluids for brakes, dyes, perfumes, waxes, paints, polishes, nylon, etc. It serves as a bioenergy resource in the reduction of greenhouse gas emissions. Biotechnological development has made it possible to create insect and disease-resistant varieties by inserting genes originating outside the organism of concern or stably expressed genes into the castor genome to express exogenous genes directly or regulate the endogenous gene expression (Jiannong et al. 2018). Castor, despite not having subspecies has important agronomic and morphological traits, and appreciable variability has been explored also for growth and development. Traditional breeding techniques like induced mutations and wide crossing are some possibilities for the genetic improvement of castor. However, there are problems associated with the successful breeding of castors such as limited information about the genetics of yields that are smaller and unpredictable as well as the susceptibility to diseases and insects. Recombinant DNA Technology protocols provide an alternative solution for an improved transgenic castor. A review of recent expansion in castor genetic improvement involves the utilization of genetic markers and the further theoretical report stated that later genetic development efforts in castor can be achieved through genetic transformation. The focus was on the development of reproducible in vitro propagation procedures, efficient enough to be applied to an extensive range of genotypes and genetic transformation (Singh et al. 2015).

1.1.1   Biotechnologies and genetic transformations

Evaluation of the genetic variation of a crop type is necessary for its development hence it is important to identify the genetic makeup characteristics of castor genetic resources for creating superior cultivars. Future castor breeding efforts will depend on understanding the genetic variety of the crop to enhance the output of oil and satisfy the constantly growing demand for industrial uses of castor oil, such as biofuel. Programs for crop advancement also essentially require the genetic makeup of germplasm. According to research on Tunisian castor, PCR amplification of DNA using specific primers produced numerous DNA fragments that could be scored in all genotypes. RAPD protocol could be employed to provide vital knowledge on present-day breeding strategies (Vivodík et al. 2019). The application of commercial heterosis in castor was successful in India after the generation of stable pistillate lines from a dominant and epistatic "S" type pistillate source. Due to the diversification of pistillate sources using NES and other new sources, it is important to identify the diverse male combiners within the current gene pool. In a study using 60 breeding lines/ genotypes, genetic diversity, and population structure were characterized using EST-SSR primers. The molecular data revealed a weak or low population form and low SSR allelic diversity. Hence, it became necessary to diversify breeding lines through mutations, as well as to cross various lines from various isolated regions to harness heterosis and create better hybrids. Knowledge of genetic diversity generated to exploit heterosis for the development of hybrids is helpful in the selection of diverse genotypes for breeding (Betha and Lavany 2019). For the identification, use, and preservation of the genetic variability of *Ricinus communis* L. species, information on the genetic makeup characteristics of castor may be useful for understanding the genetic preservation of species, germplasm management, controlling programs, and preservation strategies (Bajay et al. 2009).

The expanding importance of genetic engineering tools for crop improvement for yield and product development has been well acknowledged in scientific literature. Breeding programmes in castor have depended mostly on the primary gene pool variability available as it is monotypic. There are problems associated with the profitable cultivation of castor which includes exposure to diseases, pests, and insect attack toxicity of the press cake, restricting its use as cattle feed. The improved limitation in resistance to biotic stresses is controlled by conventional breeding techniques and in upgrading quality leading to low trait genetic variability. Protein-based markers were employed in assessing genetic diversity, while the use of molecular markers is at the primary stage of development. Fortunately, studies on castor in vitro in shoot proliferation from meristematic explants were successful, but not callus-mediated regeneration. Insect-resistant and ricin-free transgenics can be produced using genetic transformation protocols with a very low transformation frequency (Sujatha et al. 2008). It was considered necessary for elite genes from wild materials for castor breeding to be mined because of low productivity and the susceptibility to pests related diseases by castor bean varieties. The relevance of these materials for the conservation and breeding application was provided by the report of SRAP investigation of the genetic make-up characteristics of wild castor reported in South China (Agyenim-Boateng et al. 2019). Molecular diversity studies were carried out for the identification of varieties and phylogenetics among many castor genotypes and the identification of those with distinguishable profiling of DNA (Lakhani et al. 2015). According to Vivodk et al. (2015), random amplified polymorphic DNA (RAPD) markers were used to detect variation among the set of castor genotypes. A multi-omics approach gave comprehensive knowledge of the plasticity of *R. communis* in response to temperature. For maintaining the quality of castor seed, physiological, biochemical, and molecular responses of its seeds and seedlings to divergence temperatures using castor bean cultivation can also be made safer for farmers and industrial workers by the gene silencing technology (Ribeiro de Jesus 2015). Altering the molecular composition of fatty acids to create entirely new substitutes for traditional oils and fats has been a great opportunity rendered by biotechnology. Castor oil is just one target of current biotechnology research (Rural Advancement Fund International 1990). This is a great bio-resource for biotechnology. Even as a result of the unavailability of a high-density genetic map, restrictions were made to the conventional genetic and advancement in the breeding of castor beans. However, a reported result based on a recombinant inbred line (RIL) population and generation of high-quality genomic single nucleotide polymorphism (SNP) markers and the construction of a high-resolution genetic map provided novel proof that the phylogenetics of castor bean is virtually singular from other species or class in the Euphorbiaceae family (SNP), which the recent genome technology development described as the commonest manifestation of DNA sequence variation between alleles, in several species and its discovery and application brought about more knowledge of genetic makeup characteristics, and the knowledge of crop improvement. The results would be helpful for ongoing research into potential novel methods of changing the weight and size of castor bean seeds. The genetic map and state-of-the-art pseudo-chromosome scale genome form an essential basis for marker-assisted selection (MAS) of Qualitative Trait Locus (QTL) influencing significant agronomic traits, as well as the accelerated molecular breeding of castor bean in a financially viable manner (Morgil et al. 2020, Yu et al. 2019).

1.1.2   Castor genome resources

Limited work exists on the biosynthesis of ricinoleic acid at the transcriptional level, and the reference genomes are incomplete. However, the report showed significant improvement in the existing gene models of *R. communis* (Wang et al. 2019). Coverage of 4.6-fold sequencing of the castor genome draft for a member of the Euphorbiaceae was successfully reported for the first time. Furthermore, as previously believed, there are more members of the ricin gene family, in contrast to many of the major genes that play a single copy role in the turnover and synthesis of oil. Complete genome sequences of varying species suggested the possibility of an early hexaploidization occurrence that is preserved throughout the dicotyledonous lineage. The genome sequence of castor was considered one of the vital components to exploring changes in the structure of the genome, not restricted to the Euphorbiaceae family only, but rather generally in plants (Chan et al. 2010).

1.1.3   Bio-detoxification

The most prevalent toxic substance in castor seed that is highly ribosome-inactivating is lectin (ricin). Credit goes to genetic engineering for creating a detoxified genotype RNAi design for silencing ricin-coding genes in the endosperm of castor seeds. This result demonstrated that ricin genes were advantageously silenced in genetically modified (GM) crop plants (Sousa et al. 2017). A bio-detoxified valuable proteins-rich castor bean cake after oil extraction was found useful for animal feeding (Sousa et al. 2015). Castor var. Brigham seeds with low ricin content were found to have seeds that had been partially analyzed and processed by imaging of recognized bands that has ricin connected to improved biotechnological planning. These seeds can be used in castor breeding programs. The cost and time needed to detoxify ricin in a cake that was pressed would be eliminated in low ricin castor cultivars which would in turn contribute to changing human perception of castor toxicity and supplying a world agronomic trend of cultivating castor as an important biotech crop (Wettasinghe 2012). Accurate information for better screening for low levels of ricin cultivars and evaluation of cross-breeding success is needed by researchers to reduce ricin content. From a security standpoint, ricin in "white powder" preparations accurately quantified was considered a most vital application. Recent events have demonstrated that the two toxic lectins that are toxic to health, proteins ricin and *R. communis* agglutinin (RCA), found in castor seeds oil were noted as threat agents. A simple crude preparation of castor beans made by crushing based on recipes, or a highly purified powder with concentrated amounts of ricin will assist public workers on health and law enforcement agents to make informed decisions on further detoxification (Schieltz et al. 2015).

1.1.4 Improvement of oil yield and quality

For the improvement of castor bean cultivars, the high potential benefits will be derived by increasing oil content, improving high oil yield, and making more cost-competitive production. Nuclear magnetic resonance (NMR) technology was described as a successful procedure for increasing the content of oil via repeated selection relying on individual screening of seeds of castor in a larger population recommended for commercial production (Chen et al. 2018). Castor genotype variability was reported to occur in relation to its yield and its traits helped determine the extent and nature. To increase seed yield, careful consideration must be given to the length of the primary raceme and the total number of nodes that it contains. According to Rukhsar et al. (2018), however, cash outturns should also be considered for improving the percentage of oil yield. In addition to the molecular mechanisms required for the efficient transport of ricinoleic acid into the seed oil, reliable information on the metabolic pathways involved in the biosynthesis of castor oil has also advanced positively, paving the way for recombinant DNA technology protocols to produce castor-like oils from common oleaginous species. Genetic engineering strategies have enabled the domestication of the castor plant for ricinoleic acid, sources are explored to improve its availability (Maroto and Alonso 2018).

1.1.5 Production of castor seed cake

The production of a grade cake for the feed using a castor bean residue that has been detoxified and enhanced with tannase and phytase is now offered as a potential supplement. The tannase and phytase development of Paecilomyces variotii is simultaneously induced by this method; at high levels, both enzymes exhibit their activity and possess the ability to be used in feedstuffs because the anti-nutritional factors decrease. Castor oil fermentation results from sodium dodecyl sulfate polyacrylamide gel electrophoresis (SDS-PAGE), showing a decrease in ricin bands and a discernible increase in cellular viability (Madeira et al. 2011). According to some theories, the biological detoxification of castor bean residue could be used to make it useful as an animal feed-rich source of protein by combining the biotechnological preparations of these two enzymes as a potentially important procedure for feed enrichment (Madeira Jr. et al. 2011). Production of oil (Fig. 1.3a–b) from castor releases important by-products; husks and meal, because of the high nitrogen content, the latter is the most important by-product. However, the literature showed that they can be blended and used as organic manure. The main parameter of extracting protein from castor bean cake (Fig. 1.3c) is the solubilization in an alkaline medium (Lima et al. 2011).

1.1.6   Bioenergy – biofuel and lubricants

Nowadays, biodiesel production from castor oil can be achieved even by using lime (CaO) obtained from mud clam shells (Ismail et al. 2016). The biofuel potential of wild castor seed and castor bean oils was exhibited through Fourier Transform Infra-Red (FTIR) spectroscopic investigations and fuel characteristics (Atiku et al. 2014a, Atiku et al. 2014b). Oil from the seed of castor (*Ricinus communis* L.) crops produced biodiesel were characterized based on findings on the trans-esterification effects of castor seed oil utilizing methanol ethanol, and their mix on the yields and characteristics of biodiesel. Methanolysis was regarded as suitable to produce biodiesel from oils extracted either by cold or hot method (Enontiemonria et al. 2012, AL-Harbawy and AL-Mallah 2014). The most frequently used type of oil is lubricating oil, which is also known as commercial and industrial oil. The process of manufacturing lubricating oil can be differentiated into three ways: by refining crude petroleum oil, by extracting from crude vegetable oil, and by synthetic processes. Lubricating oil from the castor plant is regarded as one of the best sources of quality lubricating oil (Paguiliganand Villanueva 2005). Castor seed oil was transformed into an Estolide ester for use as a bio-lubricant (Tajuddin et al. 2014). Federico García Maroto led researchers from the University of Almeria to genetically engineered castor-oil plants for the industrial production of bio-lubricants. Transgenic acid profile castor-oil plants with good requirements for bio-lubricants can be obtained by using recognized genes for the biosynthesis of lipids (Innova 2009). The main concept was to produce plant oil with a higher concentration of monounsaturated fatty acids, primarily oleic and palmitic (Tables 1.1a–b). The only goal was to allow the fatty oils to accumulate in the seed without affecting other plant parts and thereby preventing adverse agronomic effects. GC-MS analysis recently reported in the literature, identified oleic and palmitic acids in both wild and cultivated castor bean plant oils respectively (Warra 2015b, Warra 2015c).

Table 1.1a. Oleic and palmitic acids derived from wild castor seed oil

S/N	Name of fatty acid	Molecular formula	MW	RI	SI% to T.C.
1.	Oleic acid	$C_{18}H_{34}O_2$	282	2175	86
2.	Palmitic acid	$C_{16}H_{32}O_2$	270	1968	91

Note: S/N = Serial number, M.W. = Molecular weight, RI = Retention index, SI% = Similarity index, T.C. = Target compound (Warra 2015b)

Table 1.1b. Oleic and palmitic acids derived from castor bean seed oil

S/N	Name of fatty acid	Molecular formula	MW	RI	SI% to T.C.
1.	Oleic acid	$C_{18}H_{34}O_2$	282	2175	86
2.	Palmitic acid	$C_{16}H_{32}O_2$	270	1968	91

Note: S/N = Serial number, M.W. = Molecular weight, RI = Retention index, SI% = Similarity index, T.C. = Target compound (Warra, 2015c)

1.1.7 Liquid fuel production

Bioenergy is one of the renewable energy options available today and in the near future. It was reported that the production of bioenergy from local varieties of castor oil plants could create potential impacts, minimize energy import and bring about economic development (Sánchez et al. 2018). Industrial and environmental benefits of biomass pyrolysis (thermo-chemical conversion) have been discovered. It is unavoidable to use kinetic models to best design and run the pyrolysis because waste biomass transformation by chemicals gives many products by pyrolysis reactions, which are dependent on the kinetic rates process (Sokoto and Bhaskar 2018). Bio-oil obtained from castor shells was recommended as a chemical feedstock for renewable fuel using chemical characterisation (Mohammed et al. 2014). Bioenergy production from local varieties of castor oil plants showed potential impacts in reducing the import of energy (Sánchez et al. 2018). A study uses a composite sorbitan monooleate and polyoxyethylene sorbitan monostearate formula as the best emulsifier formula biomass for castor oil (Zhang et al. 2011). An analysis of the effects of oil from castor on the pollutant's emissions in a continuous combustion chamber using a bio-blend of castor oil with hydrocarbon fuels (gas oil and kerosene) revealed that all pollutants—carbon monoxide CO, unburned hydrocarbon UHC, soot, and nitrogen oxide NO_2—had lower emissions when castor oil was used in different percentages of 5%, 7%, and 10%. Castor oil blends have demonstrated lower emissions as a result of having oxygen O_2 in the chemical structure of the castor oil, which is sufficient to seek complete combustion (Hasan et al. 2017).

1.1.8 Feedstock for biogas

As an efficient, cheap and alternative source of fuel even in rural areas, the biogas project has gained momentum. Well-known biogas plants are reportedly designed based on cattle dung. Even though, due to mechanization of farming, there exists limited availability of cattle dung which results in underfeeding of the plants. Castor cake is a non-edible cake among the de-oiled cake additives that have an effect on increasing biogas yield (Baladhiya and Joshi 2016). For ethanol and biogas production, residues from castor plants, i.e., leaves, stem and seed cake were used. To produce biogas, methane yield produced from castor stem is enhanced by alkali pretreatment (Bateni et al. 2014). Using castor cake (oil expelled). Biogas can be produced using oil-expelled castor by anaerobic digestion where blende culture can produce biogas that has been isolated and stabilized. Thus, the biogas will then contain small quantities of propane, ethane, and butane in addition to methane and carbon dioxide which constitutes the major component (Gollakota and Jayalakshmi 1983). Since methane can be used as a fuel substitute for fossil fuels in the production of heat and electricity, the anaerobic digestion process to produce methane-rich biogas offers a very diverse source of renewable energy (Lehtomäki 2006). This reduces greenhouse gas emissions and slows down climate change.

1.1.9   Lignocellulosic biomass

Plant biomass comprises lignocellulosic biomass (LC) (cellulose, hemicellulose, and lignin). Value chains for the energy production of chemicals, polymers and materials can be derived from LC. However, lignocellulosic biomass can replace fossil fuels from an energy point of view. So far, its limited importance is directed to the materials sector (wood-based and paper) (http://www.biocore-europe.org/pagee027.html?optim=what-is-lignocellulosic-biomass.

Castor stem has potential in the paper industry due to its high composition of cellulose fibers, according to a study that found it contains significantly more cellulose than other agricultural waste (Shah et al. 2015). According to research, castor stalks are a lignocellulosic by-product that, when compared to those of other annual crops, can only be used as a complement to wood to produce particle board. According to Grigoriou and Ntalos (2001), the assessment was done for the raw material used to directly replace wood in particle boards.

1.1.10   Fiber, resins and adhesives

A functional study of polyurethane resin derived from castor oil would be of assistance in its use as forest material and for building raw materials for industries. The use of bamboo charcoal in low-density composites with a polyurethane resin based on castor oil has been reported (Chen and Tai 2018). The wood-derived medium-density fiberboard (MDF) and eco-friendly vegetable oil-based polymer, a polyurethane resin made from castor oil and developed at the São Carlos Institute of Chemistry, University of São Paulo, was reported as harmless to humans, making castor oil-derived polyurethane resin a good adhesive for the production of MDF (Inácio de Campos et al. 2014). Castor oil when modified can serve as an epoxy resin curing agent (Patel et al. 2004). Modern researchers have discovered that solvent-free adhesives made from renewable sources are of high value because they are produced at a low cost and can be converted to biodegradable materials, in contrast to those whose preparation sources are derived from petrochemicals. Castor oil-derived polyurethane (PU) adhesives for wood joints that are produced commercially showed lap shear strength values that were 20% higher than those of a solvent-based adhesive used for wood. According to Silva et al. (2010), the newly developed castor oil-based PU adhesives offered an intriguing alternative to the currently used solvent-based adhesives for wood and foam substrates. A particle composite made of castor oil polyurethane adhesive and coconut fiber served as an environmentally friendly material for the manufacture of particle boards (Fiorelli et al. 2012).

1.1.11 Production of bio-based plastics

Sebacic acid, a dicarboxylic acid which, for example, can be produced from castor oil is significant in the production of bio-polyamides which are plastics that are found relevant for fiber production. The most common are Nylon® and Perlon®. Polyamides found applications in demanding the molten material injection into the mould, extruded products, hollowware, and textiles industry, materials required for decoration and technical fabrics. Bio-polyamides are biobased in part or whole. However, if the dicarboxylic acid, the diamine, or both are derived from renewable resources. Castor oil derived Polyamide 11 has had potential applications in automotive for more than 3 decades 30 is relevant for fuel lines and connectors, particularly for the high bio-ethanol (E10, etc.) and biodiesel fuels (Agency for Renewable Resources 2014). Hasan et al. (2018) reported on the synthesis of chitosan and yellow pumpkin starch (Cucurbita moschata) with the use of castor oil as a plasticizer to produce bioplastics.

1.1.12 Biopesticides and bio-fertilizer production

Plutella xylostella (L.), (Lepidoptera: Plutellidae) was reportedly tested for its effectiveness in controlling the diamondback moth using castor seed, root, leaf, fresh extracts, and oil emulsion. Experimental revelations on the effects at residual indicated a decrease in the mortality of larvae with time between the botanical application and insect release (Tounou et al. 2011). Nanoformulation has proven to be biodegradable, environment and user friendly. For breeding mosquitoes in large fields, such as Anopheles culicifacies, a nanoemulsion made from castor oil has been developed. Castor oil nanoemulsions are a good, safe, and effective substitute for controlling vector-borne diseases brought on by mosquito larvae and might be advised for field application, according to the results of the castor oil nanoemulsion formulation in different ratios composed of surfactants, castor oil, water, and by high-pressure homogenizer mixer (Sogan et al. 2018). The diamondback moth, Plutella xylostella (L.), (Lepidoptera: Plutellidae) can be controlled following the evaluation of the usefulness of castor root, leaf, seed kernel crude extracts, and oil emulsion. It is expected that the use of castor products will be suitable for reducing the P. xylostella population density due to low oviposition rates, oviposition deterrent, immature mortality, and the relatively low persistence of the toxic ricin oil, which ultimately showed that significant use of castor in P. xylostella control has favorable usage (Kodjo et al. 2011). Husks and meal are the two main by-products produced during the development of castor oil. 1.31 tons of husks and 1.13 tons of meal are produced for every ton of castor oil. Currently, an optimized mix is reportedly used as an organic fertilizer, and castor meal is the most significant by-product because of its high nitrogen content. (Lima et al. 2011). The castor bean was considered a good substrate to produce biogas and biofertilizer with satisfactory chemical qualities (Alves et al. 2012).

1.1.13   Industrial value addition from castor products

Castor value-added products (Table 1.2) have potential applications in the rubber, textile, paper, paint, electronics, pharmaceutical, and agricultural industries. Food, plastic cosmetics, perfume, telephony, lubricant, and castor, when given added value, can support the growth of the economy and industry. Ibeagha and Onwualu (2015) opined that establishing an information system for a functional market, capacity and organizations for farmers' development clusters would assist the development of its value chain. Large-scale employment and wealth creation can be accomplished through the local production of the necessary process technology, followed by maintenance requirements. This can successfully develop the value-added castor products at all value chain hierarchies. According to estimates made by the authors' strategies, castor could bring Nigeria's economy up to 25 billion Naira (105,488,032 million dollars) per year (Ibeagha and Onwualu 2015). Castor value-added products (Table 1.2) have potential applications in the rubber, textile, paper, paint, electronics, pharmaceutical, and agricultural industries. Food, plastic, cosmetics, perfume, telephony, lubricant, and castor, when given an added value, can support the growth of the economy and industry (Warra et al. 2020).

1.1.14   Knowledge explosion for environmental cleanup and bioprospection

Bioprospecting in plant species like *Ricinus communis* from Family: *Euphorbiaceae* is capable of phytoremediation, the production of phytomass for boosting the bioeconomy, and co-generation of bioproducts is a feasible proposition and one of the areas of research that have the greatest impact on biotechnological developments (Fig. 1.4). Mining, especially artisanal, conventional agriculture, urban waste products, and various industrial activities are implicated in severe environmental pollution. Phytoremediation emerged as plant-based technology to degrade (inorganic), sequester, or contain contaminants (organic) in natural resources (soil, water, air). Candidate plants for this operation belong to Asteraceae, Brassicaceae, Euphorbiace, and Fabaceae (among dicots);

Table 1.2. Potential of the castor derivatives for investors
(Energy Alternatives India 2016)

Industries	Potential
Agriculture	Organic fertilizers
Plastics and Rubber	Polyamide nylon, Plastic films, Adhesives, Coupling agents, Polyols, Synthetic resins, Plasticizers Surfactants, Additives for reducing viscosity
Food	Flavouring agents, Materials food packaging
Cosmetics and Perfumeries	Products from perfumery, Lipsticks, Hair tonics, Shampoos, Polishes, Emulsifiers, Deodorants
Paper	Flypapers, Defoamer, Additive for water proofing
Pharmaceuticals	Antihelmintic, Antidandruff, Cathartic, Emollient, Emulsifiers, Deodorants
Electronics and Telecommunications	Electronics and telecommunications polymers
Polyurethanes	Insulation materials
Paints, Inks and Additives	Inks, Plasticizer for coatings, Varnishes, Lacquers, Paint strippers, Adhesive removers, Wetting and dispersing additives
Textile Chemicals	Textile finishing materials, Dyeing aids, Nylon, Synthetic fibers and resins, Synthetic detergents, Surfactants, Pigment wetting agents
Lubricants	Hydraulic fluids, Heavy duty automotive, Greases, Fuel additives, Corrosion inhibitors, Lubricating grease, Aircraft lubricants, Jet engine lubricants, Racing car lubricants

Poaceae and Cyperaceae (among monocots) attained popularity. Members of the above families have gained significance in the phytoremediation of contaminated and co-contaminated substrates including waste dump stabilization. The reasons for this are: (a) A high tolerance to toxicity imposed by the contaminants, (b) High biomass production, (c) A high value for Biomass energy (d) An adequate knowledge of biotech tools and (e) many lab, pilot, and field works have been done on various issues. The shrinkage of natural forests has made us think about the utilization of waste and contaminated lands to produce industrial crops to generate feedstock to sustain production. Biomass energy production planning by converting arable land has a conflict of food vs. fuel with the concomitant danger of soaring food prices. Therefore, multipurpose crops capable of growing on contaminated and polluted soils with limited inputs need to be identified. In this context, castor beans emerged as a prominent environmental crop. Castor beans possess great potential to perform better in diverse habitats due to their plasticity. Though the "castor bean"' was introduced as an ornamental plant it has been much maligned subsequently. In this subchapter, it is suggested and supported by the research of the literature as a prospective candidate for environmental remediation and for revegetating contaminated/degraded sites for the development of bioeconomy.

1.1.14a   Potentialities

Despite the knowledge explosion in the study of phytoremediation (Prasad et al. 2010) to develop solutions to environmental contamination and pollution, there has been a bias to acknowledge the significance and the field performance of biological remediation (Montpetit and Lachapelle 2017). The environment in several countries rich in biodiversity remains polluted or contaminated. This has continued despite elaborate environmental laws and law-enforcing regulatory agencies (Lachapelle and Montpetit 2015). Deriving economically important phyto products during phytoremediation is an innovative strategy to enhance the economics of plant-based cleanup technologies (Prasad 2015, Tripathi et al. 2016). This would ensure the attention of environmental Biotech entrepreneurs and people's participation in phytoremediation. The possibility of using contaminated substrates to produce biofuels has tremendous scope (Prasad 2015, 2024). In India, there is widespread contamination due to industrialization and domestic activity. The costs for ex-situ remediation involve high-tech science and huge investments. As a result, several of the contaminated sites are untouched. This leads to the spread of toxic substances in soil, air, and water posing health risks to all Biota (Prasad 2011). Inappropriate municipal solid waste disposal of industrial effluents and municipal solid waste results in inorganic and organic contaminants. Environmental pollution was caused by the irrational exploration of natural resources and the exponential growth of the human population. Remediation of inorganic and organically contaminated soils has gained significant momentum and is a difficult task. One of the major industrial problems is hazardous waste disposal. Various adaptive traits of *Ricinus communis* enable it to perform better in a stressed environment producing high biomass. We noticed the enormous growth of castor beans (*Ricinus communis*) in the industrial waste-contaminated areas during our frequent field trips in and around Greater Hyderabad. *Ricinus communis* has demonstrated successful performance in co-contaminated sites, and based on several studies (Zhaung et al. 2007, De Souza Costa et al. 2012, Bosiacki et al. 2013, Yi et al. 2014), it has been suggested as a candidate for restoration of the contaminated sites. In comparison with other plants such as *Brassica juncea, R. communi,* it has been reported as a stress-tolerant crop and an efficient candidate for phytoremediation. Also, due to its fast growth, minimal inputs, and maintenance requirements for its establishment, it has been recommended for planting in wastelands (Ma et al. 2011, Rajkumar and Freitas 2008). In addition to its biomass's potential for biofuel, animal feed, and fertilizer when detoxified, it made contaminated soils usable. *Ricinus communis* L., belonging to Euphorbiaceae, is a fast-growing C_3 plant, native to tropical Africa and it is widespread throughout tropical regions. It is an erect, tropical shrub or small tree that grows up to 2–3 m and sometimes about 5 m.

The stem, stalks, and leaves are reddish to purple. Leaves palmate, 6–11 lobed, serrate. It is only one species of the monotypic genus *Ricinus*. It is cultivated and naturalized in India. It is an important non-edible oil seed crop in arid and semi-arid regions. Better efficiency of S-metabolism is witnessed in oil-yielding plants, therefore, making them exhibit tolerance to metal through the induced synthesis of phytochelatins which have been documented extensively. It is also said to be highly salt tolerant, salinity and drought tolerant, and to be a potential ameliorator for seashore saline soil due to its well-developed and massive root system, which penetrates the soil several meters deeper than herbaceous plants. It has provided a financial advantage in contemporary agriculture along with soil remediation for heavy metal-contaminated areas. Contaminated lands may not be suitable for food production due to the potential accumulation of toxic substances in the food chain. Even though plant growth would be hampered by the contaminants' toxicity, biomass production may help to further alleviate the issue when moderately contaminated land is used for purposes other than food. Additionally, due to mining activity primarily, a significant portion of the world's land is affected by diffuse contamination, byproducts of industrial processes like smelting, agriculture, and traffic, as well as by areas elevated by contaminated dredged sediments, abandoned landfill sites, and by industrial sites. During and after soil remediation, such contaminated land could be restored for beneficial and sustainable use, minimizing negative environmental, social, and economic effects on impacted communities. Research efforts to enhance plant growth in the presence of contaminants could support the future use of these contaminated areas for safety and large production biomass (Fig. 1.5).

1.1.14b Challenges ahead

All these characteristics make castor a potential multipurpose crop for phytoremediation and bioenergy production. The possibility of toxic substances accumulating in the food chain makes contaminated lands unsuitable for food production. To reduce detrimental effects on the environment, revegetating such contaminated land by remediation of soil for non-food crops could render them again into sustainable use and bring about social and economic impacts on affected communities. The exploitation of metal-tolerant species is very necessary for the successful rehabilitation of metalliferous as well as municipal dump sites. By enhancing plant growth in the presence of contaminants, future generations may be able to reuse contaminated areas for more effective and secure biomass production. With this background, the objective was to assess the potential of castor beans for use in integrated and improved phytoremediation for sustainable development and environmental security.

1.1.14c Contaminated site cleanup and cogeneration of byproducts

Economic growth, extensive development of industries, extraction of natural resources, and lack of will for cleanup have caused contamination and pollution in the environment. Large quantities of toxic waste have created contaminated sites throughout India. These pollutants can be divided into two categories: organic and inorganic. The challenge is to develop creative, and affordable solutions to remove contaminants from polluted environments. The development of cost-effective solutions for the decontamination of polluted sites is very challenging. Bioremediation is an emerging technology for environmental cleanup. Cultivation of appropriate non-food crops and energy crops for value chain and value additions appears profitable. Bioproducts and Bioenergy from contaminated substrates (soil and water), and brownfield development for smart Bioeconomy are some specific examples.

1.1.14d Approaches and results

A bibliometric method was applied to glean data on this subject. Searches using all available journals, Patent offices, conference proceedings, books, government agencies and publications were performed using major search engines (Koelmel et al. 2015). The cultivation of ornamentals on contaminated substrates, the production of bioproducts from contaminated substrates (soil and water), energy from biomass, biofuel from contaminated substrates, and the commercialization of algal bioproducts are some practical options for advancing the biobased economy. The biorefinery was conceptualized to recover value-added products from bioremediation biomass, products from bioremediation to boost the bioeconomy, and Chrysopogon Zizanioides cultivation to produce aromatic oil and a variety of other products (for more information, see Prasad, 2015, and 2016). Various strategies and approaches for phytoremediation of polluted sites are presented which stabilized contaminated and polluted sites also helped in recovering resources are presented with specific examples.

1.1.14e Conclusions

The published scientific reports unequivocally suggest that *R. communis* L. (Castor bean) is an appropriate candidate for environmental cleanup and co-generation of bioproducts (Table 1.3). It has been considered a promising candidate for phytoremediation applications owing to its adaptative and phenotypic plasticity to heterogeneous and stressful environmental conditions besides yielding numerous bioproducts as mentioned below:

- **Medicinal products:** Abd-Ulgadir et al. 2015, Jytothirmaye and Lingumpelly R. 2015, Loss- Morais et al. 2013, Ma et al. 2014, Oloyede 2012, Preeti and Verma 2014, Singh and Geetanjali 2015, and Zhou et al. 2015
  Tolerant to toxic metals: Wei et al. 2016
  Nutrient content in seeds: Akande et al. 2012
- **Biodiesel:** Al-Harbawy and Al-Mallah 2014, Anastasi et al. 2015, Atiku et al 2014, Mohammed and Abdullah 2015, Zaku et al. 2012, Bale et al. 2013, Borugadda and Goud 2014, Cabrales et al. 2011, Maria et al. 2012, Nielsen 2011, Okechukwu 2015, Ozturk et al. 2014, Perdomo et al. 2013

Table 1.3. *Ricinus communis*, as an important multipurpose environmental/
phytoremediation crop for sustainable development and environmental security

---

1. **Amendment and chelator assisted phytoremediation** – Capuani et al. 2015, Chhajro et al. 2015, de Abreu et al. 2012, Wiszniewska et al. 2016
2. **Arbuscular mycorrhizal fungi assisted phytoremediation** – Cabral et al. 2015, Bhalerao, 2013, Schneider et al. 2016
3. **Bioenergy and biodiesel** – Andreazza et al. 2013, Armendáriz et al. 2015, Atiku et al. 2014, Bauddh et al. 2015, da Silva J.A.C. et al. 2015, Onzález-Chávez et al. 2015, Lavanya et al. 2012, Liu et al. 2015, Magriotis et al. 2014, Moncada et al. 2015, Scholz and da Silva 2008, Silitonga et al. 2016, Zhang et al. 2015, Zhang et al. 2016, Zhi-xin et al. 2007, Visser et al. 2011, Sánchez et al. 2015
4. **Biomonitor of atmospheric pollution** – Bonanno, 2014, Kammerbauer and Dick 2000, Mendes et al. 2009
5. **Biosorbent for dyes and metals** – Al-Rmalli et al. 2008. Makeswari and Santhi 2013, Martins et al. 2013, Oladoja et al. 2008, Santhi and Manonmani 2009, Santhi et al. 2010, Zhang et al. 2015, Zhang et al. 2014
6. **Drought tolerance** – Babita et al. 2010, Shi et al. 2015
7. **Ecophysiological investigations** – Adhikari and Kumar 2012, Hadi et al. 2016, Huang et al. 2016 Kang et al. 2015, Li et al. 2010, Li et al. 2011, Pinheiro et al. 2008, Ribeiro et al. 2015, Rodrigues et al. 2014, Romeiro et al. 2006, Severino et al. 2013, Severino et al. 2015, Severino et al., Shi and Cai 2009, Silva et al. 2014, Singh et al. 2010, Srinivasarao et al. 2016, Srivastava et al. 2015, Sun et al. 2013, Tan et al. 2009, Vwioko et al. 2005, Wei et al. 2016, Yi et al. 2016
8. **Nanocomposites preparations** – Carreno et al. 2009, Medeiros et al. 2015, Prasad et al. 2012, Yasur et al. 2013
9. **Association with plant growth promoting rhizobacteria** – Ma et al. 2011a, Ma et al. 2011b, Ma et al. 2010, Ma, et al. 2015, Ma et al. 2016, Rajkumar and Freitas 2008, Rajkumar et al. 2012
10. **Phytoextractor** – Ananthi and Manikandan 2013, Ananthi et al. 2012, Annapurna et al. 2016, Bosiacki 2013, Coscione et al. 2009, dos Santos et al. 2012, Malarkodi et al. 2008, Niu et al. 2009
11. **Phytoremediation (Heavy metals/salinity)** – Bauddh et al. 2016, Cecchi and Manag., 2005, Costa et al. 2012, de Abreu et al. 2012, de Souza Costa et al. 2012, Hadi et al. Sci., 2015, Haung et al. 2011, Lu and He 2005, Mahmud et al. 2008, Melo et al. 2009, Melo et al. 2012, Nazir et al. 2011, Neto, et al. 2014, Olivares et al. 2013, Pal et al. 2013, Pandey 2013, Pandey et al. 2016, Rani et al., Rissato et al. 2015, Tripathi et al. 2013, 2016, Vwioko et al. 2006, Wang et al. 2013, 2016, Wu et al. 2016, Wu et al. 2012, Xiaoyi and Chiquan 2005, Yashim et al. 2016, Yi et al. 2016, Prasad 2015
12. **Phytoremediation of crude oil contaminated soil** – Baishya et al. 2015, Bauddh and Singh 2012a, b, 2014, 2015
13. **Sustainability development and bioeconomy** – Amouri et al, 2016, Bauddh 2014, Scarpa and Guerci 1982
14. **Xylem and phloem translocation and complexation of contaminants** – Aziera et al. 2015, Chen et al. 2014, Rigby et al. 1994, Pate 1976, Peuke 2009, Stephan et al. 1994, Ye et al. 2010

- **Biosorbent for metals and dyes:** Al-Rmalli et al. 2008, Makeswari and Santhi 2013, Martins et al. 2013, Oladoja et al. 2008, Santhi and Manonmani 2009, Santhi et al. 2010, Zhang et al. 2015, Zhang et al. 2014
  Sustainability assessment of castor derived biodiesel using LCA approach: Amouri et al. 2016
  Rhizobacteria enhanced lead phytoextraction has been demonstrated: Ananthi and Manikandan 2014, Ma et al. 2015
- **Seed yield and oil quality [COFAME], agro-climate zoning:** Falasca et al. 2012, Imankulov 2012, Jhonson 2007, Nangbes et al. 2013, Oswalt 2014, Salimon et al. 2010, Severino and Auld 2013
  Adaptive modulation traits for different environmental situation by applying several tools such as RAPD analysis, molecular diversity, genetic variation, genomics, SNP, traditional breeding: Chan et al. 2010, Chandrasekaran et al. 2014, Foster et al. 2010, Goodarzi et al. 2011, Goyal et al. 2014, Kallamadi et al. 2015, Lakani et al. 2015, Meena et al. 2014, Milaini and Nobrega 2013, Ovenden et al. 2014, Perea-Flores et al. 2011, Ramprasad and Bandopadhyay 2010, Saadaoui 2015, Shaheen 2002, Velasco et al. 2015, Vivodik et al. 2015, Wang et al. 2013, Wu et al. 2015; including seed and genetic variation Khan et al. 2014, Radhamani et al. 2012
- **Castor oil safety assessment:**
  Antimicrobial: Zarai et al. 2012,
  Biopesticidal value: Mohammadi et al. 2014, Nekonam 2014, Saadaoui et al. 2015, Wafa et al. 2014, Fungicidal Yekeen et al. 2014
  Biotechnology of Euphorbiaceae has been highlighted for environmental issues: Ladda and Kamthane 2014, Neto et al. 2015, Maghuly et al. 2015, Msaakpa and Obasi 2014, Rana et al. 2012
- **Drought and saline stress:** Karimi et al. 2012, Lima et al. 2015, Pius et al. 2014, Rahmati et al. 2015, Riberiro et al. 2014 and 2015, Rodrigues et al. 2014, Severino and Auld 2014
- **Photosynthesis:** Shi et al. 2014
- **Biopolyester:** Peres et al. 2014
- **Ricin:** Schieltz et al. 2015, Słomińska-Wojewódzka and Sandvig 2013, Wettasinghe et al. 2013

1.2 Physic nut (*Jatropha*)

There exist two of the commonest varieties of Jatropha, *Jatropha curcas* L. and *Jatropha gossypifolia* (Warra et al. 2019). However, different species exist in many parts of the world which include: *Jatropha cinerea, Jatropha platyphylla, Jatropha integerrima*Jacq, *Jatropha multifida* L., *Jatropha podagrica, J. neopauciflora, J. oaxacana, J. rufescens, J. ciliata,* and *J. rzedowskii* (Hernández-Nicolás et al. 2018, Soto-Leon et al. 2014, Sharma and Singh 2013, Blessing et al. 2011, Makkar et al. 2011). The genus *Jatropha* (*Euphorbiaceae,*) is a crop of the tropics and subtropics, it can grow on constrained land easy to propagate and grows rapidly, it is drought tolerant, has a high oil content with multiple uses ranging from biofuel to cosmetics (Warra and Prasad 2016). Numerous human and veterinary diseases can be successfully treated using most of the parts of Jatopha curcas (Fig. 1.6). Curcain, Jatropha, and Jatrophine are alkaloids discovered in the jatropha plant that has anti-cancer capabilities, and white latex is also a possible disinfectant. Moreover, it was effective in treating piles, ulcers, and skin conditions in domestic livestock. Children's mouth infections are treated with leaves. The latex of the Jatropha plant contains substances such as apigenin, vitexin, isovitexin, and others that make it useful against rheumatic, muscular, and malarial problems. Jatropha has been found to have antibiotic activity against a variety of microbes, including *Escherichia coli* and *Staphylococcus aureus*. Antidotes against snake venom are contained in the roots. Bleeding from gums can be checked by root extracts. Jatropha oil soap is efficient against buttons (Thomas et al. 2008). According to Patil et al. (2013), Jatropha curcas has antimicrobial, anti-cancer, and anti-HIV activity. There are opportunities for biotechnological exploitation through tissue culture, selection, hybridization, and mutation as a result of the taxonomic identity and description of the plant, requirements for its distribution and ecology, and ability to adapt to a variety of agro-climates (Divakara et al. 2009). Recently there is a series of literature on in vitro culturing technology and efficient tissue culture, specifically callus induction, somatic embryogenesis and micropropagation.

The development of in vitro regeneration of the Jatropha curcas plant was achieved for the purpose of germplasm collections, breeding programs, and mass propagation (Nahar and Borna 2013). Latex-rich Jatropha curcas is rich in high oil content for biodiesel production. The latex component is also considered rich in proteins with potentially important physiological functions and secondary metabolites potentially useful for new drugs (Yang et al. 2017). For the planning of an effective breeding program in Jatropha curcas, genetic information such as gene action and the contribution of genetic components to the inheritance of the quantitative characters is required. Against this background, combining the ability for flowering and yield traits of Jatropha (*Jatropha curcas* L.) was selected for the development of hybrid varieties in Jatropha (Aminul Islam et al. 2016).

1.2.1 Genetic analysis

Utilizing transcriptome data from high and low fruit yield, the sequencing effort, and gene function analysis to help with the breeding of high-yielding *J. curcas* cultivars, jatropha germplasm has the potential to provide a significant scientific asset (Hui et al. 2018). Understanding the genetic makeup of J. curcas populations would make it possible to have genetic material on hand for future development. The Jatropha curcas genetic diversity study at the cutting edge was described in Ovando-Medina et al. 2011. It was an impressive and newly reported strategy for gene transfer to Jatropha using an agrobacterium. The genes responsible for fatty acid production have been found. Priority should be devoted to boosting resistance, decreasing toxins, and producing more female flowers in inflorescence (Moniruzzaman et al. 2016).

1.2.2 Genetic regulation and lipid biosynthesis

The advantages of Jatropha, including its ease of genetic manipulation, rapid generation, and relatively small plant size, have made it a new model for woody energy crops. Functional genomics and transgenic analysis were used to conduct a thorough review of the regulation of seed development, genes involved in lipid biosynthesis, lipid turnover, and a group of epigenetic or transcriptional regulators on lipid biosynthesis and seed development in the Jatropha plant. In addition, methods for improving the quality of biofuel and seed oil productivity were given, along with information on how epigenetic and transcriptional regulation affect lipid biosynthesis and turnover in plant seeds. Understanding the biosynthesis of lipid/turnover and then spatiotemporal manipulation of the process in Jatropha is vital for the improvement of productivity of its seed and quality as biofuel raw material. Lipid turnover processes involve Triacylglycerols (TAGs) hydrolysis into Fatty acids (FA) and glycerol and this reaction is catalyzed by TAG lipases, these occur during Jatropha seed germination. Among the recognized lipases are the unusual patatin-like TAG lipases (PTLs), which are oil body-associated enzymes and play an important role in the initial stages of TAG degradation in yeast, mammals, and insects. Additionally, TAG lipases start the hydrolysis of TAG into glycerol and FFAs, which are then metabolized via the beta-oxidation pathway to release carbon sources for the early growth of seedlings. According to Ma et al. (2017), the knowledge produced is essential for improving agronomic traits in related energy crops and oil crops.

1.2.3 Genetic resources, transformation and products development

From testing the physical and chemical characteristics of seed oil of *J. curcas's* genetic resources there is the possibility of its being a potential feedstock for biodiesel production. One of the possible eco-friendly substitute sources of energy, for the predominantly used fossil fuels (Parthiban et al. 2011). Profit-making byproducts obtainable from Jatropha curcas made it a suitable crop for the following reasons (i) β-sitosterol, β-amyrin and taraxerol are all present in the bark. (ii) Resorsilic acid, p-OH-benzoic acid, o- and p-coumaric acid, protocatechuic acid, β-sitosterol, saponins and tannins, taraxerol, and β-amyrin are all present in the aerial parts. (iii) The flavonoids apigenin, flavonoidal glycosides, vitexin, isovitexin, and atriterpene alcohol (C63H117O9) are all found in leaves. (iv) Curcin, lectin, phorbolesters, esterases, and lipase are all present in seeds. (v) Roots contain and the curculathyranes A and B, β-D-glucoside, marmesin, β-sitosterol, jatropholone A and B, propacin, the curcusones A–D, diterpenoids jatrophol and, coumarin-tomentin, coumarin-lignan-jatropha, and taraxerol. (vi) The seed kernel contains a substantial amount of phytates, saponins and trypsine inhibitors, which also remain as a major constituent of the seed cake. Because of the toxic components of the seed, genetic transformation is required to genetically engineer the seed toxins to useful and non-toxic components (Misra and Misra 2010).

1.2.4 Toxicity and detoxification

Toxic substances like curcin (a protein) and diterpenoids (phorbol-esters), which are present in *J. curcas* seeds, must be removed before the phorbol esters can be used for industrial or medicinal purposes. Even though the primary toxin, phorbolester, is not heat sensitive. However, it can be hydrolyzed to fewer toxic substances that can be extracted either by water or ethanol (King et al. 2009, Goel et al. 2007, Usman, et al. 2009). There exist other chemical compounds including curcin (an alkaloid) in its seeds that render it not fit for consumption (Thomas et al. 2008). It was reported that not all the Jatropha species are toxic, non-toxic nature of *J. platyphylla* was confirmed, it was reported further that *Jatropha platyphylla* seeds after roasting are consumed by local communities in Mexico. The kernels of *J. platyphylla* contained ca. 60% oil and were free of phorbol esters. Trypsin inhibitors, lectins and phytate are contained in the kernel meal of this Jatropha species. However, no harm to the local people for eating roasted seeds, this is because trypsin inhibitors and lectins are heat labile (Makkar et al. 2011). The removal of toxicity from *J. curcas* seeds allows drug discovery of its active components (Gudetam 2016).

1.2.5 Fragrance development

Vanillin production is feasible when the potential of lignocellulosic biomass is utilized. The description of the chemical constituents and resource recovery potential of vanillin products were evaluated from the stem of J. curcas. It was claimed to be the first report to show the possible use of J. curcas stem as a value-addition raw material for vanillin flavor production and further indicated the possibility of using microorganisms for lignocellulosic biomass utilization for vanillin flavor production. However, optimization of the process and the evaluation of cost investment are needed (Vaithanomsat and Apiwatanapiwat 2009). Some volatile constituents were also found in the *Jatropha gossypifolia* L. essential oil; sesquiterpenoid germacrene D-4-ol (23.3%) and hexahydro farnesyl acetone (15.4%). Other important compounds were δ-cadinene (7.7%), tetradecanal (6.8%), and cubenol (6.1%). Instrumental gas chromatography coupled with mass spectrometry (GS-MS), the composition of the hydrodistilled oil was analyzed (Aboaba et al., 2015). Two forms of diterpenoids (Fig. 1.7) are detected in this plant, the diterpenoid Curcuson A and the diterpenoid Curcuson C, all possessing aromatic qualities (Director General Department of Alternative Energy and Efficiency Royal Government of Thailand, nd). This showed the potentiality of the jatropha crop in perfume and cosmetics production.

1.2.6   Bio-based products from *Jatropha curcas* crop

Bio-based derived biofuels and bio-based materials (biochemicals and bioplastics) are achievable using the jatropha crop. In integrated biorefineries, jatropha plant products are used for bio-based materials manufacturing (Table 1.4). Bio-oil from the Jatropha plant has great potential for health and the environment (Philp and Pavanan 2013). As a result of anti-nutritional factors present in the Jatropha oil that is non-edible, it is rendered undesirable to compete with the food chain and cannot be used for nutritional use as a result of the presence of compounds that are toxic such as phorbol esters. Results of the investigation of Jatropha curcas oil (JCO) showed that the bio-oil is useful in rejuvenating aged bitumen to conditions that resembled the original bitumen; the bio-oil is also beneficial from both health and environmental perspectives. In terms of storage, the rejuvenated bitumen was found to be very stable (Ahmad et al. 2017).

Table 1.4. *Jatropha* bio-based materials value chain (Philp and Pavanan 2013)

Bioproducts	Applications
1. Bioplastics	Processing for application in packaging, food and agricultural applications, automobile consumer and household appliances
2. Biolubricants	Biodegradable and unbiodegradable types
3. Biosolvents	Extraction of substances and retaining the chemical composition
4. Biosurfactants	Easy mixing of chemicals and as detergent cleaners, personal care, textiles etc.

1.2.7   Energy and biofuel potential

Due to the need for sustainable wealth in alternative energy sources, research on jatropha oil as a raw material for biodiesel development has attracted interest (Devappa et al. 2010). According to Moniruzzaman et al. (2016), jatropha is a good candidate for potential biodiesel production due to its high oil content and quick growth. It is also easy to propagate, drought-tolerant, requires relatively few inputs for agricultural irrigation, and is resistant to insects and pests. *Jatropha gossypiifolia* (JG) and *Jatropha curcas* L. (JC) species were reported as potential feedstock for biodiesel production (de Oliveira et al. 2008, Ahaotu et al. 2013). Suitable conversion technologies provide the possibility for obtaining energy products in solid, liquid and gaseous forms. Technology-based options, ranging from biodiesel to other potent energy carriers such as biogas and pyrolytic products, were reviewed (Jingura et al. 2010) in order to expand the energy carrier matrix derived from *Jatropha curcas* L. A zero-waste approach to using *Jatropha curcas* L. as an energy crop is made possible by the variety of emerging technology substitutes and *Jatropha curcas* L. processing products. Different

Jatropha curcas L. products make it possible to obtain both second-generation energy carriers like pyrolytic oils and syngas as well as first-generation energy carriers like biodiesel (Table 1.5).

Table 1.5. An energy model for *Jatropha curcas* L. (Jingura et al. 2010)

Component	Quantities (tonnes ha⁻¹)	Energy value (MJ kg⁻¹)	Total energy ha⁻¹ (GJ)	Conversion technology
Shells	1	11.1	11.1	• Combustion • Devolatilization (Pyrolysis) • Anaerobic digestion
Husks from seeds	0.6	16	9.6	• Briquetting • Gasification
Plant oil	0.9	39.8	35.8	• Combustion (direct) • Trans-esterification
Cake obtain by pressing	1	18.2	18.2	• Anaerobic digestion • Pyrolysis
Woody products	3	15.5	46.5	• Combustion
Total			121.2	

1.3 Beniseed (*Sesamun indicum L.*) crop

Beniseed (*Sesamum indicum* L.), a seed oil-producing crop grown worldwide with almost all its parts useful for domestic and industrial applications (Fig. 1.8a–h), and has a high content of oil and the genome of a small diploid, achievements have been made so far on studies concerning genes that are oil content related, biosynthesis of fatty acid and yield and it is established that many of the candidate genes responsible for oil content in sesame (*Sesamum indicum* L.) seeds encode enzymes involved in the metabolism of oil bundle of knowledge needed for breeding and improvement strategies (Warra et al. 2016a, Warra et al. 2016b, Warra 2011, Wei et al. 2015, Tiwari et al. 2011). Improvement of sesame is also observed in polygenic mutation. Obtaining haploid plants through another culture is now possible; the biotechnological protocols involve interspecific hybridization coupled with the embryo rescue technique in order to create new genetic variation. Tissue culture, molecular markers, genomics, gene isolation and transfer are also involved in successful breeding programs, the importance of the selection of genotypes in breeding and conservation in the metabolic diversity evolution in commercial sesame traits are highlighted in the literature (Laurentin et al. 2008). Genome sequencing as a giant project was achieved by the Sesame Genome Working Group (SGWG) providing sequence and assembly of the sesame (*Sesamum indicum* L.) genome a broad area of paramount importance to recombinant technology (Zhang et al. 2013). The importance of sesame in systematically reserving germplasm banks and is helpful for parents in breeding programs and is observed in a high level of genetic diversity among the genotypes through the RAPD technique (Akbar et al. 2011).

1.3.1 Transgenic technology

Limited genetic variability within germplasm has restricted traditional breeding techniques. Approaches that extend beyond breeding techniques are a requirement for new sesame cultivars to be developed further resistance to biotic/abiotic and particular characterized functions, emerging biotechnological methods have provided some promising options or more of the required protocols, like regeneration, agrobacterium genetic transformation, or by direct gene injection into plant cells which are required for these technologies to be applicable. It was reported that to date no production of a transgenic sesame plant has been successful; limited information concerning the sesame regeneration system is responsible (Suh et al. 2010). FAD_3 transgenic lines were reported by multiplication and acclimatization. However, insufficient transformation efficiency in sesame has hindered the analysis of the transgenic lines and the development of new cultivars. An increase in transformation efficiency in sesame is a possible effort to overcome this hurdle (Suh et al. 2010).

1.3.2 Genetic diversity

A report has shown that no complete research study was done on the genetic diversity of sesame at either the phenotypic or molecular level. However, in the year 2000, a sesame core collection (CC) was established in China as a basis for a theoretical guide for further collection, efficient protection, a good application, and a comprehensive investigation of sesame genetic resources as a provision for technical guidance, assessment of the genetic diversity genetic makeup characteristics of sesame CC in China was carried out using molecular and phenotypic data and by sesame mini-core collection (MC) extraction. A detail of molecular and phenotypic characterization of the genetic diversities of the sesame CC was provided by the study. Utilizing phenotypic and molecular data, an MC was extracted. The provision of a good representation of the genetic makeup characteristics of the original CC by the MC was a result of the low mean difference percentage (MD %) and the variance difference percentage (VD %), and the large variable rate of coefficient of variance (VR %) and coincidence rate of range (CR %). The MC was more genetically diverse with higher diversity indices and a higher polymorphism information content (PIC) value than the CC. To effectively select materials for the breeding of sesame and for biological studies of genotype, and as a population for association mapping of sesame, MC plays a significant role (Zhang et al. 2012). A few studies about morphological diversity have been looked into due to the lack of information on the genetic diversity in sesame, and these studies have typically been focused on regional interests. Due to this, closing this gap requires in-depth research, which was done on this background to gain a thorough understanding of the diversity in a sesame germplasm collection that includes both accessions from various regions of origin and commercial cultivars or experimental lines. The three various degrees of genetic diversity were evaluated using functional metabolites, metabolic profiles, and amplified fragment length polymorphism (AFLP) of DNA (using the effect of plant extracts on soil-borne pathogenic fungi). According to the study, all accessions include molecules that are harmful to fungi, but the effect on fungal development ultimately depends on how well those compounds are balanced with those found in the plant organ. Research showed that significant screening within a species is observed when searching for antifungal activity in plants (Laurentin 2007). There is a large variety of sesame crops in India, and the National Genebank of the National Bureau of Plant Genetic Resources maintains a landrace collection (NBPGR). Exploiting the breeding potential of this germplasm pool has been extremely challenging due to the challenge of transferring diversity into a form that breeders and farmers can use easily. A selection of 24 of the most diverse and unadapted parental lines, including one accession of the wild species *S. mulayanum*, was made as part of a core collection strategy, and these were intercrossed in various combinations to maximize genetic diversity and develop locally adapted genetic resource pools. At four chosen target sites, a weak and decentralized selection regime was kept for the offspring of 103 crosses. Based on these promising types of plants with desired plant features and a high seed yield, the range of variation in the chosen F4 progenies was evaluated and chosen. Realized genetic gains were also evaluated, particularly for variables associated with yield. In the study, a limited fraction of the existing diversity held in the gene bank was used and there was much more diversity available for large-scale genetic enhancement of sesame in the future (Bisht et al. 2004). 22 selected sesame seeds' tolerance (*Sesamum indicum* L.). An examination of genotypes from India, South Africa, China, Iran, and Pakistan was conducted in the field in the northwestern region of Iran between 2007 and 2008. Using widely used quantitative markers of stress tolerance, genotype responses were examined under optimal and restricted watering circumstances. There were significant differences between the grain genotypes at both normal and reduced irrigation circumstances and stress tolerance indices. This showed the existence of significant genetic heterogeneity among the genotypes and allowed for the identification of genotypes within the sesame germplasm which was drought tolerant. Principal component analysis revealed that the genotypes have a high amount of genetic variability (PCA). A 92.76% of the total variation was demonstrated by the first and the second principal components. In comparison to genotypes from other parts of the world, Iranian sesame was extremely drought tolerant. Stress Tolerance Index (STI), which may produce high yields under normal settings, was shown to be the best index for selecting drought-tolerant sesame cultivars after considering correlations between indices and grain production under stress and normal conditions (Gharibeshghi et al. 2016). The availability of germplasm banks, which serve as reservoirs of genetic diversity, is essential for the evolution of species. To identify duplicates, organize core collections, and facilitate the selection of parents for breeding programs, this heterogeneity must be described by genetic and phenotypic criteria. For both plant breeders and germplasm curators, the assessment of diversity in germplasm collections is vital for optimizing the use of the variability available. Various genetic markers can enable the estimation of diversity. Estimation by multivariate analysis of the 30 morphological genetic divergences of agronomic traits in 108 sesame genotypes was reported in an examined study. According to reports, the Cole-Rodgers index was used to create the dissimilarity matrices. The principal component analysis found the traits that contributed the most to the divergence and the genotypes were clustered by Tocher's optimization. Despite the limited genetic basis, the markers were effective in characterizing the genotypes and locating the clusters of genotypes that shared the greatest traits, as well as duplicate and divergent genotypes. The traits of grain yield and number of capsules per plant showed the most variation (Arriel et al. 2007). The genotype of Nigerian sesame (*Sesamum indicum* L.) and its relationship to phytochemical content were successfully characterized genetically using simple sequence repeat markers and standard and spectrophotometric techniques of the Association of Analytical Chemists (AOAC). Among the genotypes with the repeat motifs (TC12-TC25) genetic variability of a high level was observed, generated from ten highly informative primer pairs. The results of the analyzed phytoconstituents showed that sesame seeds contain phytochemicals with a potential for health and nutrition. For successful programs of plant breeding, the observed broad genetic base in the sesame germplasm found in Nigeria would give the basis for the future of genotype selection (Nwalo 2015). Due to this selection process, it was noted that this crop was being conserved in Vietnam and Cambodia. Sesame accessions from various origins were evaluated using morphological and molecular markers together with oil content and quality studies. Many sesame accessions in Southern Vietnam and Cambodia demonstrated a good potential for high seed yields and oil content, according to the results of morphological and oil content investigations. Substantial baseline knowledge about the sesame populations in Vietnam and Cambodia has been compiled, providing a road map for future breeding plans and sesame improvement in these countries (Pham 2011). For genetic relatedness studies in sesame, it was shown that amplified fragment length polymorphism markers (AFLP) were a reliable tool in determining the extent of genetic diversity as related to their origins geographically and morphologically. To distinguish between the many sesame accessions and to establish relatedness by biochemical analysis, AFLP patterns will be helpful. Morphological characteristics, geographic origins, and observations on genotype-specific amplified bands of AFLP will also help utilize their economic value and exploring the various genotypes for further classification (Ali et al. 2007).

1.3.3 Industrial, nutraceutical, and pharmaceutical value addition

For biotechnological applications, sesame is regarded as an important oilseed crop. A large number of sesame seeds contain bioactive compounds which are vital for nutrition and health. The development of varieties in sesame with high nutritional and functional value was possible by strategic genetic manipulation. The crop possesses the potential for export and subsequently may come to be one of the most significant due to having a high-value natural food advantage. The medical importance of sesame seed is attributed to the presence of microcapsules with bioactive substances with high variability composition. Phytochemicals that are health-promoting and biologically active and such as phytosterols, sesamolin, tocopherols, phytates, sesamin, PUFA, and other phenolics are present in the sesame seed. There were reported nutritional components that vary widely (tocopherols lignans and phytosterols) in the collections of Indian sesame germplasm which offer high prospects for breeding sesame. A review was made of the value addition of traditional nutraceutical, industrial and pharmacological products of sesame seeds (Table 1.6) with respect to high antioxidant value bioactive components. In improving the daily diet of the global usage of sesame seeds, important knowledge of the superior functional

Table 1.6. Some industrial, nutraceutical, and pharmaceutical potential of sesame products (Morris, 2002)

Product	Uses
	Industrial
Chlorosesamone	Antifungal
Sesamin, sesamolin	Bactericide, insecticide
Myristic acid	Cosmetics
	Nutraceutical
Lecithin	Antioxidant, hepatoprotective
Myristic acid	Cancer preventive
Fiber	Cancer preventive, cardioprotective
Sesamin, sesamolin	Fatty acid oxidation, antioxidant
Sesame oil	Prevent heart disease, Skin softener
	Pharmaceutical
Sesame oil	Drug vehicle and laxative
Flavonoids	Hypoglycaemic activity
Linoleate in triacylglycerol form	Linoleate in triglyceride form

components of sesame is significant. Despite a large stock of a sesame germplasm collection, minimal success was recorded due to limited efforts in research which expatiate on the utilization of traditional and biotechnological protocols; there was the report of developing cultivars that are superior nutritionally. Efforts in sesame value addition have resulted in the genotypes production with high antioxidant activity and subsequently would assist in free radical-related diseases prevention. Having a composition of bioactive components in sesame production with stabilized sesame oil and enhanced shelf life and better market value after modification will be possible (Pathak et al. 2014).

1.3.4   Biotechnological advancements

Sesame is an important multipurpose oil crop for future biotechnological applications; it contains high-quality plant oil, having antioxidant properties vital for human health and nutrition. There has been research on many possible strategies for the genetic manipulation of sesame for the purpose of developing new higher nutritional and/or functional value varieties. Some of these methods that exploit a sesame hairy root culture system include improving the genetic makeup of seed shattering and the production of transgenic sesame for use in molecular pharming. Establishing transgenic sesame lines expressing foreign genes to enhance the sesame seed oils composition is the current research effort. To generate industrially beneficial substances, it is expected that genetically improved sesame varieties and sesame hairy root technology may be available globally (Suh et al. 2010). A current molecular strategy in developing value-added sesame varieties was discussed in the literature; biotechnological targets within sesame seeds were the focus of these studies. Byproducts of the sesame seed, such as phytic acid, seed storage proteins and tocochromanols were presented as important biotechnological applications geared toward modulation. The recent trend in the transgenic sesame plant technology was also assessed including various recombinant DNA protocols for these two sesame seed oils' fatty acids (Suh et al. 2010).

1.3.5   Lipids composition

Among the primary fatty acids of sesame documented are linoleic, oleic, stearic, and palmitic acids. In fact, linoleic and oleic acids are components of 80% of these fatty acids. In comparison, the entire amount of sesame fatty acids, which has a very low quantity of linolenic acid, only contains 0.4-0.5% of linolenic acid. Sesame's nutritional value would be improved by increasing its linolenic acid content, making it an important biotechnological target (Suh et al. 2010). Among the 8 toco-isomers, sesame seeds are particularly high in tocopherol, therefore increasing the production of other toco-isomers in sesame that could both improve the nutritional use of the seed for humans and protect the sesame plant from oxidative stress (Suh et al. 2010).

1.3.6   Genetic improvements for storage proteins

Oleosins and storage proteins found in sesame seeds suggested that there is the possibility of removing or altering the allergenic epitopes using protein engineering, such that transgenic sesame seeds with reduced allergenicity can be generated, credit to transgenic plant technology (Suh et al. 2010). In developing sesame seeds using the gene myoinositol 1-phosphate synthase encoded, in order to create transgenic sesame seeds with decreased phytate expression, construction was built in an expression cassette of antisense, opening the path for its current use in feed crops like sesame (Suh et al. 2010).

1.3.7   Lignan content biosynthesis and modification

By the phenylpropanoid pathway, the biosynthesis involved [1–14C] conversion of tyrosine into coniferyl alcohol. Pinoresinol is created by the coupling of two coniferyl alcohol molecules, and it is subsequently transformed into "oxygen-inserted"' lignans (sesamolin or sesamin) by a piperitol molecule with a single methylenedioxy bridge (Suh et al. 2010). A study found ESTs for all but one of the enzymatic steps in the biosynthetic pathway that turns tyrosine into coniferyl alcohol (cinnamyl-alcohol dehydrogenase, caffeoyl-CoA O-methyltransferase, cinnamate-4-hydroxylase, caffeic acid O-methyltransferase, flavonoid-3-hydroxylase, coumaric acid hydroxylase, CoA ligase, and cinnamoyl-CoA reductase; tyrosine ammonia-lyase was the lone exception). A dirigent protein, which is involved in the creation of pinoresinol from coniferyl alcohol via bimolecular phenoxy radicals, was also found to be encoded by numerous ESTs, according to a report. Sesamin and sesamolin are produced from pinoresinol by a cytochrome P450-dependent enzyme that uses O_2/NADPH to create a methylenedioxy bridge and insert one oxygen atom. Despite all the progress in identifying the sesame biosynthesis lignan pathway genes, the majority of the enzymes in these pathways have not yet been described, making protein purification challenging. The potential creation of high-quality sesamin for use as a dietary supplement will aid in the clarification of the biosynthetic process involved in the manufacture of sesame lignan (Suh et al. 2010).

1.4   Moringa (*Moringa oleifera* Lamarck)

Moringa oleifera L. (Syn M. pterygosperma Gaertn) is a naturally cultivated variety of the genus *Moring of the* family *Moringaceae* (Mahmood et al. 2010). *Moringa oleifera* Lam. (*MO*) is approximately 5 to 10 m high. It is widely and globally cultivated due to the multiple utilities of its parts (Fig. 1.9) (Farooq et al. 2012). *Moringa oleifera* Lam (Moringaceae) thrives in many tropical and subtropical countries (Mehta et al. 2011). The fatty acid constituents are regarded similarly to that of the Olive plant. As a recommended seed oil in aromatherapy for massage or as carrier oil, and as a light plant oil that can be easily spread on the skin which makes it useful for skin and hair care formulations. Moringa oil brings moisturizing effects, softness and smoothness to skin and hair. It is the behenyl acid that gives the much-sought rich smoothness with no greasy after-feel (See section 4.7.) (International Flora Technologies 2008). Ojiako and Okeke (2013) described the antioxidant potential of moringa seed oil and its use in the creation of body cream. It has been claimed that the oil from the *moringa oleifera* Lam family can be used to make sunscreen and to test the sun protection factor in vitro (Kale and Megha 2011). Moreover, soap made with a native Moringa seed oil extract was described by Warra 2012. Recently, a viral vector-based screening of Moringa extracts showed inhibitory activity against the early steps in the infectivity of HIV-1 lentiviral particles (Nworu et al. 2013). The concentration of phenolic and non-phenolic chemicals in *M. oleifera* ripe seeds has been found to have high antioxidant activity, according to reported data (Adebayo et al. 2018). As an alternative, commercial livestock rations can be replaced with moringa. Its production and management are comparatively easy and encouraging due to the ease with which moringa can be sexually and asexually reproduced, as well as the plant's minimal need for water and nutrient-rich soil, especially in developing countries. Its potential as livestock fodder is largely dependent on high nutritional quality and biomass potential, especially in dry periods. It is attributed to growing as crop or tree fences in alley cropping and agroforestry systems, and even with not much value for agricultural development with high temperatures and low water availabilities. Aside from that, it does not deplete the changing patterns of land use or existing resources (Nouman et al. 2014). The prospects of the seed oil are also found in arts and for lubricating watches and other delicate machinery, and the potential in the production of perfumes and hair care products. Pressed seed cake may be used as a fertilizer. Industrially, its wood is useful in the textile and paper industries, and in the tanning industry, the bark is important (Mulugeta and Fekad 2014). It showed anti-dyslipidemic and antidiabetic activities with powdered leaf preparations verified by leaf powders and extracts in animals. The hydro alcohol and water extracts of *M. oleifera* leaves have a variety of biological effects, including antioxidant, antihypertensive, tissue-protective (heart, lungs, kidneys, and testes) and immunomodulatory effects, as well as antiulcer, analgesic, and immunomodulatory properties. The effects are caused by phenolic acids, polyphenols, flavonoids, and alkaloids as demonstrated by several research (Bashir et al. 2016). Isolation and characterization studies of *Moringa oleifera* root starch as a potential pharmaceutical and industrial biomaterial was reported (Adebisi et al. 2013). The species of moringa collected from 14 states of Nigeria were reported to have high ethnomedicinal value, including addressing close to 20 conditions, including ear infections (71.8%), malaria fever and typhoid (78.7%), blood pressure (64.7%), the lowering of blood sugar (diabetes mellitus) (65.2%), and eye infections (66.9%)(Stevens et al. 2015). According to certain research, the tree's components may help lower cholesterol and blood sugar levels. These appealing qualities have inspired researchers to hunt for other creative uses for the moringa tree, particularly as a source of anticancer medications after extracts from various moringa tree sections were tested on several cancer types both in vitro and in vivo with varying degrees of effectiveness (Khor et al. 2018).

1.4.1 Biotechnologies for agricultural and environmental applications

The high degree of polymorphism reported suggests that RAPD is very useful for genetic diversity studies in *M. oleifera*; using RAPD markers a study showed significant variation. Existing moringa accessions can be beneficial in the establishment of moringa breeding programmes to produce superior varieties (Ojuederie et al. 2012, Opare-Obuobi 2012). Environmental biotechnology using a laboratory hands-on protocol by a physical-chemical process (coagulation-flocculation-sedimentation) and micro-filtration method has made it possible to evaluate the effect of extracts of moringa seed as a natural organic polymer for getting rid of heavy metals from Landfill Leachate. The generation system has also made it possible to develop and transform Moringa using the strain that harbours the plasmid-containing enzyme genes driven by a promoter. By skipping the pre-culture step, it is also possible to make a transformation by use of the vacuum infiltration-assisted agrobacterium infection method. Enzyme assays, PCR (polymerase chain reaction), and blotting (especially Southern blot) can also be used to validate genomic integration and transgenic expression (Muyibi et al. 2003). Moringa transformation caused by agrobacterium tumefaciens-mediated has also been documented (Zhang et al. 2017).

1.4.2 Micropropagation

Nature and origin of the explant, likewise the type of cytokinin are among the dependent factors for the micropropagation of moringa to be successful. Sourced explants from in vitro-grown plant materials are preferable compared to the seedlings that are soil grown. In terms of the number of shoots/explants, frequency of shoot production, and the number of nodes/shoot, nodal segments outperformed shoot cuts (Hassanein et al. 2019). For appreciable moringa micropropagation, it was suggested that in vitro-generated nodal segments be successfully acclimated on full-strength MS media containing 0.56 mg l^{-1} of BAP, without significant verification or root growth being significantly slowed down (Hassanein et al. 2019).

1.4.3 Antioxidant potential

The new innovations in products derived from food and the increase in the selling value of processed moringa leaves in the form of antioxidant instant drinks of high nutritional value, as demanded by consumers, were provided as a piece of valued information. This is attributed to the presence of many active substances present in the leaves of moringa, quercetin considered one of them. Free radicals can be neutralized by quercetin, an antioxidant compound and prevent free radical damage to normal cells, fats and proteins. These antioxidants are known to protect cells from the harmful effects of reactive oxygen free radicals (Dwi et al. 2018). Both aqueous and ethanolic extracts of moringa leaves exhibit antioxidant activity, which could be important for solving health-related problems and in chronic disease prevention resulting from oxidative stress (Verghese. et al. 2018). *M. oleifera* leaves possess antioxidant potential with the highest free radical scavenging activity of the methanol extract (Fitriana et al. 2016).

1.4.4 Anticancer

The significant impact on the utilization and efficacy of *Moringa Oleifera* as an anticancer treatment was accounted for (Christianto and Smarandache 2019). Water concentrates of the leaf of Moringa showed anticancer action against different human cancer cell lines, including non-small cell lung cancer, extracts of the leaf of the moringa was analyzed for anticancer activity in hepatocellular carcinoma HepG2 cells in human which supported its potential as orally administered therapeutics for treating cancers of the human liver and lung (Jung et al. 2015). The aqueous extract of *M. oleifera* exhibited cytotoxic effects on Hela cells and the least cytotoxicity on lymphocytes (Nair and Varalakshmi 2011) *Moringa oleifera* and *Indigofera arrecta* leaf extract reaction with 5-fluorouracil against selected cancer cell lines showed drug interaction potentials. According to research using *Moringa oleifera* leaves which have been extracted with methanol and dichloromethane, the plant may be useful as a source of alternative innovative anticancer medications (Ndung'u et al. 2018, Charoensin 2014). The potential of *Moringa oleifera* in natural medicine, synthesis of phytonano particles and as an antiproliferative agent against cancer was reviewed (Tiloke et al. 2018). Results from an in-vitro MTT assay using a chemical isolated from the ethyl acetate fraction of Moringa oleifera flowers against the human liver cancer HePG2 cell line showed that the drug had a high anticancer potential (Rajeshkanna et al. 2017). After research on the effect of extracting Moringa oleifera leaves on ovarian, prostate, and breast cancer in human cancer cell lines, it was expected that Moringa oleifera leaf extract will prevent the growth of cancer cells (Zayas-Viera et al. 2016). *Moringa oleifera* has also been discovered to have anti-cancer properties against colon and rectal cancer cell lines, suggesting that its leaves may one day be employed as a prophylactic measure as well as a treatment for breast cancer (Poobalan et al. 2018, Al-Asmari et al. 2015). Induction of apoptosis in cancer cells by the *Moringa oleifera* tree was reported (Adebayo et al. 2017). In the presence of glucosinolate, which has been shown to have the ability to induce apoptosis in anticancer trials, M. oleifera showed chemo-preventive effects. Studies using leaf extract to stop the growth of human cancer cell lines have demonstrated the efficacy of this substance in chemoprevention. Highlights of *M. oleifera's* benefits for chemoprevention, where glucosinolates may help to halt the carcinogenesis process. *M. oleifera* with other drugs and safety have shown synergistic effects, which are essential for chemo-prevention as well as safe for consumption by the human body and is also effective. The recommendation was made for the need for extensive research as a result of the anticipated rise of cancer in coming years and to gain more information about the mechanisms involved in *M. oleifera* influence, which despite its promising evidence in chemoprevention could serve as a potential source in inhibiting several major mechanisms involved in cancer development (Abd Karim et al. 2016).

1.4.5   Food ingredients and functional foods development

It has been demonstrated by many researchers that Moringa is beneficial as a functional food product ingredient. This is attributed to its richness in macro and micronutrients like carbohydrates, protein, potassium, calcium, phosphorus, iron, vitamins (vitamins A, B, C, D, E and K.), minerals (K, Mg, Ca, Mn, Zn, Cu, and Fe), beta carotene, and other compounds that are bioactive and vital for proper body function and certain diseases prevention (Saa et al. 2019, Singh and Singh, 2019, Singh et al. 2018, Sahay et al. 2017, Alli rani and Arumugam 2017). The purpose of the moringa leaf powder is to add it to guava juice to make guava jelly. According to reports, adding moringa leaf powder to jelly boosts both the shelf life of the jelly and its nutritional value (Rizvi et al. 2018).

1.4.6   Cosmetic uses

In ancient Egypt, the oil from *Moringa oleifera* seed was utilized as a strong remedy for many skin problems. Moringa oil contains essential fatty acids, making the oil a healing humectant and emollient for rough, dry skin. Due to its stability, perfume makers admire the oil owing to its use in getting rid of unwanted odors. The fatty acid component is found to be similar to that of olives. The oil is considered a good carrier oil for massage in aromatherapy because it is light and can spread easily on the skin. The antioxidant activity of Moringa seed oil and its utilization in the preparation of a body cream was reported by Ojiako and Okeke (2013). *Moringa oleifera* Lam. (family-moringaceae) oil sunscreen cream formulation and evaluation for sun protection factor was reported in vitro (Kale and Megha 2011). Soap preparation from the oil of a native Moringa seed was also reported (Warra 2012).

1.4.7   Behenic acid

Behenic acid (Fig. 1.10) is a saturated fatty acid, a carboxylic with a chemical formula $C_{21}H_{43}COOH$. It is composed of white to cream color or powder crystals in appearance; it has a melting point of $80°C$ and a boiling point of $306°C$. It is soluble in both ethanol and ether. Behenic acid (22 carbon atoms) is the main component of Ben oil, also known as behen oil or Moringa oil (first reported by Voelcker A. in 1848). Oil extract from the Moringa *seeds* has potential in cosmetics. The seeds of this tree were harvested from the Persian Mountain Bahman where the name "behenic" was derived. Behenic acid is mostly used commercially to provide emollient properties to hair conditioners and moisturizers. It is also useful as lubricating oil; in paint removers, it is utilized as a solvent for retarding evaporation. In detergents, dripless candles and floor polishes the amide of behenic acid is utilized as an anti-foaming agent (Akoh and Min 2008, Chow 2008).

1.4.8 Water treatment

Polyelectrolyte has been identified as one of the active ingredients in the seed of Moringa. It is used globally as a coagulant. Due to its very low reduction in alkalinity; *M. oleifera* is used in softening water, providing a buffering capacity to attain the needed treatment objectives (Bichi 2013). The seeds of moringa are utilized in the water treatment method as an accrued suspension of water-soluble extract, bringing about an effective natural clarification agent for highly turbid and untreated pathogenic surface water (Lea 2014). Combination of alum with *Moringa oleifera* is found to be a potential coagulant which serves for the initial treatment in waters with humic materials and high iron content as shown by the jar test results. Unwanted pre chlorination by-products formation can also be eliminated (Kalibbala et al. 2009).

1.4.9 Production of bioenergy and biomass

Production of biodiesel and briquettes was reported as an energetic biomass application of *Moringa oleifera* (Pereira et al. 2018). The biomass production potentiality of the drumstick (*Moringa oleifera*) genotypes was reported (Savitha et al. 2014). Seeds of moringa were reported to be a reliable feedstock to produce bioethanol as it contains lignocellulosic material via simultaneous saccharification and fermentation (SSF) process by using *Saccharomyces cerevisiae* having the potential of producing fuel that is biodegradable while causing no pollution (Ali and Kemat 2016).

1.5 Sugarcane (Saccharum spp., Poaceae)

Sugarcane (Saccharum spp., Poaceae) is a major perennial crop in the tropics and subtropics worldwide. About 70% of the total global sucrose production is reliant on sugarcane. Vegetatively propagated from axillary buds on stem parts for commercial production; however, substituted micropropagation by tissue culture is a faster method than vegetative propagation by budding and is used in some nations to produce new varieties. Moreover, sexual reproduction is used to create new types of sugarcane. Complex polycrosses (melting pot crossing) are frequently used to increase the number of progeny populations in order to increase the likelihood of generating superior clones (Zhou et al. 2014). Sugarcane crops are used efficiently to produce (1G) ethanol. Also, the bagasse from sugarcane can be used to produce (2G) ethanol (Furlan et al. 2013). Genetic engineering of sugarcane for increased sucrose and consumer acceptance has been reported (Conradie 2011).

1.5.1 Genomic resources and breeding for energy utilization

Sugarcane as an energy crop is regarded among cultivated crops as one of the major crops required for presenting cultivated plants with the highest tonnage. Due to its high sugar productivity, it is considered a green alternative to petroleum, bioethanol and bioelectricity. Also, a large range of goods made from sugarcane biomass is linked to the motivation behind breeding initiatives to develop varieties with larger fiber yields and more robust and sustainable performances. Researchers thoroughly discussed the energy cane's description, types, breeding efforts, end uses, current understanding of cell wall metabolism, and bioinformatics tools and databases that were made available to the community in a recent publication on genomic resources for the development of this crop (Diniz et al. 2019). A review of the potential, problems, and genetic and genomic resources for sugarcane biomass improvement was provided to sugarcane breeders, geneticists, and larger scientific groups involved in bioenergy production (Kandel et al. 2018).

1.5.2 Industrial bio-products

Industrial biotechnology advancement has offered important opportunities for the economic utilization of residues of agro-industries such as sugarcane bagasse (Pandey et al. 2000). Sugarcane bagasse (SB) and leaves (SL) were found useful in industrial application in developing countries as cheaper sources of carbohydrates, exploration was made on their potential in products of commercial value including commercial assessment, and their use in sustainable bio-based fuel systems advancement (Singh et al. 2019). In the furniture industry, sugarcane bagasse panels were found to have great potential for use. The sugarcane bagasse panels produced exhibits comparable or superior physically and had mechanical properties to those made from Eucalyptus and Pinus (Oliveira et al. 2016). Employing the use of the cellulolysis process, sugarcane bagasse and corn stover are some potential lignocellulosic substances that are utilized as feedstocks for bioethanol production (Zakir et al. 2016). The current trend in commercialization and new business start-ups in the biorefinery value chain reported different protocols for the hemicelluloses production of sugar through pretreatment and enzymatic hydrolysis and recent developments in improvement of strain methods for use of hemicellulosic sugar, and fermentation process protocols for ethanol and value-added products (Chandel et al. 2018). Various bioproducts are obtainable from sugarcane bagasse (SCB) under the biorefinery concept (Fig. 1.11). It is now one of the finest possibilities for manufacturing biofuels due to its distinctive ability for biomass production (high carbohydrate sugar + fiber composition) and a favorable energy input/output ratio (Hoang et al. 2015). Application by-products of sugar industries, such as bagasse and press mud to soil improves the soil's physical, chemical, and biological properties and enhances the quality of crop and yield. There is an extremely large likelihood of by-products of sugarcane industries being utilized in agriculture to cut down the chemical fertilizer requirements (Dotaniya et al. 2016).

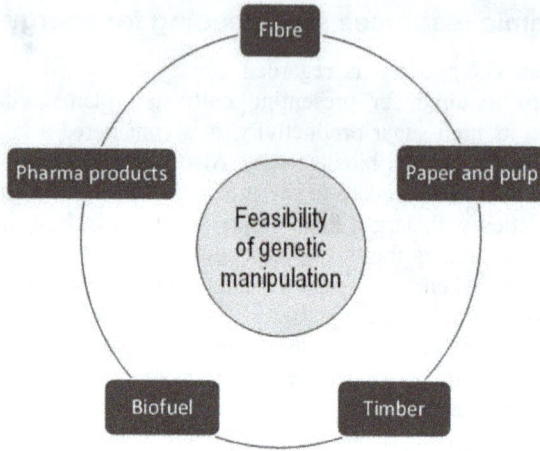

Figure 1.1. Various areas of genetic maniplulation for industrial application (Devi et al. 2019).

Figure 1.2. (a) Castor bean (*Ricinus communis* L.) plant. (b) Wild castor (*Ricinus communis* L.) plant. (c) Flowers of wild *Ricinus communis* L. variety. (d) Flowers of *Ricinus communis* L. bean. (e) *Ricinus communis* L. bean fruits. (f) Wild *Ricinus communis* L. fruits. (g) Dried unshelled wild *Ricinus communis* L. seeds. (h) Dried unshelled *Ricinus communis* L. bean seeds. (i) Fresh and dried unshelled wild *Ricinus communis* L. seeds (j) Wild *Ricinus communis* L. seeds. (k) *Ricinus communis* L. beans. (l) Castor seed powder.

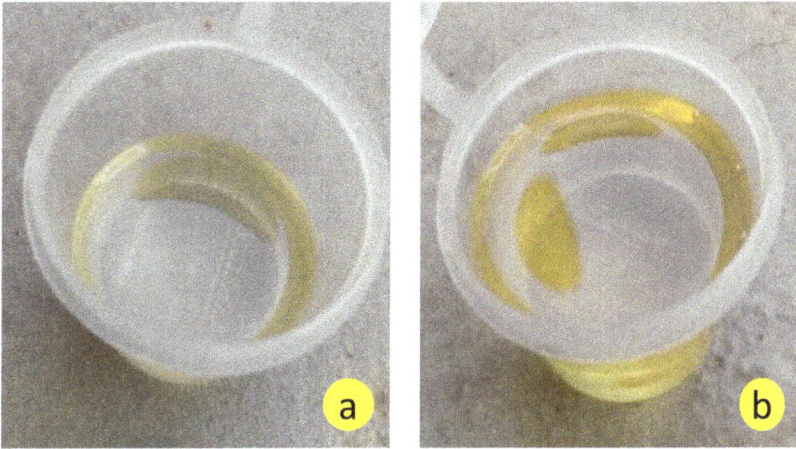

Figure 1.3. (a) Wild *Ricinus communis* L. seed oil. (b) *Ricinus communis* L. bean oil.

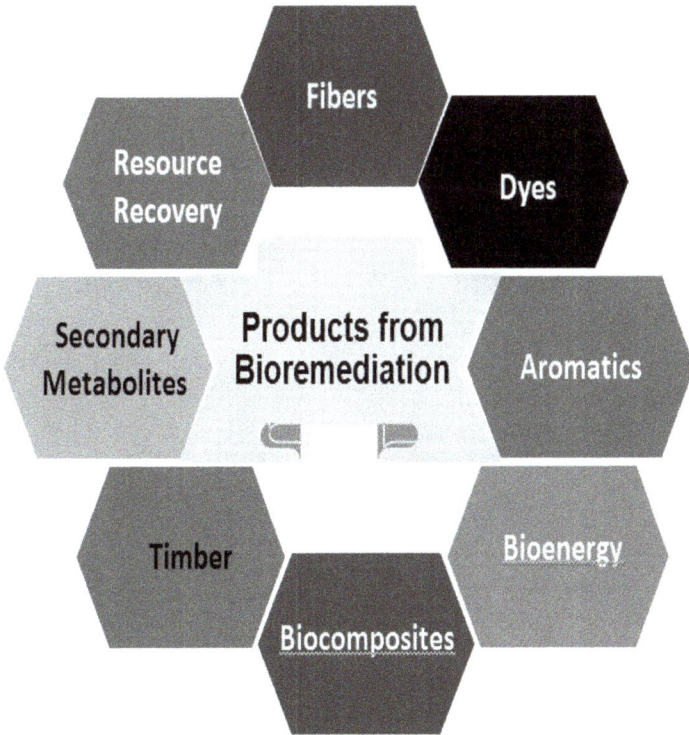

Figure 1.4. Possible products from bioresources used in bioremediation.

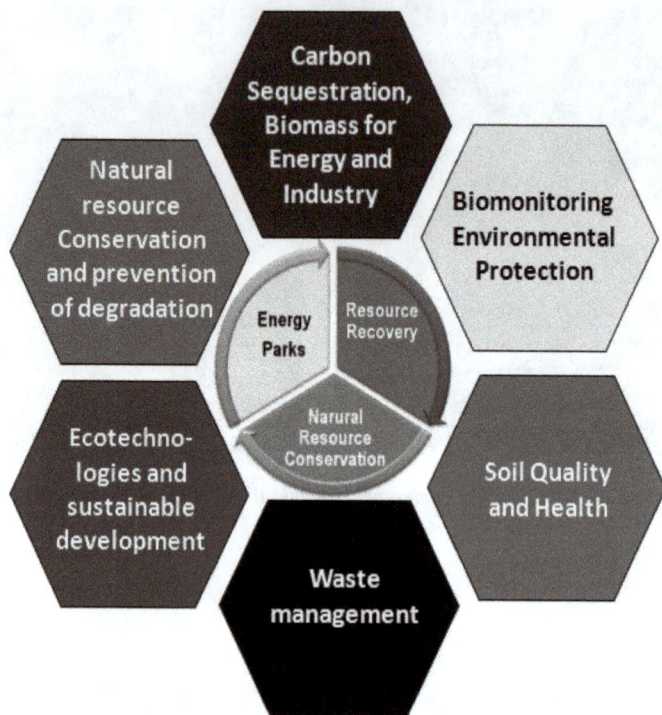

Figure 1.5. Biomass resources.

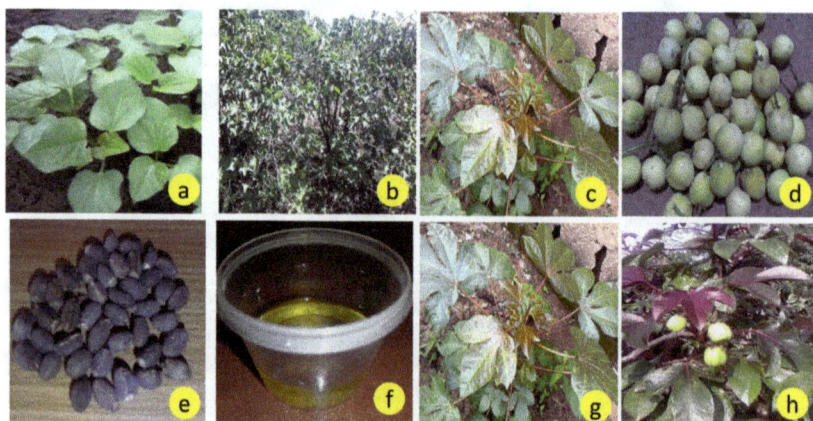

Figure 1.6. (a) *Jatropha curcas* L. plant regeneration. (b) *Jatropha curcas* breeding for plant oil and biomass. (c) *Jatropha curcas* L. riped fruits. (d) *Jatropha curcas* L. fresh seeds. (e) *Jatropha curcas* L. dried seeds. (f) *Jatropha curcas* L. seed oil. (g) *Jatropha gossypifolia* regeneration. (h) *Jatropha gossypifolia*, genetic resources breeding for bioproducts.

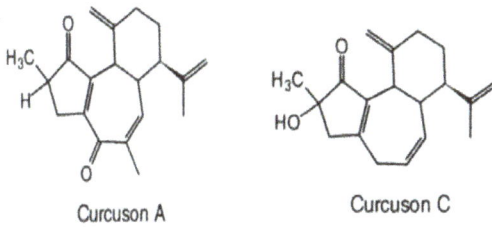

Figure 1.7. Two forms of diterpenoids in *Jatropha curcas* L. The diterpenoid Curcuson A and the diterpenoid Curcuson C.

Figure 1.8. (a) *Sesamum indicum* L. plant. (b) *Sesamum indicum* L. flower. (c) Matured *Sesamum indicum* L. crop. (d) *Sesamum indicum* L. plantation. (e) White *Sesamum indicum* L. seeds. (f) Brown *Sesamum indicum* L. seeds. (g) Brown *Sesamum indicum* L. seed oil. (h) White *Sesamum indicum* L. seed oil.

Figure 1.9. (a) *Moringa oleifera* plant regeneration through nusery. (b) Growing *Moringa oleifera* plant. (c) Grown moringa plant. (d) *Moringa oleifera* flower. (e) Fresh *Moringa oleifera* pods. (f) Dried *Moringa oleifera* pods. (g) *Moringa oleifera* seeds. (h) *Moringa oleifera* seed oil.

Figure 1.10. Chemical structure of Behenic acid.

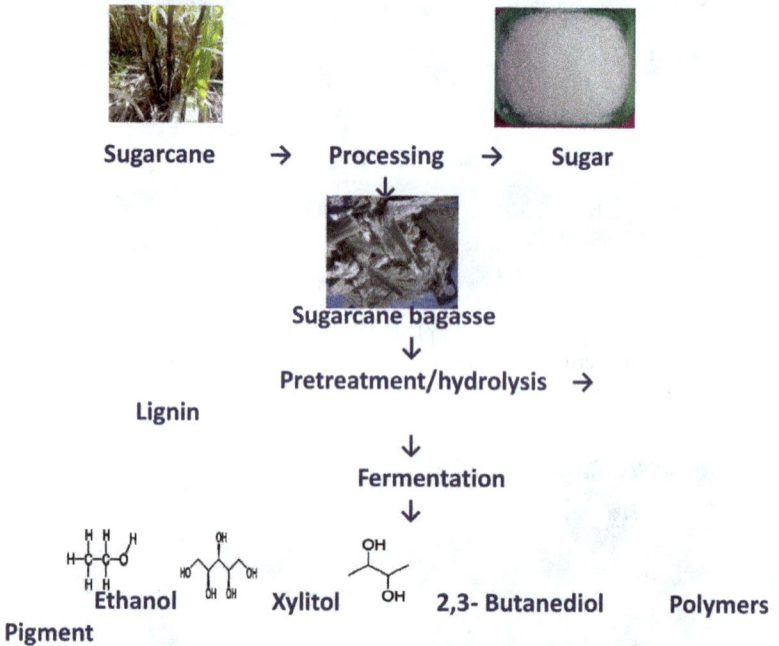

Figure 1.11. Conversion of sugarcane bagasse (SCB) into ethanol and value-added products by a biotechnological process (Hoang et al. 2015).

Fiber Crops

Fiber crops are crops that have a high concentration of cellulosic material from the parts of plants which are mostly bast tissue and largely used for the industrial production of fiber. Various types of fibers were identified in plants, the most mentioned of these are wood fibers, bast fibers, leaf fibers, fruit fibers, and seed fibers. Fiber crops are sources of raw materials in the textiles industry, parts for automotive, composites for building, the paper industry, packaging materials, and weaving or thatching.

2.1 Cotton (*Gossypium hirsutum* L.)

Cotton, a fiber plant, globally considered one of the most important agricultural cash crops, is a major feedstock in the production of textiles and receives a lot of attention in research and development (Noman et al. 2016). According to reports, there are roughly 33 million acres of cultivated land spread throughout more than 80 nations in the tropics and subtropics, with an annual production of 19 to 20 million tons of bales. The principal cotton-producing nations include India, China, Argentina, the United States of America, Australia, Greece, Pakistan, Uzbekistan, Brazil, and Egypt. These nations produce around 85% of the cotton used in the world. Cotton was a key component of more than 25 different industries as a raw material. Bt cotton, a short version of the common soil bacterium Bacillus thuringiensis, is the subject of research. During the stationary phase of its growth cycle, this spore-forming gram-positive bacterium forms parasporal crystals. The highly toxic 'endotoxins' are synthesized crystalline proteins harmful to some insects. While acting on the epithelium tissues of the midgut, they kill insects like caterpillars (Mayee et al. 2018).

2.1.1 Development of transgenic cotton

Moreover, transgenic fiber crops may be helpful in the phytoremediation of heavy metals. According to Ludvková and Griga (2019), cotton (*Gossypium hirsutum* L.), a crop used for making fiber, is a good chelator of heavy metals like cadmium. The introduction of Bt cotton to control bollworms in cotton has made it possible to provide growers with modern equipment. This is the emerging technology's additional purpose for the industry of cotton growers worldwide at many economic, societal, and environmental levels. Benefits are directly included, such as a reduction in the use of pesticides, improving the effectiveness of crop management, production costs reduction, improvement in the effectiveness in crop management, improvement in yield, and profitability; in the areas of severe pest infestation it brings about farming risk reduction in risk and opportunity to improve growth cotton farming (Mayee et al. 2018). Even though there are several risks associated with the sustainability of cotton production due to challenging pest problems (Fig. 2.1a–d), biotechnological developments like insect-resistant cotton (*Bt Cotton*), using desired genes e.g., chimeric cry2AX1 Gene overcome these problems. Bt cotton is developed from bacteria (Agrobacterium) which has natural resistance against some insects and weeds increasing in yield from conventional or desi cotton. Embryogenic calli of cotton can be cultivated with the Agrobacterium to achieve the desired result (Dhivya et al. 2016, Jadhav et al. 2015, Rashid et al. 2016). Molecular analysis using the particle bombardment method showed the presence of the crylAb gene in the genome Bt cotton indicating virus—a resistant genotype of *Gossypium hirsutum* L. (Majeed et al. 2000). Sometimes an insecticidal gene with targeted expression is used to develop insect-resistant cotton lines (Bakhsh et al. 2016). Early identification of cotton tolerance to salt stress using random amplified polymorphic DNA (RAPD) analysis also provides a resource to recombinant technology while on the other hand, random amplified polymorphic DNA-polymerase chain reaction (RAPD-PCR) allow exploration of genetic diversity and identification of some cotton genotypes. Marker-aided selection (MAS) is a functional genomic way that can provide drought resistance to cotton, paving the way for increased breeding. Translocation lines are promisingly useful in the identification of some monosomic lines, providing details of morphological characteristics of cotton monosomic stocks geared towards the identification of specific chromosomes for collection using *a priori* chromosome-associated DNA markers. DNA markers accurately enable possible differentiation of cotton hybrids from their parental lines for determining hybrid seed purity (Ghatage and Das 2016, Dongre et al. 2004, Selvam et al. 2009). In response to salt stress, small RNAs identification and elucidation of functional significance widen our understanding of post-transcriptional regulation of genes; against this background, analysis of cotton small RNAs and their target genes, by comparison, was reported (Yin et al. 2017). Nowadays, researchers focus on providing background for future researchers to induce the regeneration of plantlets from the callus of cotton plants so that new genotypes can be obtained (Memon et al. 2010; Abdellatef et al. 2008). Asexual methods of generation of cotton (*Gossypium hirsutum* L.) were reviewed by Ahsan et al. (2014). Genetic diversity analysis of high-yielding genetically modified cotton genotypes suggested utilization for future breeding programs (Baloch et al., 2015) (Fig. 2.2). New biotechnologies that emerge to improve yields of cotton, value-added products and services that feed into the whole value chain of agro-processing, such as bacterium Bacillus thuringiensis (Bt) mediated transgenic cotton, developed to create a chemical that can naturally harm only a small fragment of insects like bollworm, when adopted could increase yields and enhance competitiveness (Abro et al. 2004). This is in line with what a policy advisor Getachew Belay, also a senior biotechnology, told Zimpapers Syndication as reported by The Herald on 4th March 2018. Recombinant DNA technology is regarded as a crucial technique in cotton breeding that empowers conventional approaches to increase net yield and related parameters. Biotechnological advancements over the past few decades have contributed significantly to bettering Bt cotton breeding. Analysis of the performance over the last two decades provides a good picture of the output and current state of Bt cotton in relation to expectations (Norman et al. 2016).

2.1.2 Value additions – Cotton fiber improvement

For improvement of the growth and germination of cotton seedlings by the evaluation of treatment of biofield energy and its impact, improved cotton fiber can be produced using simple sequence repeat (SSR) markers for analysis of polymorphism. From primitive times, fiber crops have been associated with the improvement of human lives; food, clothing, and housing are some of the potential uses. There are thousands of different species of fibrous plants encompassing cotton. Cotton is one of the most popular transgenic crops for commercial purposes (Trivedi et al. 2015).

2.1.3 Cotton textile supply chain and environmental impacts

The cotton/textile supply chain has enabled employment opportunities for millions of the populace in industries allied with the supply of inputs of agricultural equipment and machinery production, crushing of cottonseed, and the manufacturing of textiles, affecting global social impacts and political debates. A systemic supply chain structure has contributed to the understanding of the impacts of the environment on the cotton textile chain. The broad range of environmental impacts assessed include the cotton planting process, processing, and production of cotton textiles, cotton textile products consumption, and textile waste disposal (Pan et al. 2008).

2.1.4 Biofuel

An investigation has found anaerobic treatability and potential of methane generation from stalks of cotton, seed hull, and oil cake of cotton wastes. A promising amount of methane was generated which can be utilized as a source of renewable energy (Ischia and Demirer 2007). The biodiesel was confirmed as a source of renewable energy by High-Performance Liquid Chromatographic (HPLC) studies, which showed that cottonseed oil's triglycerides (triacylglycerols) were almost entirely transformed into fatty acid methyl ester (FAME) (Fan et al. 2011). For the development of cotton seed oil methyl ester, research into the transesterification process has been published. Since the biodiesel produced under ideal conditions had similar qualities to mineral diesel, cotton seed oil methyl ester might be suggested as a mineral diesel fuel substitute for compression ignition (CI) engines in both the transportation and agricultural sectors (Singh et al. 2015). Cotton stalks and seeds have the potential to be utilized on an industrial scale production of biodiesel (Uyan et al. 2020).

2.1.5 Industrial cosmetic products from cottonseed oil

The gross value of the total cotton product contains about 12% of the cotton seed oil as a by-product. The plant oil, which previously competed fiercely with the more extensively used animal fats, is now in ninth place in the production tables, behind three land animal fats (tallow, lard, and butter) and five vegetable oils (soybean, palm, rape/canola, sunflower, and groundnut). Compared to other commodity vegetable oils, it has a comparatively high concentration of palmitic acid (usually 23%), oleic acid (17%), and linoleic acid (56%). Identified as β′ promoter when it is in the sn-1 or sn-3 positions, palmitic, the major saturated fatty acid in cottonseed, is stable in the β′ crystal form, this has made it suitable for many products due to its stability which promotes a smooth, workable consistency, usually referred to as plasticity. Cottonseed oil is thought to have a non-random distribution of fatty acids, with saturated fatty acids predominantly located in the sn-1 or sn-3 locations and unsaturated fatty acids in the sn-2 position. Most triacylglycerols have a combination of linoleic, oleic, and palmitic fatty acids because they make up over 90% of cottonseed oil's fatty acid composition. Almost 30% of the triacylglycerols contain only unsaturated fatty acids with no molecules completely saturated (O'Brien et al. 2005).

2.2 Cannabis (*Cannabis sativa* L.)

Cannabis sativa L. belongs to the family Cannabaceae. Whether the cannabis genus contains one or many species has been a subject of discussion for a very long time. The particular class of terpenophenolic that the plant produces is called cannabinoids. 120 of the 565 cannabis components that have been identified from Cannabis sativa are phytocannabinoids. The plant's possible therapeutic properties date back to ancient times and has acquired prominence recently due to its current use as a source for modern pharmaceuticals to treat a number of human disorders; not just because of specific rules associated with its use as a drug (Chandra et al. 2017). According to information provided by member states on the sources of cannabis resin confiscated, it has recently been revealed that Europe, North Africa, and the Near and Middle East remain the principal markets for cannabis resin, which is still produced mostly in Morocco and Afghanistan (World Drug Report 2016). Farmers in Africa have cherished the cannabis plant for millennia (Fig. 2.3). Morocco has been regarded as the world's top producer of cannabis resin (hashish) for more than 50 years, for a variety of reasons (Blickman 2017). *Cannabis sativa* is a variation of the same plant species as industrial hemp and marijuana. Hemp, on the other hand, differs genetically and can be identified by its application and chemical composition. The production of industrial and other goods is associated with non-drug use. The world market for cannabis was estimated to have consisted of over 25,000 products from hemp. It can be grown as a crop for fiber, seed, or other dual-purpose. The potential of its fibers is justified by the large range of derived goods, which include fabrics and textiles, yarns and raw or processed spun fibers, paper, carpeting, home furnishings, building and insulation materials, auto parts, and composites. The internal stalk (hurd) has a variety of uses, including animal bedding, raw materials, subpar papers, and composites. The food and beverages industry utilizes seed and oil cake; it can also be used as an alternative source of food protein. In a wide range of body care products, hemp seed oil serves as an ingredient and is also found useful in nutritional supplements as well as industrial, cosmetics and personal care, and pharmaceutical oil (Johnson 2013).

2.2.1   Fiber production

Fiber from hemp is an ancient bast fiber possessing good quality, now grown in many places for a variety of purposes. Industrial hemp fiber has potential in the production of particle boards, textiles, paper, and building materials. Currently, it found a viable commercial application in composite car parts (Ngobeni et al. 2016). Monoecious cultivars, dioecious cultivars, as well as fast-retting phenotypes bearing distinctive morphological markers in low tetrahydrocannabinol (THC) plants, are being bred for the production of high percentages of primary fiber components. THC, CBD, and other cannabinoid-free plants are all included in the selective cross-breeding for cannabinoids. There aren't many cultivars that have been specifically bred for seed output (Grassi and McPartland 2017).

2.2.2   Bioenergy Production

Biomass from the *Cannabis sativa* L. plant has the potential to provide bioenergy (Kołodziej et al. 2023). Seed oil from industrial hemp, an industrial and nutritional crop, a variant of (*Cannabis sativa* Linn), can be used to produce biodiesel via a transesterification process. Hence, oil from hemp seeds provides a good alternative feedstock for biodiesel production. Besides kinetic viscosity and oxidation stability parameters, superior fuel quality is experienced in hemp seed oil biodiesel, in which chemical additives inclusion will yield an improvement. As a prohibitively expensive plant, hemp remains a "niche" crop in the food supply chain, rendering it a primary feedstock in biodiesel production. The perception and legal aspect in the way of wide-scale hemp biodiesel production remain the major challenges (Alcheikh 2015).

2.2.3 Genetic diversity and Conservation of transgenic cannabis (hemp)

Historical crop selection produced the genetic diversity that hemp bioprospectors appreciate for commercial purposes. The African colonial and post-colonial administrations excluded cannabis from agricultural development programs, devalued the crop, implemented cannabis controls earlier than most other countries, and excluded cannabis as an extension of drug-control policies by public agricultural institutions. Private enterprises continue to be in charge of maintaining the genetic variety of the psychoactive cannabis crop while ignoring the inherent intellectual property rights in local landraces. Initiatives to decriminalize cannabis should prompt analysis of its functions in African agriculture to preserve its genetic diversity globally (Duvall 2016). The strategies offered for plant selection are becoming more sophisticated as a result of the recent development in the area of molecular genetics. With the help of recombinant DNA technology, a specific gene encoding for a desired trait is taken from an organism's DNA, copied, and put into the genetic code of a plant. When hemp is grown for fiber, the key selection criteria are high yield, fiber content, and fiber quality. Moreover, research has been done to generate a hemp variety having this property in order to boost oil output. Recently, hemp has been chosen with a focus on these two needs. Moreover, specific selection programs for "drug variants of hemp" have been carried out in order to create hashish or marijuana in addition to medical compounds. Creating decorative hemp is another instance (Berenji et al. 2013).

Other uses of cannabis

Pharmaceutical and medical uses of the plant are exploited besides its known application in making ropes, textiles, shoes, etc., by extraction of Δ9-tetrahydrocannabinol and cannabinoids. The hemp plant's leaves and fluorescence, which only have up to 0.3% of these chemicals, are used for the extraction. For creating a thick cream for use in cosmetics, cannabidiol (CBD) is mixed with oil (Dölle and Kurzmann 2019).

2.3 Bamboo (*Adansonia digitata* L., Malvaceae)

Bamboo species have great industrial and environmental applications due to their important impact on ecology and economy and their vital role in establishing sustainable development. In exploring desirable traits for breeding purposes, the determination of genetic differences and morphology among bamboo genera and species is very important. Some emerging technologies that have been found useful in the advancement of the taxonomy of bamboo are molecular fingerprinting and sequencing protocols (Konzen et al. 2017, Gupta and Ranjan 2016).

2.3.1 Genetic breeding characterization

For genetic diversity and characterization, a road map for creating a biotechnology database for common bamboo species based on their molecular traits is crucial. The documentation of genetic diversity assessments among the known bamboo species based on DNA fingerprinting profiles, either alone or in conjunction with physical features, was published by several researchers (Yeasmin et al. 2015). The Random Amplified Polymorphic DNAs (RAPD) method has been helpful for breeding programmes in assessing genetic diversity, identifying species, and genetic divergence taxa bamboo (Gami et al. 2015, Nayak et al. 2003).

2.3.2 Uses of bamboo by-products

Feedstocks for the development of bio-briquette fuel such as bamboo fiber and sugarcane skin as waste biomass materials are found suitable and the bio-briquette produced is useful as high-quality fuels (Brunerová et al. 2018). Bamboo is useful in the production of Biofuels such as bamboo charcoal, bamboo pellet biofuel, ethanol, and methane also useful in construction. In the construction industry, culms of bamboo and bamboo byproducts are mostly applicable in load-bearing and non-load-bearing structures, It is also good for walls and partitioning in housing, useful as a flooring material, roofing, furniture, handicrafts, plastic paper, textiles, and bio-based composites (Akinlabi et al. 2017, Viel 2013). Shoots of bamboo are regarded as potential health foods owing to their richness in proteins, carbohydrates, vitamins, fibres, minerals and very low fat contents (Nongdam and Tikendra 2014). According to some research, bamboo shoots were recommended as a production requirement for the categorized standard table of the nutrient composition of baobab products that is helpful in formulating balanced diets, focusing on biological characteristics in which the baobab trees grow (phenotypic/genetic, age), and environmental factors (soil, climate, care, the season of harvest, etc.) (Habte et al. 2019).

2.4 Kenaf (*Hibiscus cannabinus*)

Kenaf is cultivated worldwide for its fiber; also attracting increasing attention recently is the medicinal properties of this plant (Zhao et al. 2014). In the high growth rate production of xylitol, kenaf could be used as a sustainable feedstock. According to the reported research, using kenaf as the fermentation feedstock may help the growth of sustainable xylitol production (Shah et al. 2018). Kenaf has evolved from its known use as a crop plant for making cordage (rope, twine, and sackcloth) to its many new commercial uses, which include absorbents, building materials, paper products, and feed for livestock. Efforts range from the use of fundamental agricultural production techniques to the promotion of kenaf products (Webber et al. 2002). Since kenaf plants have so many different uses and occupy so many different positions in society, it is important to take into account the several factors that affect the composition of the plant as well as its constituent parts. Novel kenaf products have developed as a result of the efforts of numerous local and advanced companies, reinforced by ongoing agrarian research and studies to provide a new variety. With the continued growth of kenaf as a financially successful industrial crop, these can serve as encouragement in expanding its bright future (Saba et al. 2015a).

2.4.1   Biotechnological approaches for improvement

Utilizing the information obtained for gene segments of DNA (putative) involved in cellulose and lignin-biosynthesis in kenaf was considered relevant biotechnological approaches useful in modifying the desirable traits in mutants. The transcriptome data set is a potential reference and is useful as a resource for molecular genetic studies in kenaf. This will also serve as information dissemination concerning these genes to make them useful in modifying kenaf plants' contents for economic value addition to the cultivar (Lyu et al. 2020). According to a study, Random Amplified Polymorphic DNAs (RAPD) analysis is a powerful method for identifying different kenaf varieties and successfully tracing their genetic links through variations in agronomic morphology. In general, the known pedigrees or origin information can provide a good explanation for the pattern of genetic diversity of the kenaf variants as revealed by RAPD research. The information acquired will help breeding programs so that the introduction of agronomically advantageous genes to some particular types can be effectively led through appropriate germplasm (Cheng et al. 2002). Bast (phloem) fibers are sourced from widely cultivated kenaf; the genome of kenaf (*Hibiscus cannabinus* L.) was reported to provide insights into bast fiber. Recently, two population bottlenecks in kenaf were shown from genomic analysis providing suggestions for systematic domestication and improvement that have rendered a high yield in fiber biogenesis. This genome, which is chromosome-scaled, provides a significant framework and toolkit for sequence-directed genetic improvement of fiber crops (Zhang et al. 2020). Fiber yield and quality genetic control in Kenaf were reported. It showed that for appreciable improvement in fiber yield, it is important to select the segregating generations (Behmaram et al. 2014).

2.4.2   Fiber production and uses

Due to its high strength, elasticity, stiffness, low density, low cost, and eco-efficiency as well as its minimal health risks, renewability, and possession of useful thermal and mechanical properties, kenaf fiber is found to be an excellent reinforcement in polymer composites. It is also biodegradable. An overview of current advancements in the field of kenaf fiber and composites, including their chemical, microstructural, mechanical, and dimensional stability as well as their thermal stability, product development, and application, was published (Adole et al. 2019). There have been studies on the Kenaf bast fibers properties in the textile industry, for the production of textiles. There is research designed to soften kenaf fibers, woven fabrics, production of non-woven fabrics characterization, and materials products development such as underlays of furniture, the backing of the carpet, and covering of walls. When kenaf is combined with cotton fibers and given the correct care, the resulting products have a higher value, making kenaf a viable textile fashion fiber (Ibrahim and Ogunwusi 2017). Recent developments have sufficed in kenaf fibers and composites production (Sapuanetal 2018). In some markets, many bast fiber crops, like hemp, jute, and flax; likewise, wood or wood wastes, compete with kenaf such as those for wall panels and pulp, and paper applications. Jute and related fibers' demand and development have decreased as a result of traditional gunny sack markets switching to affordable man-made synthetic fibers based on fossil fuels during the previous few decades. The systemic transition from a petroleum-based economy to a biobased economy is expected to increase demand for these types of cellulose resources. Increased kenaf cultivation and the growth of kenaf-based companies, especially in areas with less wood supply, present a wealth of potential (Steef et al. 2013).

2.4.3 Bioenergy potential

Industrial kenaf (*Hibiscus cannabinus* L.) as agricultural biomass was successfully evaluated resulting in high prospects for this crop to be utilized as a renewable and raw material for the sustainable development of bio-energy. In countries like Malaysia, it is considered as a possible resource for energy (bioethanol, biohydrogen, bioenergy) supply sustainability in the years to come (Saba et al. 2015b).

2.5 Jute (*Corchorus capsularis*)

Jute is a kind of fibrous plant that is widely cultivated and has essential physiological traits like biomass, a deep root system, and resistance to metal stress. Jute fibers, which are silky, are one of the most popular natural bast fibers in the world (Majumder et al. 2018). Native leafy vegetables called Corchorus species have also demonstrated the ability to phytoremediate many hazardous metals, including lead, cadmium, zinc, and copper (Pb). Jute has considerable potential to tolerate metal-contaminated soil and is regarded as a phyto-accumulator of a large number of metals in its body parts (Saleem et al., 202). Jute is one of the best fibers for completed goods production that has been widely promoted and sold as an eco-friendly raw material, especially for textiles (Biswas et al. 2019). Yet, in the majority of jute-producing nations, an enhanced jute variety created through conventional breeding is commercialized. Nonetheless, the progress made by transgenic technology, including in Bangladesh and India, is being assessed. As with other genetically modified (GM) crops, recently developed jute, fungus-resistant (FR) transgenic jute, and herbicide-tolerant (HT) transgenic jute may face certain regulatory problems. Yet, Bt jute may have a higher chance of acceptance because, similar to Bt cotton, all regulatory difficulties have been resolved (Majumder et al. 2020).

2.5.1 Biotechnologies for improvement

There was a reported proposal on the future potentialities and special emphasis on jute breeding; these include the fiber quality, production by genetic improvement and mechanization, genomics, and germplasm innovation as a combination of biotechnologies. Marker-assisted selection through tissue culture, transgenic technology and molecules contributed greatly to jute breeding in addition to other conventional breeding methods (Zhang et al. 2019).

2.5.2 Jute derived products

Jute is an important feedstock for the production of products such as cloth, ropes, sacks, hessian, etc., as packing materials. Emerging technologies have made bulk use of Jute to overcome the declining market of these new products, to produce high-value addition and price competitive intermediaries; it serves as a raw material for final products. New innovative products like jute textiles, pulp for paper, jute composites, handicrafts, etc., have been developed with high-value addition. Textiles for home, geotextiles, fashion accessories, pulp and paper, shopping bags, jute-reinforced composites, particle boards, handicrafts, etc., are among the various diversified jute products having the potential for wider use and application (Khaled and Abu Darda 2016). In the production of nanocellulose, fibers of jute were reported to be the raw material employing the ball milling technique of high energy; it has potential usage as fillers in biodegradable nanocomposite plastics used in automotive, packing, and agriculture applications (Abbasi and Baheti 2018). Jute can be applied in artificial fibers made geotextile that serves as one of the desired tools for overcoming geotechnical problems due to its suitability in ensuring effective control over soil-related distresses, and most important, it is environmentally friendly (Sanyal 2017).

2.6 Sisal (*Agave sisalana*)

Sisal has been an important resource for the development of bioproducts because it has high cellulose and hemicellulose contents that generate fermentable sugars capable of producing compounds of industrial value (de Medeiros et al. 2020). Sisal leaves provide quality fibers for the manufacture of ropes, anchors, cordage, and handicrafts. It is regarded as an excellent green material for application in areas such as marine, automotive, renewable energy, and construction because of its superior engineering properties (Srinivasakumar et al. 2013). Due to its fiber durability and versatility, it is regarded as an important input material for different applications; it can easily be raised on wastelands in different types of agro-climatic conditions. As a source of employment opportunities and income generation, sisal fiber and its related activities like cultivation, extraction of fiber, and development of value-added products are significant in rural development thus leading to sustainable development (Nayak et al. 2011).

2.6.1 Biotechnological development

Diverse ecosystems have driven the search for new biotechnological microbial sources of significance. Soils that have exhibited high microbial biodiversity with various enzyme profiles in Brazil, like proteinases, xylanases, and cellulases, may possess important prospects for the discovery of emerging strains with novel characteristics. It was claimed that cellulose-degrading enzymes were produced on sisal and other agro-industrial leftovers using a new Brazilian actinobacteria strain called Streptomyces sp. (Macedo et al. 2013). A cutting-edge procedure for a modified soil burial test (SBT) was used to create systematic knowledge on the biodegradation profile of sisal fiber components (Saha et al. 2015).

2.6.2 Prospects of sisal fibers

Among fiber plants, sisal fiber represents 2% of the global production (plant fibers provide 65% of the fibers globally) occupying 6th place. The fiber has a high content of cellulose and hemicelluloses of the three grades of sisal fiber which is the lower grade relevant to the paper industry. The fiber utilized in making binders, twine, ropes, and baler in the cordage industry is medium-grade. For marine, agricultural, and industrial applications, twines and ropes are widely used. The higher-grade type of fiber is used by the carpet industry after being processed and converted into yarns. Wall tiles and furniture constructed of echoed sisal are two examples of sisal-based items that are currently in development. Sisal composites that are reinforced and used in package holders, inside car linings, car door sides, wheel wells, panels, ceilings, etc., offer the benefits of reducing vehicle weight to lower fuel consumption. Recently it is found useful as a replacement for asbestos as a strengthening agent, as a substitute for silk fiber, and in the automobile industry as an environmentally friendly component. In polymer (thermoplastics, thermosets, and rubbers) composites, sisal fibers are good reinforcements; because of their low density and high specific properties, they may have relevant applications in the automotive and transportation industry. The fibers are also useful feedstock in the development of plastics and packaging, tying, and gardening. The sisal wastes are useful as fertilizer or animal feed. Cat scratching posts, hot pads, spa products, jewel boxes, pet toys, slippers, cloths, lumbar support belts, rugs, disc buffers, handbags, etc., are other sisal fiber byproducts (Shanghai Exhibition Centre 2017). When sisal (*Agave sisalana*) pulp hydrolysis liquor was considered as a potential substrate for biosurfactant production, surfactin production was successful, making the liquor from sisal pulp hydrolysis a workable and sustainable substrate (Marin et al. 2015).

Figure 2.1. (a) *Gossypium hirsutum* L. (cotton) crop breed. (b) Cotton (*Gossypium hirsutum*) field. (c) *Gossypium hirsutum* L.) yields. (d) *Gossypium hirsutum* (L.) under attack by the African bollworm pests causing losses by farmers.

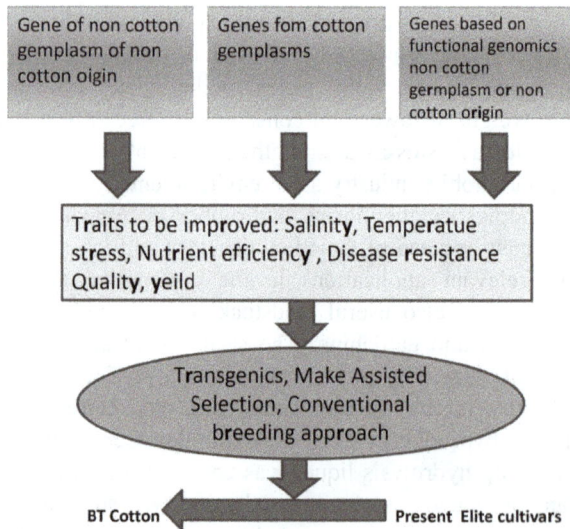

Figure 2.2. Flow sheet displaying probable future target for breeding of transgenic cotton.

Figure 2.3. Cannabis (*Cannabis sativa* L.) plant.

Leguminous Crops

Legumes are vital in the improvement of soil structure, organic matter and fertility, contributing significantly to the conservation of bioresources and the prevention of soil erosion. Leguminous crops are made up of pulses, the main crops belonging to the Leguminosae family. They are annual and perennial crops of which food and feed are the potential uses.

3.1 Groundnut (*Arachis hypogaea* L.)

Groundnut crop is the source of nuts for consumption in raw form, either roasted, boiled, turned into confections and flour, or crushed by mechanical means for edible industrial oil production. The shell can be used as fuel, animal feed, filler in feed, fertilizer for particle boards, and to produce alcohol and acetone after fermentation. As a legume, it also provides soil with organic matter and nitrogen (100–152 kg/ha N) for the roots. After the oil is extracted, the groundnut cake is used as fertilizer, weaning food for kids, and animal feed. Refining is done to free groundnut cake and oil produced from contamination by aflatoxin (Nigam 2014). Groundnut products are derived from crop parts (Fig. 3.1) In most West African countries, groundnut oil is used as a dietary component in domestic and industries. In some countries like Nigeria, Gambia, and Senegal, the extraction of oil from the nuts has been an important rural cottage industry for many years (Nautiyal 2002).

3.1.1 Genetic breeding for yield improvement

It was reported that groundnut can be subjected to extensive mutagenic treatments for the induction of variability. Yield improvement was experienced in two groundnut cultivars (GPBD4 and TPG-41) via induced mutagenesis by the use of electromagnetic spectrum (EMS) and gamma rays. The superior mutants isolated were found to be of good oil quality and stability coupled with pod yield and other agronomic features superior or comparable to the parents. These top-notch mutant lines can be used in recurrent mutagenesis and hybridization programs to combine desirable features. The findings amply illustrate the potential of mutation breeding for generating broad genetic diversity for a variety of quantitative and qualitative traits, as well as for creating advanced mutant breeding lines that are varied and desirable within groundnut cultivars (Kavera and Nadaf 2017). For attracting the consumers' attention and high demand, groundnut genotypes with large seed sizes, seed weight, and improved seed quality were developed. To facilitate the choice of the most efficient breeding and selection procedure, knowledge of the genetics system controlling expressions of these traits is important. A study showed that genetic recombination for seed quality traits will be achieved through hybridization (Kabbia et al. 2017).

3.1.2   Value addition by-products

Groundnuts comprise plenty of functional compounds like protein, fiber, and polyphenolics as by-products of groundnuts are useful as functional ingredients when incorporated into processed foods (Zhao et al. 2012). Some global applications of most groundnuts are in the production of oil, butter, roasted peanuts, snacks, confectionaries, and in meat product formulations as extenders. Poly and monounsaturated fatty acids are some good sources obtainable from groundnut. Large quantities of by-products are produced in the process of harvesting and extracting groundnut and groundnut oil respectively. A large part of groundnut skins, hulls, meals, and vines that are rendered as agricultural wastes can be mixed in some food preparations as supplementary and emergency foods due to their high nutritive value, especially, as they are a very good protein source (Dhanesh and Kochhar 2013). Groundnuts can help reduce a considerable amount of protein, energy, and micronutrient deficiencies; 100 g (3.5 oz) of groundnut seed has 585 calories (Nigam 2015). Potential uses for groundnut shells in industry and commerce have been discovered. Groundnut shells can be used to produce a variety of bioproducts, including biodiesel, bioethanol, nanosheets, enzymes, and hydrogen (Duc et al. 2019). A study demonstrated that groundnut shells can be successfully utilized as an adsorbent which is low cost for the Eriochrome black T (EBT) dye removal from aqueous solutions, showing the adsorption performance of agricultural waste material (Bouchet et al. 2016). The synthesis of activated carbon from groundnut shells as dye adsorbents for wastewater treatment was also reported (Wu et al. 2019).

3.2   Soya bean (*Glycine max* L. Merr)

Soybean (Glycine max (L.) Merr) is a commercial crop with various uses as food, feed, and feedstock for industries (Khojely et al. 2018). The cultivated soybean, *Glycine max* (L.) parts and products are of industrial importance (Fig. 3.2a-e). A plant oil crop and source of animal protein feed worldwide, 80% of the world's soybean output is obtainable via mechanical crushing to get the meal oil. A minute proportion of the total crop is directly used by humans (Pawar et al. 2011). Collaborative research efforts are much needed to develop the economic sector of this crop. The many potential uses of soy, which is seen as a greener alternative to non-renewable resources, are still untapped. The most popular source of vegetable protein utilized globally is soy products. Soybean oil is a high-quality vegetable oil when processed and a source of protein for livestock (Newkirk 2010).

3.2.1 Genetic modification

Scientists have modified and chosen specific characteristics and traits from the genes in soybeans to increase yield, increase disease resistance, and profitability for farmers, provide health benefits for people and animals, and bring varieties to market that are specifically relevant for industrial uses. One of the success stories of soybean biotechnology is the production of cultivars that are resistant to herbicides like glyphosates. Herbicide-resistant first-generation characteristics varieties helped farmers cut costs and increase production effectiveness. The creation of variants that increase the nutritional content of the soybean for consumption by humans and animals is one example of second-generation features in soybean breeding. The techniques used include lowering anti-nutritional components like lectins or trypsin inhibitors as well as creating high-lysine, high-oleic, or low-phytic acid variations. Several seed companies are developing soybean varieties strong in omega-3 fatty acids in response to rising customer demand for these nutrients in food. The third-generation soybean research focuses on green energy and industrial use. Soybean can be used in the production of biodiesel, medicines, plastics, textiles, waxes, hydraulic fluids, lubricants, adhesives, and other products (Newkirk 2010). Through genetic modification, soybeans production has been improved via traditional crop breeding, making the species sexually compatible (Gianessi and Carpenter 2000). Improvement of soybean cultivars is attributed to the discovery and genetic modification of novel desirable genes for adaptation to new habitats, new management techniques, and new end applications (Scaboo et al. 2010).

3.2.2 Domestic, agricultural and industrial utilization of soya bean and products

Soybean (*Glycine max* (L.)) is one of the sources of protein globally. Soybean oil and phytoconstituents promote health, nutrition and livestock feed. Soybean improves soil health due to its deep root system and ability to fix atmospheric nitrogen. It also remains an important world commodity for industrial utilization. Production and uses of soybean and its derivatives for food, animal feed, pharmaceutical, and industrial applications have been one of its great potentials. Due to the expansion of the livestock industry, there is an increasing need for high-protein feeds such as soybean meal. The oil or fat derived from soya is of relatively low iodine number as a result its primary use is in the edible market. The principal uses of oil in the edible market are shortening, margarine, and salad dressings (Ali 2010, Goldberg 1952).

3.3 Faba bean (*Vicia faba* L.)

The faba bean (*Vicia faba* L.) possesses high nutritional value, medicinal properties, and good biological nitrogen fixation. Many regions in the world grow faba bean crops as a legume (Fig. 3.3a-c). Integrating faba beans in cropping systems is expected to provide diverse ecosystem benefits (Etemadi et al. 2019). Due to its ability to withstand different climatic and soil conditions, it is described as the probable third most important feed grain legume and one of the good candidate crops under situations of global warming and climate change (Singh et al. 2013).

3.3.1 Genetic resource and development

Producing simple sequence repeat (SSR) markers from expressed sequence tags (EST) was a significant method for assessing the genetic diversity of a population. In evaluating the faba bean germplasm, a study demonstrated the high effectiveness of the EST-SSR marker (Gong et al. 2011). To successfully breed faba beans for disease resistance, traditional breeding techniques such as recurrent mass selection have been well-established. Future faba bean cultivar development has several potential applications for molecular breeding approaches that combine the most recent discoveries in genetics and genomics with conventional breeding methods (Gnanasambandam et al. 2012).

3.3.2 Potential applications

A novel 15-kDa Bowman-Birk type trypsin inhibitor (termed VFTI-G1) isolated from faba bean seeds (*Vicia faba* cv. Giza 843) has potential medicinal applications (Fang et al. 2011). Based on the demonstration that broad beans have a relevant prospect as nutraceutical with potential nutritional value and promising health-promoting factors. Analysis using GC/MS showed the presence of α-tocopherol, phytol, phytosterol, stigmasterol, campesterol, and fatty acids, with promising therapeutic uses (Pasricha et al. 2014).

Recently, there has been a lot of concern about a newly emerging method for the utilization of waste byproducts from the agriculture industry in the development of food additives or supplements. A study explained the tentative identification of more than 140 phenolic and other phytochemical compounds in the extract by characterizing the phytochemicals of faba beans pods extracted using methanol and higher performance liquid chromatography (HPLC) connected with quadrupole time of flight tandem mass spectrometry. This study represented a new development in the understanding of broad bean pods. Consequently, it is the first time that broad bean pods have been found to contain more than 90 phytoconstituents, including phenolic acids, flavonoids, iridoids, lignans, and derivatives of terpenoids. The information acquired showed that vegetable waste from the food industry might be utilized to great effect as a prospective source of bioactive ingredients to create new nutraceuticals and functional meals with a valuable future market (Abu Reidah et al. 2017).

3.4   Jack bean (*Canavalia ensiformis* L.)

Jack bean (*Canavalia ensiformis* (L.) DC.) is rich in protein, carbohydrates and minerals even though there is a limitation to its use in ruminant and monogastric diets due to much anti-nutritional presence (concanavalin A, canavanine, and canatoxin). However, in the diet of ruminant animals, dried jack beans can be used. Various parts of the plant (Fig. 3.4a-b) are useful for domestic and industrial utilization.

3.4.1   Molecular technologies

The biotechnological potential of *Canavalia ensiformis* is related to the expression of a particular set of proteases—active and thermostable enzymes in particular organs with much activity and stability, rendering this legume a potential source of proteases (Gonçalves et al. 2016). The proteins urease and canatoxin, a subtype of the jack bean urease, are found in *Canavalia ensiformis* (jack bean) seeds. The discovery of a family of urease genes in the jack bean and the identification of a cDNA encoding of a new member of this gene family came from the cloning of another isoform of urease dubbed JBURE-II (Pires-Alves et al. 2003). Jack bean (*Canavalia ensiformis* (L.) DC.) genetic variability and character association studies suggested that the selection of early genotypes would be beneficial in boosting the pod length, pod weight, number of pods, and eventually pod production per plant (Lenkala et al. 2015).

3.4.2   Industrial utilization

The guidelines for the possible utilization of Jack bean seed flour in good functional foods for nutrition and food formulation have been provided (Marimuthu and Gurumoorthi 2013). As a result of an increase in the cost and supply of wheat and the limited demand for the production of wheat flour biscuits, attention is nowadays given to the potential of indigenous grains to completely or partially replace wheat in bakery products. Fortification with fermented Jack bean flour in the preparation of shortbread biscuits was reported (Friday et al. 2017).

3.4.3   Phytoremediation potential

Jackbean was reported as a potential plant for phytoremediation of multi-element contaminated soils. It showed a high potential to phyto remediate soils contaminated with several chemical elements like those occurring in regions affected by mining activities (Silva et al. 2018). According to research, *C. ensiformis* crude extracts may be used to stabilize and immobilize heavy metals in contaminated mining waste to stop them from dispersing further into the environment (InHyun et al. 2016).

Figure 3.1. (a) Groundnut (*Arachis hypogaea* L.) crop. (b) Undeshelled *Arachis hypogaea* L. (c) Deshelled *Arachis hypogaea* L. (d) *Arachis hypogaea* L. oil.

Figure 3.2. (a) Soya bean (*Glycine max* (L.) crop farm. (b) *Glycine max* (L.) yielding flowers. (c) Freshly matured *Glycine max* (L.) beans. (d) Matured *Glycine max* (L.) beans ready for harvest. (e) Dried *Glycine max* (L.) beans.

Figure 3.3. (a) Fresh faba bean (*Vicia faba* L.). (b) Dried un-deshelled *Vicia faba* L. bean. (c) Dried *Vicia faba* L. bean.

Figure 3.4. (a) Fresh jack bean (*Canavalia ensiformis* L. (b) Dried *Canavalia ensiformis* L. beans.

Conifer Crops

Conifers are plants known as Pinophyta, they are also called Coniferophyta or Coniferae, from which the common name 'conifer' was derived. All conifers are woody, mostly trees and occasionally shrubs. Conifers appear globally as the longest-living, tallest, and most massive trees. These trees flourish largely in the northern hemisphere and produce needle or scaly foliage and bear cones. Conifers are predominantly evergreen trees, although there are a few exceptions. Most categories of conifers are medium to very large trees with some taking the shape of shrubs. Conifers are very easy to grow as ornamental trees.

4.1 Pine

Pine is a conifer shrub or tree species from the *Pinus* genus of plants—there exist more than 120 species of pine in the world (Myers 2020). Plants of the genus Pinus can thrive in a wide range of rough or harsh environmental conditions. Pine forests serve to absorb more carbon from the atmosphere, contributing to greenhouse effect mitigation. Pinus can be used to produce a variety of goods, including cellulose, wood and non-wood products used in the food, chemical, and pharmaceutical industries as well as for biorefineries (Rodrigues-Correa et al. 2012). A significant tree crop, pines produce useful goods with significant industrial value. One of the most popular and widely used non-wood products is resin. Some industrial applications of terpenes from resin include chemicals, agrochemicals, food additives, pharmaceuticals, and bioenergy. It has recently been found to be useful in the manufacture of green polymers and biodegradable batteries. In addition to resin, pine trees also produce bark, needles, and cones that can be used for landscaping, as well as substrates for plant culture, biofuels, and bioherbicides. Its use in metal biosorption composites and nano-fibrillated cellulose medium-density fiberboard components are a few of its other uses (Neis et al. 2019).

4.1.1 Biotechnologies

The development of molecular and biotechnological techniques has made it possible to produce oleoresin by screening trees with the use of appropriate DNA-based markers in order to indirectly select pine genotypes exhibiting the desired phenotype. To better understand the biochemical and physiological foundation of the development of oleoresin, information about their genes and their regulatory patterns for the biosynthesis of terpenes in commercial forests will be crucial. The selection and establishment of super-resinous woods will be aided by the link between DNA-based markers and the knowledge of the molecular underpinnings of terpene production. The improvement and application of clonal propagation procedures along with genetic transformation techniques for pines will promote the development of trees that produce sufficient amounts of high-yielding, high-quality oleoresin (Rodrigues-Correa et al. 2012).

4.1. 2   Useful oleoresin products from pine

The potential of oleoresin derivatives is found in cosmetic, pharmaceutical, food, and chemical industries for the development of products, like insecticides, adhesive paints, varnishes, and disinfectants. Biotic and abiotic variables that impact the formation of oleoresin can be used to increase yields by boosting particular signalling and biochemical defense mechanisms (Kreps et al. 2017). There is a great demand for alternative fuels because petroleum-based products must be replaced and the use of personal vehicles and air travel are becoming more common, particularly in developing nations. Terpenes from plants, have been suggested as a renewable alternative feedstock for biofuel. It was noted that pine trees have biological and economic benefits for the chemical industry, particularly in terms of their contributions to enhancing food safety, replacing petroleum-derived chemicals and fuels, and boosting carbon storage. Protection of stored food and insecticides is also obtained from pine oleoresin-derived terpene (Rodrigues-Correa et al. 2012).

4.1. 3   Other industrial prospects

There are many uses for pine nuts and pine nut products (Table 4.1). Pine nut oil is a resource for cosmetics as an oil for beauty products. Also, it offers possibilities for wood finishing, using paintings as paint bases, and treating delicate skins in the leather sector. The result of pine nut oil pressing is pine nut

Table 4.1. Industrial uses of pine products (Sharashkin and Gold 2004)

Industry	Uses
Bakery Industry	Breads, cookies, cakes, pastries. For cakes, pine nuts are made into flour.
Confectionary Industry	Serves as an ingredient in chocolates and bars.
Beverage Industry	In making pine nut beverages and cream.
Oil Industry	Source of vegetable oil in dressing for salads.
Cosmetics	Potential use in cosmetics and soaps.

flakes used in granolas, chocolates, and crunch bar manufacture. The flakes still have up to 30% oil that can be used to make flour or pine nut meal, which has a variety of culinary applications. It is a high-end alternative to wheat or rye flour that gives pastries, pancakes, and other baked goods a rich, nutty flavor. The meal becomes a dairy-free milk-like beverage with a sweet, creamy nutty flavor when combined with water (Sharashkin and Gold 2004).

4.2 Spruce

Species of spruce (Picea spp.) are extremely vital for the stability of the world's ecosystem and biodiversity. They are the source of most of the global wood and fiber supply as well as feedstock, renewable and other industrial biomaterials (Ralph et al. 2008).

4.2.1 Genetic transformation for improvement

A new method for growing transgenic white spruce was reported (Picea glauca) which utilized Agrobacterium tumefaciens. In 15 separate transformation events, the method described was effectively used to produce 1200 transgenic white spruce plants that included insecticidal genes from Bacillus thuringiensis. Through the application of recombinant DNA technology and genetic manipulation, it is intended to facilitate physiological investigations and improve white spruce development (Le et al. 2001). Conifers like spruce whose embryogenic tissues are transformed can play the role of the right choice of plant raw material for insecticidal or fungicidal transgene efficiency testing. Using somatic embryo explants, stable *Agrobacterium*-mediated transformation of spruce embryogenic tissues was reported (Pavingerová et al. 2011). Reducing the amount of lignin in a tree or changing its composition to enable delignification during pulping is an excellent reason for tree breeding, in transgenic angiosperm tree species, as described. For the first time, it was demonstrated that a transgenic method may successfully modify the content and composition of lignin in a conifer. Five-year-old transgenic Norway spruce plants (Picea Abies [L.] Karst) expressing the spruce gene for cinnamoyl CoA reductase (CCR) in antisense orientation have altered lignin quantity and composition (Wadenbäck et al. 2008). One of the models for conifer genomics has been constructed using the gymnosperm white spruce (Picea glauca). Innovative technologies for the assembly of very large genomes, as well as the conifer genomics resources produced in this process, were detailed. The draft genome assemblies of two genotypes of white spruce, PG29 and WS77111, were also described. Comprehensive annotations were also provided for the white spruce mevalonate, methylerythritol phosphate, and phenylpropanoid processes. These analyses brought to light the extensive gene and pseudogene duplications in a conifer genome, especially for genes of secondary (i.e., specialized) metabolism, as well as the potential for function gain and loss for defense and adaptation (Warren et al. 2015).

4.2.2   Potential byproducts

Potential medicinal compounds possessing antioxidant properties were reported from the waste biomass of spruce. It was discovered that the bark can serve as a substitute for antioxidants and medicinally useful chemicals that can replace manufactured medications. Bark, however, is a waste biomass product produced by the wood industry every year producing quantities in the millions of tons. Sterols and terpenes found in silico analysis were the most prevalent substances found in hexane extracts of bark. Their distribution, metabolism, excretion, and potential forms of biological activity, as well as their absorption and metabolism, were projected. Abietic acid (Fig. 4.1), dehydroabietic acid, and -caryophyllene oxide, all of which were derived from the bark of Norway spruce, which most likely each possessed antihypercholesterolemic, mucomembranous protective, and anti-cancer activities, respectively (Kreps et al. 2017). Spruce bark is the industrial source of condensed tannins. But, the content of free, glycosidic, and polymeric sugars in the raw extract may need to be gotten rid of before industrial use (Kemppainen et al. 2014).

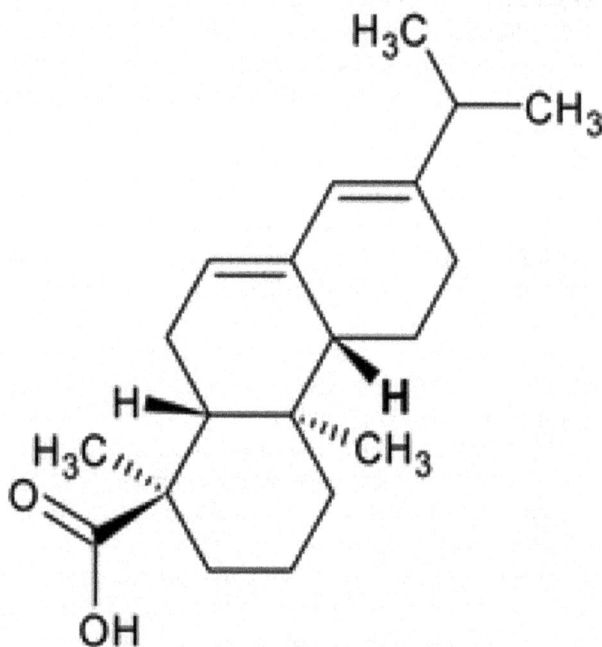

Figure 4.1. Abietic acid.

Crucifer Crops

Crucifers are various plants in the Brassicaceae family. The name "Crucifer" is derived from the former taxonomic classification for the family, Cruciferae. It is also commonly referred to as the mustard family.

5.1 Camelina (*Camelina sativa*)

Camelina sativa is a crufter of oilseeds from the Brassicaceae (mustard) family. It can withstand a range of climatic and soil conditions. Due to its many valuable agronomic attributes and its seed quality, there is a renewed interest in camelina. Camelina matures early and possesses a high-yield potential. It has a high drought tolerance and resistance to heat and many common pests and diseases. It has a unique unsaturated fatty acid profile largely (>90%) and the seed meal is relatively low in glucosinolates when compared with other crucifers (Eynck and Falk 2013). *Camelina sativa* (L.) Crantz of the Brassicaceae family is a significant oil crop with numerous advantageous agronomic characteristics, such as the minimum input of fertilizer and water, strong resistance, and adaptation. The camelina oil plant has been used as a model to research the regulation of lipid metabolism and genetic improvement because its life cycle is short and straightforward genetic transformation, as well as the availability of genome and other "-omics" data. Camelina has a unique ability to quickly modify its metabolism in order to produce and accumulate significant amounts of uncommon fatty acids and modified oils in its seeds, making them more stable and eco-friendlier. Such genetically modified camelina is discovered to be a very useful resource for the development of high-value chemicals, biofuel oil, and foods and medicines that promote health (Yuan and Li 2020). *Camelina sativa* L. Crantz (gold-of-pleasure, large-seeded false flax) (camelina) is a promising oilseed crop for making edible oil, seed meal for animal feed, and/ or feedstock for biodiesel. Camelina doesn't rival agricultural crops for food, hence, it does not require prime agricultural land and requires limited irrigation and nitrogen inputs (Lohaus 2019). The potential of camelina oil is seen in various areas like pharmaceuticals, cosmetics, animal feed etc. Camelina seeds have a high fatty acid content due to the oil's 50–60% unsaturated, 35–40% omega 3 and 15–20% omega 6 concentration. The plant can withstand drought and is frost resistant to disease and pests. The seed has a high oil content and produces exceptional seed yields, especially when managed with minimal input and in constrained conditions. *Camelina sativa* is a promising oil-seed crop with multiple applications including environmental uses. It has been determined that animal feed is a crucial component of environmental sustainability. Due to this, numerous investigations concentrated on analyzing the environmental performance of Camelina sativa production in various parts of the world, including the United States of America, Spain, and France, by performing a life cycle assessment. *Camelina sativa* can be integrated into the future circular economy aside from other utilities (Pîrvan et al. 2020).

5.1.1 Genetic improvement

To successfully characterize the genetic diversity and population structure of the *C. sativa* species, genotypic data was used. This allowed researchers to speculate about how natural selection and plant breeding may have influenced the formation and differentiation within the species' natural populations and how the genetic diversity of this species might be applied to future breeding efforts (Luo et al. 2019). A report on the successful genetic modification of camelina's agronomic traits, particularly the fatty acid profiles of its seed oils, supported the possibility of using camelina as an industrial crop for the creation of novel biotechnology products (Lu and Kang 2008). It has been advocated for use as the optimum crop for the creation of biodiesel and bioproducts. Triacylglycerol (TAG) synthesis will demonstrate the crop's profitability by improving the flux of carbon from increased photosynthesis into seed oil output. Camelina was genetically altered to co-express the Arabidopsis thaliana (L.) Heynh in order to boost oil output, this was under the guidance of a seed-specific promoter, for the genes for triacylglycerol acyltransferase 1 (DGAT1) and yeast cytosolic glycerol-3-phosphate dehydrogenase 1 (GPD1) (Chhikara et al. 2018).

5.1.2 Prospects

In addition to retaining increases in tissue n-3 LC-PUFA contents, Camelina, when genetically modified to contain high levels of EPA and DHA, can replace fish oil in feeds for European sea bass without harming growth or feed efficiency. Transgenic Camelina sativa expressing algal genes was also used to produce oil containing n-3 LC-PUFA to replace fish oil in salmon feeds. The oil has no negative effects on the growth, metabolism, or nutritional value of farmed fish (Betancor et al. 2021, Betancor et al. 2015). It was looked into if camelina seeds may be used as a source of meal for one-portfolio products, sugars for ethanol, and oil for domestic biofuel production. After harvest, camelina residues (straw) can serve as a useful source of raw material for green sugars. As high-value-added goods, seed meals and glycerin were discovered to be a reliable source of income and can generate an extra $1/kg of generated oil (Mohammad et al. 2018, Moser 2010). The production of a novel biological replacement for jet fuel through the accumulation of medium-chain, saturated fatty acyl moieties in seed oils of transgenic *Camelina sativa* was demonstrated in a paper to be possible using the California bay gene (Hu et al. 2017). Current methods for modifying oil yield or altering the endogenous lipid profile in camelina have had some success, and can currently provide products made from camelina seeds that are suitable for the market, such as omega-3 L-PUFA-enriched oil (Faure and Tepfer 2016). A thorough analysis of recent developments and difficulties in using molecular markers, genomics, transcriptomics, miRNAs, and transgenesis for improvement in biotic and abiotic stresses, carbon assimilation capabilities, seed yield, oil content and composition in camelina for biodiesel fuel properties, nutrition, and high value-added industrial products like bioplastics, wax esters, and terpenoids was conducted (Saingera et al. 2017).

5.2 Cauliflower (*Brassica oleracea* var. *botrytis*)

Cauliflower flourishes in India and the *Brassicaceae* is a family of cauliflower leaves. Its leaves are rich in calcium, iron and b carotene, even though it has a higher waste index. Due to the fact that it can come in the waste products category, cauliflower leaves are also used in the treatment of anaemia disease and micronutrient deficiency as a value-added product (Pankar and Bornare 2018).

5.2.1 Resource utilization

A study reported the valorization of cauliflower as an appreciable industrial application, to (*Brassica oleracea* L. var. *botrytis*) byproducts as a resource for antioxidant phenolics using two developmental extraction protocols, indicating that the cauliflower byproducts are an inexpensive source of antioxidant phenolics and interestingly enough the industrial angle and the possible utilization of ingredients to functionalize foodstuffs (Llorach et al. 2003). Development of the biological extraction process of kaempferol derivatives from cauliflower byproducts, as well as of the production of their unique metabolites using fungi as a potential candidate was reported. These could be valuable substances for medical, cosmetic or food demands (Huynh 2016). Cauliflower flour can be utilized as a substitute for 5% to 10% of wheat flour in Balady bakery without adverse effects on the consumer's acceptability of the product showing the possible use of the cauliflower (Brassica oleracea L. ssp. botrytis) stem flour in improving Balady bread quality (Hegazy and Ammar 2019). Recycling Cauliflower and Romanesco wastes could potentially be used in ruminant feeding (de Evan et al. 2020). It was discovered how to create acetone-butanol-ethanol (ABE) utilizing cauliflower waste and Clostridium acetobutylicum NRRL B 527. Using drying kinetics and evolving bioprocess, this waste is used to produce sustainable biobutanol. This is a road map for investigating the economic development of biofuel(s) using various second-generation resources to meet the increasing worldwide need for fuel (Khedkar et al. 2017).

5.3 Rape (*Brassica napus*)

The most productive oil crops are those of *Brassica napus*, sometimes known as rapeseed or oil seed rape. Rape is primarily used in Europe to produce biodiesel; in fact, rapeseed is the source of over 65% of the biodiesel developed in Europe (Pavlidou 2016). It is also a plant crop that gives much to the feed industry and contributes a lot of proteins (Nesi et al. 2008).

5.3.1 Resource recovery

Brassica napus growth in combination with amendment with chars from manure waste served in assisted phytoremediation of mining soils (Cárdenas-Aguiar et al. 2020). A review provided an emerging trend and the future development in fatty acid profiles and modification of rapeseed oil for use as food, and industrial raw material for biodiesel production. Aside from rapeseed oil uses dating back from early civilizations, erucic acid (C22:1) and the presence of glucosinolates has transformed a reduction from the mid-nineteenth century to its popularity. As a result, several efforts have been made to create cultivars free of toxins. Breeders previously reported success in growing '00'-grade rapeseed, often known as 'Canola'. Such an achievement was guaranteed by the targeted mutagenesis of the Brassica napus fae-1 and fae-2 genes. Thereafter, 'canola' maintains its momentum in the market as a good vegetable oil. The FAD2 alleles responsible for desaturating oleic acid (C18:1) to linoleic acid have also been chemically mutated to produce high oleic acid rapeseed lines that contain 86% oleic acid (C18:2). Rapeseed oil with high erucic acid content has recently remained popular for biodegradable, emollient polymers used in the cosmetics industry as well as for biodiesel. Thus, breeding strategies were explored; regrettably, they were unsuccessful in raising the percentage of erucic acid in seed oil above 50%. Rapeseed genotypes over-expressed with Ld-LPAAT individually and Ld-LPAAT-FAE chimeric construct jointly was attempted, but the erucic acid content was not able to exceed 60%. Then, a coordinated effort between conventional breeding and transgenic techniques is used to get through three hypothetical bottlenecks, which, as previously mentioned, limited the level of erucic acid to about 60%. In the end, genotypes of rapeseed containing 78% erucic acid were effectively developed (Nath et al. 2021).

5.4 Mustard (*Brassica juncea* L.)

According to a paper, *Brassica juncea* has a great ability to absorb and accumulate Cd, making it a successful plant to use for the phytoextraction of Cd-contaminated soil. The report revealed Cd buildup in all plant sections (roots, stems, leaves, and shoots). Brassica juncea is an excellent plant to use for the phytoextraction of Cd-contaminated soil because it has the ability to adapt and accumulate Cd, according to the phytoextraction procedure. The report revealed Cd buildup in all plant sections (roots, stems, leaves, and shoots). The phytoextraction protocol is ideal for the remediation of cadmium from Cd-contaminated soils with the aid of the Brassica juncea plant (Goswami and Das 2015, Bhadkariya et al. 2014). Indian mustard (*Brassica juncea*) was found useful as a biofumigant in managing Rhizoctonia solani. Rhizoctonia solani is a fungal pathogen that affects various crops, including common beans, and has a significant negative impact on production. This has made it possible to use *B. juncea* as a defatted seed meal in addition to using it as ordinary green manure to treat root rot fungus (Abdallah et al. 2020).

5.4.1 Mustard by-products utilization

Co-pyrolysis of paper waste composed of packing paper made locally, newspaper-based supermarket food packs, printing paper, and mustard press cake in a semi-batch pyrolyzer allowed for the optimization and characterization of bio-oil. The study's projected findings might be applied to the generation of energy for the management of trash in large urban areas. Although pyrolysis is the prelude to thermo-chemical gasification, the information and knowledge gained from the study may also be used to anticipate the behavior of the co-gasification of these raw materials, namely paper waste (PW) and the mustard press cake (MPC) (Sarkar and Chowdhury 2016). Mustard powder and mustard cake prepared from naturally available mustard seeds were effectively determined to be used as coagulants for treating synthetic municipal wastewater using the coagulation-flocculation process (Bhargav and Joga 2018). To meet the bioaccumulation, toxicity and biodegradability potential standards that reduce the adverse effects on the aquatic atmosphere compared to conventional mineral-based lubricants. Nowadays, environmentally friendly lubricants are prepared to meet this need. A study was conducted in an effort to intensify the feasibility of using the oil from mustard as a lubricating agent for commercial purposes (Vignesh et al. 2021). It has been suggested that mustard (*Brassica juncea* L.) oil could be used as a feedstock for the production of biodiesel (Bello et al. 2019). Considering a lot of pressure being placed on the available capacities for the disposal of plastic waste in addition to the unnecessary use of plastics the need for plastics that are biodegradable and plastic waste biodegradation has become very important in recent years. Production of bioplastic from mustard oil was considered relatively low-cost, readily available, included in vegetable oil and having fewer volatile characteristics. A report concluded that mustard oil is also a low carbon source for bioplastic development as it is readily available in South Asia and its utilization can lead to the production of bioplastic at the commercial level (Javed and Jamil 2015). The demand for the generation of fuel from alternative sources was motivated by the depletion of fossil fuel resources as well as high crude oil prices. To successfully replace fossil fuel, such fuel must be both economically appealing and performance-capable. Brassica campestris, an Indian mustard plant, produces used mustard oil waste that is typically dumped in the trash after being used for cooking. A study looked into the viability of using this used mustard oil to make biodiesel (Singh et al. 2010).

CHAPTER

6

Forage and Fodder Crops

Forage is the edible part of the plant, aside from the separated grain, which can serve as a feed resource for animals. This material is harvested and used up by the livestock themselves to meet their nutritional requirements while fodder is coarse grasses, such as sorghum, corn, pennisetum, millet, lablab, cowpeas, grain sorghum, etc. These are harvested with the seed and leaves, and then completely cured and fed to the livestock.

6.1 Genetic manipulations

Despite the slow nature of regulatory steps for the approval of genetically modified crops and the politics involved, along with complications in forage grasses and legumes, a certain bioresource is associated with the development of forage and fodder crops mainly in forage, turf and breeding for bioenergy crops. Forage that is grown most widely, turf and species for bioenergy (e.g., switchgrass, white clover, alfalfa,) are said to be highly unable to be fertilized by their own pollen and out-breeding. They possess a high ability to transfer their genes to adjacent plants compared with inbreeding species (Wang and Brummer 2012). The study of the forage crops gene pool through formation, analysis, and documentation by mobilizing the world collection of cultivars and wild specimens, which are sources that are valuable for selection, remains in a good direction as highly suitable and having a significant impact on the development of selection – and genetic science with an anticipated effect on the breeding of new varieties. On the food safety front, these varieties will have an ecological and social output. Researchers have assembled about 9500 specimens of the gene pool of forage crops, of which 59.44% are the collection of alfalfa and 32.36% is wheatgrass (Fig. 6.1) (Meirman et al. 2013).

6.2 Industrial and environmental benefits

It was proposed that forage legumes have the potential to contribute significantly to environment-friendly agricultural land and sustainable intensification of livestock production in the tropics, thus providing ecosystem services (Schultze-Kraft et al. 2018). A report has highlighted how value-added feed and fodder are relevant in alleviating the antinutritional effects of tannins in animals and upgrading the tanniniferous biomass feeding value (Bhat et al. 2013).

6.3 Miscanthus (*Miscanthus sinensis*)

Miscanthus crop is found useful in the conversion of biomass to liquid fuel and biorefineries systems in the production of chemicals and liquid fuels; it has attracted significant attention most recently. Its yield, carbohydrate, elemental, and lignin composition are highly substantial to be assessed for the future development of biofuel (Brosse et al. 2012). Possible applications of forage miscanthus are making pulp and paper, energy production, production of fermentation products, manufacture of construction/building materials, light natural sandwich materials (LNS) preparation, thatching and bioremediation (Fowler et al. 2003).

6.3.1 Biotechnological transformations

A genetic structure for *Miscanthus sinensis* from RNAseq-based markers showing new tetraploidy was reported. The genetic structure will be a potential biological discovery and breeding attempt to improve this prosperous biofuel crop and also supply a valuable resource for a better understanding of genomic responses to tetraploidy and chromosome fusion (Swaminathan et al. 2012). It has been reported that M. sinensis was subjected to drought stress and high-throughput transcriptome sequencing, considerably enhancing the present genomic richness accessible. A plethora of potential genes implicated in drought tolerance regulatory networks was discovered by the comparison of DEGs under various drought stress conditions, which will facilitate future genetic advancement and molecular research of M. sinensis (Nie et al. 2017). For the investigation of the environmental performance of the miscanthus-based value chains, a Life Cycle Assessment was conducted. The study showed how important it is to consider all environmental effects when choosing the best utilization pathways. The study's showing of the beneficial environmental performance of marginal land for the production and utilization of miscanthus biomass was another important finding. It was noted that this is a fantastic chance to increase biomass development without competing with food crops for the development of the European bioeconomy with a continually growing demand for biomass (Wagner et al. 2017).

6.3.2 Biomass, bioenergy and biofuel production

Due to its high yield and nutrient-efficient characteristics, miscanthus is an essential feedstock for the manufacture of bioethanol. Despite the difficulties that must be addressed to establish the commercial production of bioethanol from lignocellulosic material. The pre-treatment step of this process, where the feedstock is thermo-chemically treated to maximize the release of fermentable sugars from the cell walls, is one of the most obvious opportunities for improvement. By using plant breeding to improve cell wall composition, pre-treatments can be made less chemically intensive, which in turn lowers costs and boosts the production efficiency of second-generation biofuels. In addition to cell wall composition, plants from the OPTIMISTIC multi-location trials from the second year, an international collaborative effort that aims for the optimization of the production of bioenergy and bioproducts from miscanthus, were examined for saccharification efficiency. Based on a model created with data obtained the year prior and the use of near-infrared spectroscopy (NIRS), glucose conversion was anticipated. A fresh cross-validated model also produced accurate predictions of this characteristic. The results of this study improved the predictive models for cell wall composition and saccharification efficiency and offered intriguing new information about how these attributes varied in newly established miscanthus plants. This is one of the most intriguing possibilities for producing sustainable and renewable liquid fuels (Martínez 2015). The complete results of the EU-funded research project OPTIMISM were described. This study looked into ways to maximize the production and usage of miscanthus biomass. The potential of miscanthus as a crop for marginal locations was shown, and knowledge and technology for the commercial application of miscanthus-based value chains were provided through the investigation of miscanthus bioenergy and bioproduct chains (Lewandowski et al. 2016). Growing the perennial biomass crop miscanthus for bioenergy in the UK was reported to have environmental costs and benefits (McCalmont et al. 2017). During phytoremediation, a biomass such as miscanthus which is produced can be used as a biofuel. It is used for heating homes which are located close to where it is produced. Consideration for alternative energy technologies for Miscanthus and other plants, including liquid fuels such as ethanol, methane from anaerobic digestion, and pyrolysis (thermal processing), and retreatment alternatives to convert cellulose to glucose are reviewed (Zhao et al. 2021). According to a paper, Miscanthus sacchariflorus, a biofuel grass, induces blooming as a quantitative short-day response, but delayed flowering during long days boosts biomass accumulation (Jensen et al. 2013). Literature reported divergent bioconversion routes for biomass quality of miscanthus genotypes (Iqbal 2017). Miscanthus has recently become a very effective energy substitution crop, in large part due to its minimal input needs and high potential dry matter output. Due to its comparatively high yield potential and better biomass qualities, including low ash, Ca, Si, and Mg concentrations, M. sacchariflorus can be chosen for the development of solid biofuels. Inter-annual variation in biomass yield and composition between 2004 and 2010 was investigated in a trial that was of a multi-genotype variety, planted in south Germany (Iqbal and Lewandowski 2014). When biomass composition is optimized, ash-related issues including slagging, fouling, and corrosion as well as emissions (NOx, SOx) during combustion can be reduced. Also, it was reported on the biomass composition and ash melting characteristics of several miscanthus genotypes in Southern Germany (Iqbal and Lewandowski 2016). It was investigated how much greenhouse gas (GHG) was released during the quick pyrolysis of miscanthus to produce bio-hydrocarbons. The following are the results of a survey conducted by the National Institute of Standards and Technology (NIST) on the effectiveness of the NIST's e-learning program. Miscanthus transport, miscanthus cultivation, and system upgrades all made minor contributions to GHG emissions. The rate of SOC in the Miscanthus cultivation subsystem in particular had a significant impact on net GHG savings. More than 60% of emissions were saved using the bio-hydrocarbons created from the two upgrading processes to replace the equivalent of fossil fuels, which is the standard required by the EU rule for new biofuel plants (Shemfe et al. 2016). Miscanthus sp. was investigated as a potential lignocellulosic feedstock for biorefineries based on hot-water extraction as a pretreatment. Hot Water Extraction (HWE) or autohydrolysis is focused on extracting xylans for further use – acid degradation to furfural or fermentation to ethanol, with partial removal of lignin from native biomass. It showed that a rapid-growing, perennial grass such as miscanthus can be an effective source of bio-based products, which can be harnessed through an environmentally-friendly process of hot water extraction, without the use of harsh chemicals and the associated costs of chemical recovery (Nagardeoleka et al. 2017). DTXs can be used as an insecticide for the biological control of pine wilt disease through Monochamus Alternatus (MA, Japanese pine sawyer) termination, according to a report on a highly systematic, advanced process for producing destruxins (DTXs) from Miscanthus (MCT). The report also included a plan of application (Kim et al. 2019).

6.3.3   Potential for phytoremediation

The establishment of miscanthus and its application when growing in contaminated soil, plant selection and breeding are reviewed. Since the knowledge of how to obtain beneficial results by adding soil amendments has advanced significantly, the effects of soil amendments on the fate of contaminants and plant growth are included. Improved results have been reported for miscanthus production when plant growth regulators have been added (Davis et al. 2021).

6.4   Alfalfa (*Medicago sativa*)

Alfalfa was described as a major protein source for ruminants (Veronesi et al. 2010). Green technology production of industrial enzymes is another potential of Alfalfa. Transgenic alfalfa that produces a variety of microbial enzymes has been created and field-tested by a multidisciplinary research team at the University of Wisconsin, Madison. They include four distinct cellulases, alpha-amylase from Bacillus licheniformis, phytase from Aspergillus niger, and manganese-dependent lignin peroxidase (Mn-P) from Phanerochaete chrysosporium. The manufacture of the animal feed enzyme phytase, which was expressed at commercially viable levels (up to 2% total soluble protein) under field circumstances, has been the most successful to date (Austin-Phillips and Ziegelhoffer 2001).

6.5 Switchgrass (*Panicum virgatum*, L. Poaceae)

The lignocellulosic perennial grass known as switchgrass (*Panicum virgatum* L.) has great potential for use in the production of bioenergy. Because lignocellulosic bioenergy crops are largely resistant to cell wall breakdown, they produce less biofuel than they actually could. In comparison to non-transgenic controls, transgenic plants produced up to 18% more ethanol while retaining comparable growth and biomass. Switchgrass FPGS1 is inhibited, which enhances the production of lignocellulosic biofuels (Mazareietal, 2020). Switchgrass is a bioenergy raw material that is relatively abundant; however, its difficulty in managing is one of the economic obstacles to developing biofuels. Switchgrass's lignin content and S/G ratio were decreased at the cellular level by the caffeic acid O-methyltransferase (COMT) gene deletion, and transgenic lines outperformed wild-type switchgrass in terms of fermentation yield with Saccharomyces cerevisiae and wild-type Clostridium thermocellum (ATCC27405). Thus, indicates that the advantages of a raw material modified can be coupled with a modified consolidated bioprocessing microorganism as expected (Yee et al. 2014).

6.5.1 Bioenergy and biomass potential

The structure and division of organic compounds in the switchgrass biomass, were described. The disposition of carbohydrates and lignin concentration fluctuation was investigated in switchgrass biomass considering switchgrass (Panicum virgatum L.) as a multifunctional energy plant. Switchgrass biomass at seed filling in the second harvest year had a high lignin concentration of 105 g kg^{-1} dry matter (DM), indicating that it was highly relevant for the manufacture of solid biofuels. A sizable amount of CH_2O (693–742 g kg^{-1} DM) demonstrated that switchgrass biomass is suitable for the synthesis of second-generation bioethanol at this stage. According to Butkut et al. (2013), switchgrass in the second harvest year produced relatively high NSC yields at heading (an average of 28.4 g plant-1) and low lignin outputs (an average of 19.3 g plant-1) that are complementing properties of resources for biogas production. Also described was the biochemical and thermochemical conversion of switchgrass to biofuels (Balan et al. 2012). Switchgrass management for summer grazing and biomass production was described as dual-use (Richner 2013). The modification of cell walls to lower lignin content is necessary for lignocellulosic feedstocks such as switchgrass to reduce the resistance of converting biomass into biofuels. Growing biomass may be equally important for bioenergy feedstocks. In the biochemical process, sucrose and uridine diphosphate (UDP) are transformed into

UDP-glucose and UDP-fructose. UDP-glucose is produced by sucrose synthase (SUS), and cellulose synthase uses it to make cellulose for the production of cell walls (Poovaiah et al. 2015).

6.5.2   Value-added bio-products

Because of its high productivity, possible low demand for agricultural inputs, and advantageous effects on the environment, switchgrass is an excellent raw material for applications requiring value-added products. Switchgrass has the potential to store carbon, recover nutrients from runoff, improve soil, and provide habitat for grassland birds, among other environmental benefits. Some other potentials for the value-addition of switchgrass are gasification, newsprint development, and reinforcement of fiber in thermoplastic composites (Keshwani and Chen 2009). It was suggested that the biochar produced in this work might have a better liming effect and improvement of soil fertility and crop growth as a soil conditioner, and lead to double wins in saline soil improvement and a new approach for switchgrass utilization. Features and prospective values of bio-products (bio-oil, syngas, and biochar) derived from switchgrass grown in saline soil using a Fixed-Bed slow pyrolysis system were reported (Yue et al. 2017).

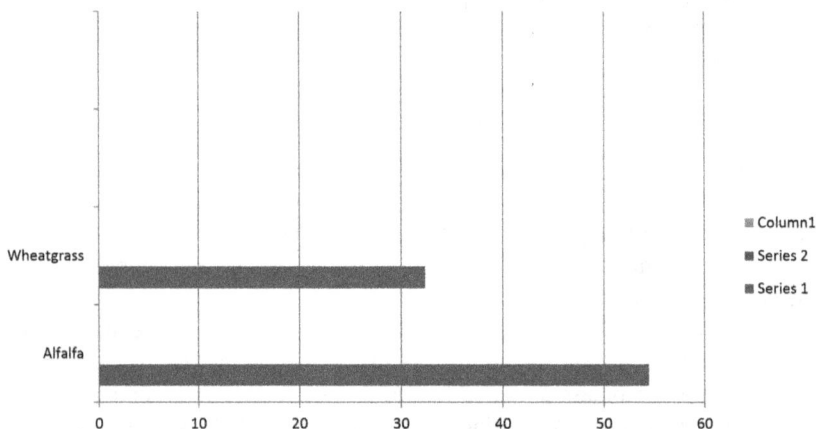

Figure 6.1. Gene pool of some forage crops (Meirman et al. 2013).

Succulent Crops

7.1 Aloe vera (*Aloe barbadensis*) crop

Aloe vera L. (Syn.: Aloe barbadensis Miller; Hindi: Ghikanvar; AV) of the family *Liliaceae (*sub-family of the *Asphodelaceae)* is a cactus-like perennial plant. In an experimental animal model, *Aloe vera* gel showed antihyperglycemic potential (Lanjhiyana et al. 2011). It is a medicinal plant with biological qualities that may be used in clinical studies, including anti-cancer, antibacterial, anti-inflammatory, anti-diabetic, and immunomodulatory capabilities (Zagórska-Dziok et al. 2017, Goudarz et al. 2015). It attracts demand in the world market owing to its nutraceutical, medicinal, and other uses. Aloe vera and its extracts are known to have significant markets in Australia, Europe, and the US. Aloe vera has the best business possibility among the numerous therapeutic plants because of the rapidly expanding demand for it on the international market. Aloe vera and other potentially valuable medicinal plants can only be grown in a few countries, including India, because of their unique geographical characteristics (Major and Ali 2016). *Aloe vera* has enjoyed great popularity in use in the scope of household remedies and cosmetics (Cseke et al. 2006). Aloe vera gel represents an alternative source of natural antimicrobial substances in the prevention of infections (Udgir and Pathade 2014). A lot of the health potential of *Aloe vera* has been linked to the polysaccharides contents in the gel of the leaves. Aloe vera exhibits a variety of biological qualities, such as antifungal, anticancer, wound healing, hypoglycemia or anti-diabetic effects, anti-inflammatory, immunomodulatory, gastroprotective, and skin permeability enhancing capabilities. There have also been documented important pharmacological uses, like the use of dried *A. vera* gel powder as an excipient in long-acting pharmaceutical dosage forms (Hamman 2008). The potential of Aloe vera constituents is shown in Table 7.1.

7.1.1 Propagation of *Aloe vera*

The protocol for the rapid propagation of *A. vera*, which is possibly applicable, was reported as a part of in vitro germplasm conservation. The report indicated

Table 7.1. Potential of Aloe vera constituents (www.amb-wellness.com)

Aloe vera constituents	Potential uses
1. Aloe vitamins	Helps in maintaining normal metabolic functions.
2. Aloe minerals	Serve as regulatory and structural elements, helps the body to neutralize toxins.
3. Aloe amino acids	Source raw material in making other cellular products, such as hormones and pigments. Play role in the formation of neurotransmitters enzymes (chemical messengers), antibodies and nutrient transporters
4. Aloe fatty acids	Needed for growth, reproduction and good health.
5. Aloe phytosterols and chromones	Agents responsible for the anti-inflammatory action.
6. Aloe flavonoids	Antioxidant, anti-inflammatory and anticancer.
7. Tannis	Possess antioxidant properties that can help prevent cellular damage.
8. Aloe enzymes	Provides some vegetable enzymes, basics of human nutrition.
9. Aloe saponins	They increase the concentration of antibodies in the blood, lower cholesterol and normalize blood sugar concentration.

that by transferring to the same medium, each divided shoot explants, explant multiplication could be continued even after a year. Successfully transferring regenerated plantlets to the soil could be possible where they grew well within 4-6 weeks with 83% survival (Dwivedi et al. 2014). The need for high-quality *A. vera* planting material is rising globally. Commercial plantation could be useful by micropropagation through shoot tip explants of *A. vera*. and is the practical application of plant biotechnology which is widely accepted and has attracted the multibillion-dollar sector globally, and these techniques can be employed to satiate A. vera need industrially. It was also stated that effective techniques for A. vera plantlet regeneration without an intermediary callus phase had been developed (Haque and Ghosh 2013, Khatun 2018). The aloe vera plant is essential for medicine and is utilized in the beauty and pharmaceutical industries all over the world. Even so, aloe grows organically through the vegetative division. However, the pace is too slow to supply the demand for high-quality planting material for commercial horticulture. The axillary branching method of micropropagation Aloe vera for elite selection employing shoot tips as explants were standardized. Microshoot rooting was completely achieved on phytohormone-free MS media. Hardened plants that had been regenerated were transplanted to soil and demonstrated an 85% survival rate. The regenerated plants resembled control plants morphologically (Kumar et al. 2011). A report highlighted the protocol involving in vitro propagation of *Aloe vera* Linn from shoot tip culture that assists in the propagation of thousands of Aloe vera plants from a few initial mother plants in a year period (Surafel et al. 2018).

7.1.2 Composition, properties and industrial potential

Processed leaves of aloe vera (Fig. 7.1a-c) create a fraction of liquid and bagasse. Most of the resulting bagasse is thrown away as garbage. For the pharmaceutical and agro-food sectors, both the bagasse and the liquid fraction may include intriguing metabolites with biological activity potential. A. vera gel and liquid fraction, ethanolic extract, EE-B, and phytopathogenic fungi can be controlled naturally in industrial crops during the pre- and post-harvest stages (Flores-López et al. 2016). The natural gel obtained from aloe vera coated with citric acid was reported to have a potential application for the shelf life and quality maintenance of mango fruit after harvest, making the aloe vera coated with gel a promising bio-preservative and a utilizable substitute to the synthetic preservative. A relatively convenient and safe measure to effectively extend the shelf life of post-harvest fruits and render them unharmful to the environment. It may provide an alternative to the more harmful synthetic coatings and other post-harvest chemical treatments. Aloe gel antimicrobial coating has also been identified as a suitable protocol to extend the shelf life of papaya fruits. Dipping mango fruit into an aloe vera extract and combining it with citric acid enables reducing the respiration rate of mango fruit and was effective at reducing the enzymatic activities typically considered to be responsible for the quality decay (Atlaw 2018, Marpudi et al. 2011). Sapota (Manilkara zapota) coated with edible Aloe vera gel was found to be a very good option for maintaining the fruit quality characteristics and shelf-life extension (Padmaja et al. 2015). Enzymatic processing is being more widely used to produce aloe vera concentrate or powder, which is considered appropriate in reducing the gel viscosity, thereby increasing the gel condensation rate, and making its drying economically efficient (Izadi et al. 2015). The development of a cloning protocol of A. vera to provide propagation material with superior quality to the private sector in southern Brazil, i.e., the A. vera juice industry. The described method appears to be a suitable technology for recombinant clonal multiplication of A. vera on a large scale, affording propagation material with high genetic and sanitary quality, the adoption of this biotechnological strategy could, in any extension and taking into account the specific conditions of each enterprise, be applied as a model ("tailor-made") to increase the incoming of Aloe-related companies, as a short-term goal, since it seems to be both technological and economical attainable (Campestrini et al. 2006).

7.1.3 Genetic characterization, genomics, and sequencing

Despite all its medicinal and industrial potential, there exists limited available genomic information on *Aloe vera*. Aloe vera's comprehensive functional genomics was studied by creating a transcriptome sequence database using the Illumina HiSeq technology and gradually annotating it with regard to the plant's metabolic specialization (Choudhri et al. 2018). Novel seed-setting soma clones in Aloe vera by morphological and genetic characterization were reported which may serve as a new genetic resource for biotechnological improvement (Das et al. 2015). High genetic variation among genotypes which is important in the conservation and exploitation of Aloe vera genetic resources was reportedly evaluated (Kumar et al. 2016). The genetic diversity of this plant species can be adequately described using RAPD and ISSR analyses, which have also been found to be the most useful for analyzing the genetic relationships between Aloe vera accessions (Rana and Kanwar 2017, Bhaludra et al. 2014). Geographical factors are among those that affect genetic variety, which ultimately changes the phytochemical and therapeutic qualities of A. vera. Traditionally, the variation in phenotypic or qualitative features has been used to estimate diversity. The ability of morphological markers to estimate genetic diversity in plants has been shown to be decreasing, with their heavy reliance on the environment for expression being the main cause. Despite this, they can still serve as a foundation for genetic variation. By directly examining variation that is regulated by genes or the genetic material, biochemical and molecular markers bypass many of the environmental influences that have an impact on characteristics. The examination of the origin and domestication of species has shown that molecular approaches are valuable for providing information on evolutionary relationships (Rana et al. 2018).

7.2 Cactus (*Opuntia ficus-indica* L.)

The cactus pear (*Opuntia*) belongs to the *Cactaceae* family. Its high taxonomic complexity has made its phenotype exhibit high variability based on the prevailing environmental conditions, and possesses common polyploidy; its reproduction is either by sexual or asexual, or with several interspecific hybrids. The term "cactus" (nopal in Spanish) is used to describe the entire plant, while "cactus pear" (tuna in Spanish) is used to describe the fruit and "cladode" (nopalito in Spanish) refers to the sensitive cladode and "pencato" to the plump mature cladodes or 'leaves' is used to describe the stems (Food and Agriculture Organization of the United Nations (FAO) (2013). The prickly pear (Fig. 7.2a-b) cactus (Opuntia sp.) plant genus is found majorly in arid and semiarid lands, owing to its abundance, it is useful to the human diet and in animal feed. It is useful also in control of desertification (Gonzalez-Cortés et al. 2018).

7.2.1 Natural and industrial resource

The processing and preservation of the fruit and cladodes are through different technologies and practices. Cactus pear is used in the preparation of traditional foods. Some products derived from the fruits of cactus pear are nectars, juices, syrups, jams, dried fruit, juice concentrates, and liquors. In the formulation and enrichment of new foods, many functional compounds can also be extracted and utilized. These are among others colorants, and natural additive gums with good potential in industries such as pharmaceutical, food, and cosmetics. Fiber-rich products can help in the control of diabetes and obesity. The plant acts as the cochineal insect's host and, as a result, serves as the foundation for making highly valuable wine. The cactus pear plant's variety of uses makes it a resource for starting small- and large-scale, artisanal, and/or commercial companies. The various plant parts can be used in a wide variety of ways. Depending on the maturation stage, different Opuntia species produce fruit with a variety of colors and cladodes with a variety of potential (either for food or feed). Consuming these foods has a number of positive health effects: veggies have been demonstrated to increase the nutritional value of diets. The plant will flourish where others fail in locations with a lack of access to food. Also, it's possible to cultivate cochineal on cladodes and gather these insects for dyeing. Functional chemicals have a wide range of potential applications in the food supplement and cosmetic industries. Opuntias, in contrast to other plants, have the advantage of having a variety of uses in the agricultural industry. There is also evidence of the nutraceutical potential and economic utility of cactus pear as a new source of fruit oils and functional foods due to the presence of lipid classes, fatty acids, fat-soluble vitamins, sterols, and b-carotene (Food and Agriculture Organization of the United Nations (FAO 2013, Ramadan and Morsel 2003). Opuntia cactus pear by-products have the potential for industrial use as a good and affordable source of minerals, unsaturated fatty acids, and natural antioxidants. However, further research must be done to assess the toxicity, in vivo activity, and bioavailability of cactus pear by-products before contemplating their inclusion as a dietary supplement or as a natural food antioxidant, (Ghazi et al. 2015).

7.2.2   Cactus pear seed oil

Cactus pear seeds are typically seen as a waste product after the manufacturing of pulp, juice, and jam. Nonetheless, it is possible to extract a small amount of edible oil from the seeds. If enough seeds are available to make the method commercially viable, the tiny amounts of edible oil extracted from the seeds will be more practically helpful. If costs are competitive, cactus pear seed oil can compete with other edible oils based on culinary application, neutral flavor, and composition. It shares some qualities with grape seed and maize oils, for example (Sáenz and Sepúlveda 2013). The commercialization of the oil makes the utilization of the entire pulp of economic value, reduces seed waste, and obtains previously untapped materials for technical and dietary uses (Ramadan and Morsel 2003).

7.2.3   Potential of Opuntia spp. fruits in food products

Industrial use for food is mostly in food preservation systems, fermented products, concentrates, and fresh fruit Juices, and frozen products, dehydrated products, and confectionery (Sáenz, 2013).

7.2.4   Cactus Pear fruit as a source of betalains colorants

The red and purple cactus pear fruit contains pigments called betalains, that are largely used in the culinary sector. It derived its name 'betalains' because the pigments are mainly extracts of beetroot (*Beta vulgaris* L.). The whole fruit is utilized in making colorants including the peel. However, it was reported that there are currently no reports of industries extracting colorants from cactus pear fruit or juice concentrates. Further study was advised to determine the best plants, varieties, and extraction methods because the pigment content of fruit and peel varies greatly, making it essential to choose raw materials with the highest pigment content (Sáenz 2013).

7.2.5   Production of hydrocolloids (mucilage)

Medical applications are an area where mucilage or hydrocolloids products are found useful. Cladodes and the fruit peel of the cactus pears can be used to extract mucilage. Despite the fact that the procedures now in use are expensive, complex, and have low yields, yet, employing extracts to treat stomach mucosa and other conditions is currently attracting industry interest (Sáenz 2013).

7.2.6   Uses of cactus cladodes

Cactus cladodes are used to make a variety of cosmetic goods, including soaps, creams, lotions, shampoos, and several kinds of face masks. They are employed as adhesives in the building sector, have the potential to stop corrosion, and are added to paints and adobe. They can be used in farming to improve the soil (Sáenz 2013). Cactus cladodes are useful for bioenergy production (biogas, biodiesel, or ethanol). Energy can be produced directly by burning crops developed expressly for energy usage, which produce 40 tonnes per hectare per year, or pruning debris from fruit plantations, which produces 10 tonnes per hectare annually. The cladodes are gathered, sun-dried, and crushed before being used in direct burning or coal-fired cogeneration; their calorific value ranges from 3,850 to 4,200 kcal per kilogram (Varnero and Homer 2017). Single-cell protein development (SCP) from lignocellulose biomass shows future promising technology aimed at providing protein supplements for both human food and animal feeds. In arid and semi-arid areas, one lignocellulosic feedstock with the potential to produce SCP is the cladodes of Opuntia ficus-indica (cactus pear) (Gabriel et al.2014).

7.2.7   Genetic resources and molecular characterization
of cactus pear

It was suggested that it would be possible to characterize the genetic resources of cactus pears and to launch genetic strategies for improvement and selection that serve as a guide for developing a conservation strategy and managing the genetic diversity of prickly pears. The study appears to be the first to employ molecular markers to analyze the genetic diversity of Moroccan prickly pear trees. To start a breeding program and a genetic selection, it could also be paired with agro morphological features (El Finti et al. 2013). For upcoming breeding initiatives, the gene pool for cactus pears needs to be characterized and evaluated.

A study looked at the developments in the use of molecular markers for the characterization of germplasm as well as the possible use of functional marker-based molecular tools to evaluate the germplasm's agronomically significant features. The Tunisian cactus accessions were studied genetically to determine their links to one another and to study polymorphisms using the random amplified microsatellite polymorphism (RAMPO) approach. The degree of diversity across genotypes of cactus pears was investigated, along with the connections between the most popular cultivars, wild accessions, and related species. The study also clarifies some of the ambiguities that resulted from classifying Opuntia germplasm based on the morphological characteristics (Zarroug et al. 2015). Evaluation of the level of genetic diversity by the application of the recent knowledge deriving using molecular markers technique was successfully reported. The effective use of the germplasm collected in designing the program of crop improvement has been one of the major utilizations of the collection. The report described recent activities aimed at introducing the program in other countries and promoting the use of cactus pear (Chessa 2010).

Figure 7.1. (a) Aloe vera (*Aloe barbadensis*) early development. (b) Matured Aloe vera (*Aloe barbadensis*). (c) Aloe vera (*Aloe barbadensis*) gel.

Figure 7.2. (a) Prickly pear cactus (Opuntia ficus-indica). (b) Peruvian apple (Cereus repandus) cactus varieties.

Root and Tuber Crops

8.1 Cassava (*Manihot esculenta*) crop

Cassava (*Manihot esculenta* Crantz) is a root vegetable grown in the tropics that belongs to the genus Manihot that can provide a staple food for an estimated 800 million world population. For smallholder farmers who are low-income, it is typically grown in peripheral areas with poor soils and erratic rainfall as one of the few basic crops that can be grown on a small scale without the need for machinery or expensive inputs (Fig. 8.1). The annual production of cassava has expanded globally since 2000 by an estimated 100 million tons (Food and Agriculture Organization of the United Nations 2013).

8.1.1 Current biotechnological advances and challenges

The importance of cassava as the fourth-largest source of calories required globally helps to develop this crop biotechnologically. Advances and new difficulties should be assessed on a regular basis. With the use of plant biotechnology, cassava has a number of potentials to improve as a crop in the face of ongoing global change. Cassava cultivars are currently subjected to genetic manipulation; with potential implications for future breeding. The first transgenic cassava plant's historical development provided the framework for studying molecular aspects of somatic embryogenesis and the creation of friable embryogenic calluses. Complicated plant-pathogen interactions were studied in order to benefit from this information and aid cassava in battling bacterial infections. Candidate genes that may be involved in resistance to viruses and whiteflies—the two most crucial characteristics of cassava—were also looked at. The literature is reviewed for analyses of key developments in transgenic-mediated nutritional enhancement, mass production of healthy plants through tissue culture and synthetic seeds, ideas on employing genome editing, and difficulties related to climate change. The use of biotechnology has led to significant advancements in cassava over the last 30 years, but they must expand the proof of concept to cassava producers' farms (Chavarriaga-Aguirre et al.

2016). The leaves, roots and stems of cassava contain potentially hazardous quantities of cyanogenic glycosides principally linamarin with the greatest percentage of lotaustralin. Acyanogenic cassava cultivars are required for cassava to develop into a consistently secure and acceptable cash crop and to lower the amount of cyanide waste generated by cassava-processing facilities. It has been shown that CYP79D1/CYP79D2 antisense plants had linamarin levels that are lowered by up to 94% and 99%, respectively, in the leaves and roots. According to the findings, young plants may carry linamarin from leaves to roots. These acyanogenic cassava plants promise a more nutritious and commercially viable food source as well as a way to examine how cyanogens affect crop yield and protection from herbivory (Siritunga and Sayre 2003). Putative transgenic lines of cassava variety H-226 were successfully developed which were PCR-positive for *hph* gene and *Rep* gene indicating the integration of transgenes of interest against the cassava mosaic virus responsible for cassava mosaic disease (Arun et al. 2016).

8.1. 2 Antioxidant potential

Cassava has a good pack of antioxidants, making the plant a natural protective against stress-related diseases. Isolation and quantification of the protein and rendering the sugar content of the cassava low together with the antioxidant potential analysis, which also included the inhibitory values on free radicals using pulverized extract cassava roots was reported (Mehran et al. 2014). Lately, experimental biotechnology techniques have been used by BioCassava Plus, a Bill and Melinda Gates-backed venture, to solve several limitations facing African cassava. There are numerous advantages of cassava for small-scale farmers, and its usefulness for industrial applications was examined. The effects of biotic and abiotic factors on production output root quality, nutritional sufficiency, marketability and acceptance, and commercial procedures are also discussed (Gbadegesin et al. 2013).

8.1. 3 Feedstock for bioethanol production and other value-added products

One of the main bio-based goods is bioethanol, which is mostly made from starchy plants. Various starchy crops are found utilizable for bioethanol development by means of fermentation protocols by microorganisms used in the fermentation process. Starchy substrates are utilized as feedstocks for bioethanol production which cannot be fermented directly to ethanol by specific microorganisms. This is due to the fact that yeasts, which are employed for fermentation, cannot consume starch molecules made up of lengthy chains of glucose molecules; therefore, simpler fermentable sugars must be hydrolyzed into simple glucose molecules prior to fermentation. One of the starchy crops, cassava, can be used to produce sugar raw materials, which are necessary for the manufacturing of ethanol and other goods with added value. It is a potential alternative to other starchy crops. For a higher yield of ethanol, it is necessary to do saccharification and fermentation simultaneously. Due to high production and comparatively less demand for use as food and fodder, preparation of this renewable fuel, especially from starchy tuber crops like cassava hold the exceptional ability to meet the future energy demand. Because of their widespread availability, high starch content, and affordable processing, tuber crops offer a possible alternative for the production of bioethanol to other substrates like lignocellulose and algal biomass, which have significant processing and maintenance costs. It has been crucial to produce products like ethanol from various starchy feedstocks in order to lessen our need for fossil fuels as well as the contribution of petroleum derivatives to climate change and air pollution. Using the right methods, starchy biomass can also be converted into additional value-added products outside biofuel, such as fermentable sugars, organic acids, solvents, and drink softeners (Jagatee et al. 2015).

8.2 Potato (*Solanum tuberosum* L.)

The genetic diversity of potatoes, with its ability to adapt to changing and diverse settings, particularly the more frequent harsh conditions produced by climate change, is still underutilized for the production of new varieties which can cope with these changes and situations, hence the need to educate farmers on conserving the genetic diversity (Ortiz and Mares 2017). Due to introgression, interspecific hybridization, auto- and allopolyploidy, sexual compatibility among many species, a mix of sexual and asexual reproduction, potential recent species divergence, phenotypic plasticity, and the resulting high morphological similarity among species, potato taxonomy is thought to be quite ambiguous. New researchers have improved the taxonomical links among potato species by using molecular methods to identify the genes governing different types of resistance. For potato breeding, ancient forms of domesticated and wild cousins of the potato have been employed as sources of desirable features, such as resistance or tolerance to diseases, pests, environmental challenges, and tuber quality. At numerous gene banks across the world, potato germplasm with relevant alleles for diverse breeding goals is kept. These resources, which contain priceless knowledge, are available for use in study and breeding. For the germplasm collection to be successfully used, accurate species identification based on evolving taxonomy is required (Machida-Hirano 2015).

8.2.1 Genetic resources preservation

Successful cryopreservation techniques have been developed for a significant number of plant species, which is turning into a highly essential tool for the long-term preservation of plant genetic resources. Due to their allogamous character, practical protocols developed for utilizing in vitro tissue culture can be a quick and trustworthy alternative to managing vegetative propagation in gene banks for the preservation of potato genetic resources. Cryopreserved materials are essential for providing a long-term backup of field collections against the loss of plant germplasm. With optimal cryopreservation techniques that offer high regrowth, the genetic diversity that occurs in tissue culture cells during protracted subcultures can be avoided, paving the way for a methodical and purposeful cryo-banking of plant genetic resources. The preservation of genotypes that are difficult to preserve using other techniques is made possible by cryopreservation treatments for potatoes, which effectively complement field and in vitro conservation (Niino and Arizaga 2015). A report on the status of potato conservation in significant gene-banks and some countries, as well as the many types of protocols that can be used to protect potato genetic resources, was published (Muthoni et al. 2019).

8.2.2 Genome and genomics

Potato (*Solanum tuberosum* L.) is a great global non-grain food crop and is central to world food security. It is highly heterozygous propagated by cloning, autotetraploid, and suffers acute inbreeding depression. According to reports, the potato genome sequence provided a platform for the genetic improvement of this significant crop (Xu et al. 2011). The use of genomics in potato research has also revolutionized the crop's production by giving potatoes new traits to combat abiotic stresses and post-harvest problems, establishing them as the primary crop for particular non-food uses, and arming them with the resources to face challenges in the future. The non-scientific community needs to be aware of the potential of research derived from genomics to address a variety of challenges, for advancements in potato research (Wendt and Mullins 2011).

8.2.3 Industrial prospects

Industrial potential for potatoes includes the development of biodegradable plastic that can be used as a substitute for polyacrylate. A feasible replacement for petroleum-based products is the goods produced from potatoes and developed from the waste products of potatoes treated for starch and/or protein extraction (Wendt and Mullins 2011). The probability of utilizing potatoes for the development of new materials, such as proteins, antioxidants, and bioplastics, instead of their use traditionally for food production was reported. Potato components, extraction technologies and summed up possible directions for the production of new products, considering the potential of processing waste as a raw material were studied (Priedniece et al. 2017). Potato peel waste can produce products like biofertilizers, biosorbents, biofuels, dietary fiber, antioxidants, and food additives through various processes like fermentation, extraction, and other treatments. The use of potato peel utilization in food and nonfood purposes for example extraction, utilization of bioactive components, biotechnological application, and livestock feed, etc. was reviewed (Javed et al. 2019).

8.3 Sweet potato (*Ipomoea batatas* Lam.)

Sweet potato (*Ipomoea batatas*) contributes importantly to food security and is a vital root crop globally, following potatoes and cassava (Wadl et al. 2018). It is the seventh most significant food crop having a lot of significance; it can tolerate varied environmental conditions and has a good nutritional value. Due to the rising food demand and the need to preserve agricultural and genetic resources, it is important to assess the genetic diversity of this important crop (Rodriguez-Bonilla et al. 2014). It was reported that sweet potato helps to avoid periods of food crisis and is a staple food during times of scarcity in Northern Benin (Adjatin 2018). Varieties of both solid and liquid organic waste products can be derived from the cultivation and processing of sweet potatoes. Derived solid waste includes peelings and trimmings from the sweet potato root and sweet potato leaves and vines. Large amounts of nutrient-rich wastewater are produced by liquid waste. Materials from sweet potato waste contain macro- and micronutrients, proteins, phenolic compounds, pigments, and carbohydrates that could be extracted or used for a variety of downstream processes and products (Akoetey et al. 2017).

8.3.1 Genetic diversity

The blending of two divergent techniques of characterization, genetic markers and agronomic traits has been successful in differentiating or clustering the sweet potato genotypes in agreement with their geographical origin or phenotypic descriptors. This knowledge could be applied by both breeders and farmers to discover and safeguard commercial cultivars, and hence for traceability purposes (Palumbov et al. 2019). A study for the establishment of ways of released identification, elite and local sweet potato genotypes available in the fields for farmers in Nigeria which could help breeders in the genetic diversity study of this crop and for its improvement was reported. The three morphological clusters and four from the analysis of isozymes showed some peculiar agronomic traits of the genotypes, which could be utilized for population improvement of the crop (Efisue 2015).

8.4 Yam (*Dioscorea* spp., Dioscoreaceae)

Yams (*Dioscorea* spp., Dioscoreaceae), are significant crops that are grown for their starchy or medicinal properties in the tropics and subtropics. Yams make an important contribution in the areas of food security and medicine in developing countries. Recent years have seen advances in our knowledge of the phylogeny of Dioscorea and the phylogenetic relationships within the genus. Additionally, cutting-edge molecular technologies have been created for genome analysis, including cytogenetics, genetic mapping, germplasm characterization, functional genomics, and tagging. Major genes controlling resistance to anthracnose and the Yam mosaic virus have also been linked to candidate random amplified polymorphic DNA markers. These markers could be converted to sequence characterized amplified regions and utilized for marker-assisted selection for resistance to diseases. To create expressed sequence tags for gene discovery and as a source of additional molecular markers, a preliminary cDNA library has been created. Complementing conventional breeding approaches, genetic engineering offers a powerful tool for yam improvement (Mignouna et al. 2008, Mulualem et al. 2018).

8.4.1   Genetic resources and genomics

Yam (Dioscorea spp.) is a popular root crop that is used as a significant industrial raw material, food source, and feed. For the creation of variety and conservation strategies, knowledge of the genetic diversity found among yam genetic resources is required. According to reports, some carefully chosen genetic resources can serve as a source of novel genes for yam breeding and variety development (Mulualem et al. 2018). Yam genomics was reported as supporting West Africa as a major cradle of crop domestication. It investigated how the African yam (*Dioscorea rotundata*), a significant crop in early African agriculture, came to be domesticated. With the help of a computer program, they were able to identify the source of the problem and provide a solution. It was for this reason that the Niger River basin saw an increase in the production of African yams (Scarcelli et al. 2019). An international team of scientists from the United Kingdom, Japan, and Germany developed a novel high-quality genomic sequence for the white Guinea yam (*Dioscorea arotundata*), a common tuber crop that is extremely important for millions of people's subsistence and sociocultural lives, primarily in West and Central Africa. "Knowing this trait and having access to the white Guinea yam's genomic resource will be invaluable in breeding a better yam, one that will increase food security in West and Central Africa and the standard of living for the region's smallholder farmers." According to a phylogenetic study, yams are a distinct monocotyledon lineage from the Poales (rice), Arecales (palm), and Zingiberales (banana). "This is an important breakthrough. The development, which began with rice in recent years, means that yam has now joined those crops with a complete DNA sequence, according to co-author Dr. Robert Asiedu, R&D Director for the International Institute of Tropical Agriculture-West Africa in Ibadan, Nigeria." The entire DNA sequence will make it much easier to understand how genes regulate important qualities like flowering, illnesses, and other traits like quality, which will speed up and improve the accuracy of new variety breeding. Professor Ryohei Terauchi, a co-lead author from Kyoto University and the Iwate Biotechnology Research Centre in Japan, said, "This will help to get rid of some of the many issues affecting yam producers in Africa and other parts of the world. These include pests and diseases, post-harvest losses, and the need for more sustainable agricultural farming systems," according to the statement (ScieNews 2017, Tamiru et al. 2017).

8.4.2   Value-added products

Value addition in yams could be helpful to promote consumer preferences enabling the producers to increase their income. A study was conducted to evaluate the nutritional diversity in three dioscorea species kept at the Kalyani Center, BCKV, All India Coordinated Research Programme on Tuber Crops, and to describe the value-added yam products with a view to creating a good market inside and outside the country. Value-added products like fried yam balls, yam chips, yam flakes and yam pickles popularization will help to erase the label of poor man's food, promoting it to find a place in the diet of the middle and upper-middle-class (Mitra and Tarafdar 2011).

8.5   Tiger nut (*Cyperus esculentus* L.)

Tiger nut (*Cyperus esculentus*), also called 'yellow nutsedge', is a plant of the regions of the tropics and the Mediterranean. The delicious, almond-like tubers have enormous potential for health and nutrition since they are rich in fiber, carbohydrates, and proteins. They are rich in potassium, glucose, phosphate, vitamins C and E and oleic acid. With the exception of minor research on its oil and the manufacturing of tiger nut milk, the tiger nut has received very little attention from science and technology. In addition to being prevalent in cooler areas, *Cyperus esculentus* is common in tropical and temperate zones. It grows in the wild and is used as a weed in addition to being used as a crop plant. The ecological plasticity of *C. esculentus* results in its exceptional variability and diversity of morphotypes (Sanchez-Zapata et al. 2012, Warra et al. 2017, De Castro et al. 2015).

8.5.1   Molecular analysis and phylogenetic

The molecular study sheds new light on C. esculentus' evolutionary trajectory. C. esculentus' phylogenetic position and biogeography were studied with the aim of adding new information to better understand its evolutionary history. The findings have a number of taxonomic and phylogenetic ramifications, including a theory on the origin and phylogeography of this species, which most likely evolved in Africa during the late Cenozoic and repeatedly reached the Americas without the aid of Columbian trades (De Castro et al. 2015). A prereading effort is suggested as a prerequisite for enlarging the genetic basis and ultimately realizing the full potential of tiger nuts in a phenotypic characterization of tiger nuts (Cyperus esculentus L.) that paves the way for the nuts' improvement (Asare et al. 2020).

8.5.2   Potential products

Agricultural products like tiger nut tubers are found useful for the development of various products. Tiger nut flour, tiger nut milk, and tiger nut oil are the three main usable products that are primarily produced from the processing of tiger nut tubers. These three products can then be used to create a variety of food and non-edible goods. Despite being used for food and feed since ancient times, tiger nut products are also used to make gluten-free bread, beverages, and snacks (Yu et al. 2022). Researchers have carried out a great deal of study to create additional beneficial products from tiger nut tubers (Fig. 8.2) (Maduka and Ire 2018).

Figure 8.1. (a) and (b) Cassava (*Manihot esculenta* Crantz) plant varieties. (c) and (d) Varieties of cassava (*Manihot esculenta* Crantz) tubers.

Figure 8.2. (a) Brown variety tiger nut (*Cyperus esculentus*) tubers. (b) Yellow variety tiger nut (*Cyperus esculentus*) tubers. (c) Yellow variety tiger nut (*Cyperus esculentus*) tuber oil. (d) Brown variety tiger nut (*Cyperus esculentus*) tuber oil.

Fruit Crops

9.1 Oranges (Citrus)

Citrus is one of the most important economic fruit crops in the world. The group of fruits that belongs to this category includes oranges, lemons, limes, grapefruits, and tangerines (Fig. 9.1) (Olife et al. 2015). Citrus essential oils (EOs) which are extracted from their peel, leaves and flowers are of high industrial value. The trash from the processing of citrus contains enormous amounts of these oils as a byproduct. Due to the qualities of citrus EOs, such as the bactericidal, virucidal, fungicidal, and medicinal properties associated with their fragrance, they are a potential alternative to chemical-based antimicrobials for food, agriculture, pharmaceutical, and sanitary applications as well as a valuable component of cosmetics and perfumes (Palazzolo et al. 2013). Orange is the world's large cultivated fruit studied. The major products of industrial processing is essential oil and juice, d-limonene, pectin and citrus pulp as by-products, the pulp is dried, and stored as a product for commercialization (Zanella et al. 2013). The monoterpene limonene was the most prevalent substance in the sweet orange's volatile component. Aldehydes were the primary constituents in the oxygenated fraction, followed by alcohols and esters. The most important ones were decanal, linalool, and decyl acetate. Low concentrations of sesquiterpene hydrocarbons were discovered (Ojeda de Rodríguez et al. 2003).

9.1.1 Genetic improvement

The development of acceptable techniques for the assessment of unintended consequences in cultivated genetically modified lines indicates the potential of genetic modification as a tool for future fruit tree enhancement. Several procedures are used in citrus genetic manipulation experiments to get the same result. In traditional citrus breeding, two DNA sets—one from the mother and one from the father—are mixed to create hybrid embryos through a process known as meiosis. These embryos create hybrid seeds that are rarely fertile. A significant number of seeds that have been properly cultivated and include both the mother's and father's DNA were assessed for identification of progeny showing industry-relevant characteristics (Grosser 2015). According to the report, long-term agricultural production of GM citrus does not have any unanticipated consequences on transgene expression, which is stable and does not change crop attributes. Based on the use of reliable and effective assessment techniques, it was demonstrated that the transgenic citrus trees expressing the selectable marker genes that are typically used in citrus transformation were highly comparable to non-transformed controls in terms of their overall agronomic performance. This suggested that subsequent investigations into the potential pleiotropic effects brought on by the integration and expression of transgenes in field-grown genetically modified oranges might concentrate on the newly inserted trait(s) of biotechnological interest (Pons et al. 2012).

9.1.2 Sweet orange (*Citrus sinensis*)

Citrus sinensis orange fruit peels were steam distilled to obtain orange oil, which was then tested for its antibacterial capabilities against a number of medically significant pathogens. The findings suggested that orange oils have antibacterial capabilities and could be employed in local therapies to treat illnesses brought on by the microorganisms examined (Obidi et al. 2013). In another development, optimizing the cryopreservation of sweet oranges using a modified aluminum cryo-plate technique, could be useful for the subsequent formation of plants from embryogenic citrus lines (Souza et al. 2017)

9.1.3 Mandarin orange (*Citrus reticulata*)

Mandarin refers to a class of delicious oranges with a loose peel that are also known as "kid-glove" oranges. One of the most widely consumed and economically viable fruits in the world is the loose-skinned mandarin orange (*C. reticulata* Blanco) (Das et al. 2014). It is suggested that oil from tangerine (*Citrus tangerine*) which is an orange-colored citrus fruit consisting of hybrids of mandarin orange, or called a mandarin variety, can be used as a natural inhibitor to inactivate the growth of A. niger on rubberwood surfaces (Jantamas et al. 2013). Citrus peels may be able to absorb heavy metals, according to research studies. To bio-adsorb nickel, cadmium, and lead from polluted soil and vegetable coriander, which is grown and eventually consumed by people, tangerine peel, one of the most prevalent fruit waste materials, was used (Sobhani and Ziarati 2017).

9.1.4 Lemon (*Citrus limon* L.)

The use of the GC-MS (Gas Chromatography-Mass Spectrometry) technology revealed that the essential oil (EO) of lemon peel exhibit showed substantial antioxidant and antimicrobial properties both in vitro and in barley soup as a food model (Moosavy et al. 2017). The chemistry and gel-forming properties of pectin from lemon peel have enabled this naturally occurring biopolymer to be utilized in pharmaceutical health promotion and treatment (Bagde et al. 2017).

9.1.5   Industrial potential of oranges

The sweet orange peel's essential oil has antifungal properties that can be used as a food preservative (Abdel-Fattah et al. 2015). Therapeutic and commercial utilization of *Citrus* peel essential oils showed potential for phytomedicine and as an antioxidant agent (Javed et al. 2014). The extraction of bio-oil from orange peels was reported (Meshram et al. 2015). From the findings of chemical constituents of essential oil of orange (*Citrus sinensis* L.) peel it was proposed that the oil could be utilized as a source of industrial feeds for chemical synthesis, as preservatives also as a flavor and fragrance. This will also serve as a new window for research in anti-TB drug development and the use of essential oil in the production of drugs and foods (Egharevba et al. 2016). Due to the citrus fruit's value as a food nutrient and flavor, as well as the fact that their waste peel is a valuable source of essential oil for cosmetics, pharmaceuticals, and other commercial and domestic uses, there are additional ways to evaluate the citrus fruit's underlying economic value. One of these is by investigating the citrus fruit's essential oil. Processing citrus peels into essential oils is a great way to turn these wastes with a high potential for environmental pollution into a resource with significant uses for economic prosperity as well as for securing the public health benefits of a safer and healthier environment, likely to be obtained from the indirect waste management alternative so suggested (Ezejiofor 2011). According to test results, tested sweet orange oil shows a lot of promise for preventing the isolation of organisms from spoiling leather and leather products. Orange seed oil is a naturally occurring organic material that degrades through biodegradation, posing minimal to no environmental risks. Therefore, it is advised that the oil be used as a preservative against the fungus that might cause the deterioration of these items in some steps of leather manufacturing, in shoe polishes, and in other shoe treatment agents (Olufunmi 2010).

9.1.6   Waste resources

Orange fruits produce wastes from orange peel mostly from human consumption which if not well managed could bring about environmental pollution. Orange fruits produce waste mostly from the orange peel from human consumption which if not well managed could bring about environmental pollution. Regarding waste recycling and circumventing littering and degradation of waste that are related to the environment, the research studied as a dominating component one of the terpenes of the orange peel among others identified in the structure of limonene, therefore validating it (one of the terpenes), that were found in modest quantities. Limonene is an essential oil having several commercial and home uses. Exploring essential oil is a further way to ascertain the citrus tree's underlying economic value because citrus has value as a food nutrition and flavoring as well as a source of essential oil that may be utilized in cosmetics, pharmaceuticals, and other industrial and home applications. Citrus peels are regarded to have a significant risk of contaminating the environment, but, turning them into essential oils might make them into a resource that has a great potential for commercial success as well as securing the advantages of a safer and healthier environment for the general public (Ezejiofor et al. 2011). Pectin, which can be used in biological conversion processes such as the production of polyhydroxyalkanoates, has been found to be abundant in the solid portion of leftover oranges. Most pectin is made up of neutral sugars and galacturonic acid (PHAs). Pectin was isolated and hydrolyzed to be used as a substrate for Cupriavidus necator's cell development after these wastes were successfully analyzed. It was confirmed that these wastes include a considerable number of soluble sugars (almost 40 g.100 g^{-1}) and pectin. The hydrolyzed extract showed to be a useful source of carbon for the growth of C's cells. This is a practical method for converting leftovers into high-value products. With a high concentration of fibers and carbohydrates (mostly soluble sugars), the waste from oranges and passion fruit holds great potential as a substrate for the biological conversion processes. The hydrolyzed orange wastes demonstrated higher growth rates, enabling their application as a less expensive carbon source for C. necator cell development (Locatelli et al. 2019). Industrial orange waste potential as organic amendments in citrus orchards were reported (Maksoud et al. 2015).

9.2 Grape (*Vitis vinifera* spp. sativa)

One of the many fruit crops that humans domesticated in the past was the grape (Fig. 9.2). The ability of grapes to produce jellies, beverages, and other goods has made them one of the most economically significant plants in the world. Of the several different substances which make up the berry's complex phytochemistry, the majority of which have been shown to have medicinal or health-improving effects (Georgiev et al. 2014).

9.2.1 Genetic engineering

Verifying the genotypes and their potential in future grape breeding programmes is important. The use of two simple sequence repeat (SSR) markers were reported to be adequate enough to determine the variation among all the grape genotypes (Singh et al. 2013). Genetic Literacy Project in its August 2018 edition reported a recent development on how grapes genetically resistant to mildew rot could cut pesticide use and French wine prices at the same time (Engadget 2018). This would be a biotechnological treasure to the agro-industry and market.

9.3 Pomegranate (Punica granatum)

Pomegranate trees are durable plants, capable of growing up to 9 meters in height as shrub-like plant (Fig. 9.3), within this family, there is only the Punica genus. There are two species of the Punica genus, *P. granatum* L. and *P. protopunica* Balf (Kahramanoğlu and Usanmaz 2016). Since ancient times, the pomegranate (Punica granatum) has played a significant role in traditional medicine, religious symbolism, and art. According to the literature, studies began as early as 1821. However, vigorous exploration didn't start until about 20 years ago. Many of the pomegranate's recently discovered health advantages are supported by science due to its capacity to reduce oxidative stress, increase nitric oxide production, and regulate inflammatory pathways (Mackler et al. 2013). Their seeds, fruits, peels, flowers and juice of pomegranate have been utilized as foods and medicines for thousands of years. They contain dietary fiber, antioxidants, healthy and saturated fats, minerals and other nutrients (Smith 2014). Pomegranate is a strong antioxidant that can compete with or even outperform red wine and green tea. Also, its potential as a therapeutic or adjuvant for the prevention and treatment of several forms of cancer and cardiovascular disease is suggested by its anticarcinogenic and anti-inflammatory qualities. Pomegranate may aid in preventing infection by pathogenic E. coli O157:H7, antibiotic-resistant organisms, and dental pathogens due to its antibacterial characteristics such as Methicillin-resistant Staphylococcus aureus (MRS) (Jurenka 2008).

9.3.1 Genetics

The restricted genomic resources have limited further elucidation of genetics and the evolution of these interesting traits (Yuan et al. 2018). Yet, another study has given crucial information for developing a pomegranate core germplasm collection without duplicating plant material, sustainably managing pomegranate breeding programs, and designing conservation methods for maintaining local pomegranate genetic resources (Çalişkan et al. 2017).

9.4 Banana (*Musa* spp.)

Two species of the genus Musa, banana plants (*Musa paradaisica*) and banana peels (*Musa sapientum*), are cultivated all over the world and used in food and medicine. Flavonoids, carbohydrates, reducing sugar, tannins, saponins, anthraquinones, steroids, glycosides, phytosterols, phenols, and terpenoids were among the phytoconstituents included in the extracts (Kibria et al. 2019). Banana (*Musa* spp.) is among the most essential economically tropical fruit crops. It ranks as the fourth-largest food crop after maize, rice, and wheat. Over a surface of 4.84 million hectares, it is produced annually in 95.6 million tons in more than 100 different countries. A nation's food security depends significantly on bananas since they are a low-cost, readily available source of energy. Many vitamins A, C, and B6 are present, as well as beta-carotene. Mood-enhancing elements like serotonin, melatonin, and tryptophan are also present in bananas. The crop is the primary source of revenue in rural regions and a substantial source of income in many developing countries.

Banana has an essential role in poverty mitigation (Deepika et al. 2018). Banana (Fig. 9.4a-c) is one of the fruit crops that is widely grown and consumed. According to reports, the fruit, peel, and banana stem are all high in total carbs, fiber, and minerals, particularly potassium. The banana stem is used to create different ropes, fabrics, papers, and handicrafts. It is also effective in treating a number of diseases, including dysentery, diarrhea, intestinal colitis, discomfort, inflammation, and snake bite. Polyphenols or antioxidants, such as gentisic acid, cinnamic acid, ferulic acid, caffeic acid, catechin, protocatechuic acid, and are regarded as potential compounds present in the stem (Dixit 2019). To understand how genomes other than M. acuminata and M. balbisiana may have had a role in the development of cultivated bananas, the taxonomic status of bananas as described in numerous research works, including numerical taxonomy, cytological studies, and molecular markers, was studied (Rekha 2016).

9.4.1 Genetic manipulation

The first two successful reports of the occurrence of genetic modification in bananas were in the year 1995. By using Agrobacterium-mediated transformation and particle bombardment, great effort has been made to generate novel cultivars with strong resilience to biotic and abiotic stressors and with higher nutrient levels. With an overview of significant developments in numerous programs being run throughout the world, a review highlighted developments in banana genetic engineering. The issues of intellectual property embedded in the technology, public perceptions toward the adoption of genetically modified bananas, as well as various regulatory barriers that prevent the technology development from moving forward, have been identified as major barriers to realizing the full potential of genetically modified bananas for human consumption (Pua et al. 2019). Banana genetic engineering is viewed as a great alternative for improving sterile cultivars or cultivars that are resistant to conventional breeding methods. Numerous successful attempts have demonstrated the power of this technology in creating transgenic banana varieties disease-resistant to abiotic stress and abiotic stress tolerance. Only a few of the genetically engineered bananas have qualified for field studies and some are currently undergoing nutritional human trials. In the near future, GM bananas may significantly improve food security by increasing nutritional value and productivity (Ghag and Ganapathi 2017).

9.4.2 Tissue culture and propagation

The banana is probably the only crop that can be eaten both as a fruit and as a starchy food. In much of the world, it thrives. Develop disease-free plants via tissue culture. By using this technique, banana plants are better able to grow into whole plants from a single meristem at the shoot apex. And can be replicated in several thousand plants in less than one year. The majority of organisms do not become extinct when the initial tissue explant is removed (Thomas and Reddy 2011). Banana in vitro multiplication has drawn more attention due to its ability to produce genetically consistent, disease- and pest-free, true-to-type plants. The production of disease-free plants can be achieved through in vitro micropropagation of bananas using shoot tip culture. The ability to grow elite clones with better growth and increased stress tolerance capability while also propagating banana plants in vitro has the potential to yield enormously profitable results (Deepika et al. 2018). The in-vitro banana plants are better than the ordinary suckers owing to their lively development, intelligence and higher yields (Saravanan et al. 2017).

9.4.3   Bioproducts development

It has been suggested that banana pseudostem sap is a valuable waste plant resource for creating thermally stable cellulosic substrates. The flame retardant property of cellulosic material was added using banana pseudostem sap, an environmentally beneficial waste plant product (Basak et al. 2015). In producing composite material, the fiber of a banana that has considerable strength is possibly easily combined with the fiber of cotton or synthetic fiber. In the process of extraction of the fiber, the generation of a large number of lignocellulosic wastes creates problems in the adjacent area when disposed of. The generated waste in the extracted banana fiber is utilizable in handmade paper capable of mitigating the problem of pollution as well as provision opportunities for employment in the rural populace (Arafat et al. 2018). By turning the banana center core, a biological waste in banana plantations, into flour, which could be used as a food product in bakery preparations and soup mix, the banana center core can be efficiently employed as a source of food material. A process for the drying and powdering of the banana central core was developed in an effort to add value to it. The research showed that the banana center core could be dried and made into flour which could be effectively utilized as an ingredient in a bakery and for the preparation of soup based on the physical and functional properties (Ambrose and Naik 2016).

9.5   Avocado (*Persea americana* Mill.)

The fruit avocado contains significant amounts of monounsaturated fat as well as a comparatively high quantity of vital lipid-soluble substances like vitamin E, beta-sitosterol, and carotenoids. Avocado (Fig. 9.5a-d) is a potential source of dietary protein. However, it was found that the high concentrations of antinutritional components (alkaloids, tannin, phytic acid, etc.) which render it useless for human and animal nutrition are the restriction to the full utilization of

Persea americana seeds. However, the concentrations of these antinutrients in the raw seeds were decreased by processing techniques such as soaking and boiling (Qin and Zhong 2016, Talabi et al. 2016).

9.5.1 Genetic improvement

A large genetic foundation exists for the crop's enhancement through breeding and selection, as evidenced by the accessions' high genetic variety (Abraham and Takrama 2014).

The avocado community now has a new genomic method that can be used to evaluate the genetic diversity of avocado germplasm globally and to optimize breeding and selection programs for avocados by combining traditional breeding methods with molecular approaches, thereby enhancing the effectiveness of avocado genetic improvement (Kuhna et al. 2019).

9.5.2 Potential uses of by-products

The avocado industry makes extensive secondary products and provides a vital feedstock for food and non-food uses, Due to the presence of bioactive components like polyphenols—a wonderful treasure for their antioxidant and anti-inflammatory properties, seeds and peels can be recycled in foods and cosmetics. Both seeds and peels can be recycled to make carbonaceous materials, which has a substantial impact on lowering environmental pollution by purifying water. Avocado wastes can be used to produce biofuel and for photocatalysis (Colombo and Papetti 2019). Techno-economic and environmental evaluation of avocado fruit suggested an attractive opportunity for a biorefinery for complete uses as well the importance of the level of integration (Dávila et al. 2017). Colorants and oil extracted from avocado waste have great prospects in the industries for making soaps and cosmetics (Hennessey-Ramos et al. 2018).

9.6 Mango (*Mangifera indica* L.)

Mango is a very significant fruit crop of the tropic belonging to the family Anacardiaceae, its cultivation dates back 4000 years. Almost 34.3 million tons of fruit are produced annually through the cultivation of mangoes in over 100 different nations. The literature outlines some of the most recent research and development advancements in mango breeding, genomics, and rootstock creation. An important achievement in the past few decades has been the perfection of propagation techniques such as veneer grafting, softwood grafting and epicotyl grafting in mango. The development of plant biotechnology tools and methods offers trustworthy, repeatable processes for somatic embryogenesis, regeneration of better cultivars, micropropagation, and germplasm preservation of commercially valuable cultivars (Bimal and Singh 2017, Ian and Bally 2011). Composition of mango fruit (Fig. 9.6a-c), and its great nutritional content and health advantages are due to the inclusion of minerals and phytonutrients, as well as the changes that these go through during development and postharvest.

The nutrients found in mangoes can be divided into three categories: phytoconstituents (colors, polyphenols, phenolics, and volatile constituents), micronutrients (minerals and vitamins), and macronutrients (fats, carbohydrates, lipids, amino acids, proteins, and organic acids). Mango fruit also contains structural carbohydrates such as pectins and cellulose. Lysine, cysteine, leucine, valine, phenylalanine, arginine, and methionine are some of the essential amino acids. Particularly the omega-3 and omega-6 fatty acids increase in the lipid composition as the fruit ripens. Chlorophylls (a and b) and carotenoids are two of the mango fruit's most significant pigments. Malic and citric acids, which give the fruit its acidity, are among the most significant organic acids. The fruit's aromatic profile is largely influenced by the volatile constituents, a heterogeneous group with a variety of chemical roles (Maldonado-Celis et al. 2019).

9.6.1 Recombinant DNA technology

Information on the biotechnological studies of mango that presented some practical biotechnological fixes for issues related to increasing mango output was evaluated (Hugo and Zuazo 2008). Biotechnology can expedite mango improvement programmes and supplement conventional breeding. Studies utilizing in vitro culture and selection, micropropagation, embryo rescue, genetic transformation, marker-assisted characterization, DNA fingerprinting, etc. are being conducted at numerous research institutes throughout the world and are being conducted to complement traditional breeding with biotechnology and accelerate mango improvement. Fruit ripening-related genes were cloned, and efforts were made to introduce these genes into plants. Numerous research facilities are also used in conducting genetic works on the genetic diversity of Mangifera species and mango cultivars (Krishna and Singh 2007).

9.6.2 Waste to wealth

Bulk residues like peels and kernels of mango are rendered as waste after industrial processing. They can be used to create wealth; it was claimed that mango starch ester could be successfully produced from the kernel using solubility tests, FTIR, 1H, and 13C NMR. It was physically a mixture of mango starch and vinyl laurate, and was thermally more stable than its antecedents, probably due to the hydroxyl groups that are lower in number, a higher number of covalent bonds and higher molar mass of the products after esterification. This industrial waste bio-based derivative is a great choice as an additive for oil drilling fluids to prevent fluid loss since it can help with better wellbore cleaning and guard against problems with fluid loss to rock formation (Marques et al. 2019).

9.7 Cashew (*Anacardium occidentale* L.)

In 2011, 4.27 million tons of cashew nuts were produced worldwide. The largest producer of raw nuts is Vietnam, while India leads the world in processing and exporting nuts. The high growth of production in some regions, like West Africa, is expected to give the cashew market a boost. For instance, over the past five years, Nigeria's production has increased by 40%. High-value by-products are also being used more and more, particularly those derived from cashew nutshell liquid (Dendena and Corsi 2014).

9.7.1 Genetic resources and diversity

The analysis of genetic relationships in cashew using morphological traits and Random Amplified Polymorphic DNA (RAPD) Banding data can be used to improve plants, describe new varieties, and assess the purity of a variety in plant certification programs (Samal et al. 2003). An evaluation of the cashew genetic diversity (Fig. 9.7a-d) would facilitate the choice of desired high-yield germplasm for the production of nuts of high quality and market value in the future (Chipojola et al. 2009). The many varieties of cashew might be distinguished using specific markers, which were identified, in the future (Thimmappaiah et al. 2016).

9.7.2 Potential of cashew by-products

Cashew nut and its by-products are safe for use in feed for animals. They can substitute corn and soy in diets for animals. The by-products are rich in nutrients (carbohydrates, fats, minerals and proteins). They enable the animals that eat them to perform in a manner that is comparable to that achieved with corn and soybeans. Along with the nutritional and financial benefits, cashew nuts and their by-products are used in livestock production (Guy Marcel et al. 2011). The protein concentrate from cashew kernels can be used in food formulation as a protein ingredient. The ability to produce protein concentrate from cashew kernels allows for the use of generated broken kernels in the cashew nut manufacturing process (Lima et al. 2021). Cashew nut generates various by-products via industrial processing such as cashew apple, cashew apple bagasse rich in organic nature and a source of lignocellulosic material for the production of bioethanol, Cashew nut shell liquid (CNSL), which is enriched with a variety of phenolic chemicals, may be useful when creating coatings and laminates. These by-products are important for industries because they can be transformed by further processing into several bioactive compounds, polymers and other products (Sharma et al. 2020).

9.8 Apple (*Malus domestica* Borkh)

An introductory knowledge, of selected titles in agricultural genetics in areas such as selective breeding, and plant science can be provided through sensory assessment of apples (Goldman 2012). Many varieties of brown apples (Fig. 9.8a-f) after cutting and breeding for reduced flesh browning have been extensively researched, coupled with an investigation of increasing vitamin C content. There are many compounds of interest in apples, but a particular one (phloridizin) has received priority in the medical community, due to its significant role in the regulation of glucose metabolism and diabetes (Brown and Maloney 2015).

9.8.1   Genetic breeding for improvement

Knowledge of the genetic form and morphology of apples provides significant help for germplasm selection and maintenance of required material superior breed cultivars. Detailed analyzed apple cultivars are utilizable for the conservation of genes and for designing a systematic breeding strategy towards beneficial cultivars of apples with increased adaptation to specific climatic conditions and fruit (Ganopoulos et al. 2018). By creating a genome variation map through the genome sequencing of 117 different accessions, a study emphasized the genetic foundation of apple domestication and evolution and provided useful information for marker-assisted breeding and apple improvement. Apple history is revealed by this genome re-sequencing, which also helps a two-stage hypothesis for fruit expansion (Duan et al. 2017). Fruit texture is a complicated character that includes mechanical and auditory qualities that depend on the alteration in the cell wall during fruit production and ripening. Two complementary quantitative trait locus (QTL) mapping methodologies were employed in-depth to determine the genetic regulation of fruit texture in Apple, which has a substantial diversity in fruit texture behavior that directly influences both the consumer's enjoyment and post-harvest performance. It is possible to choose apple types with valuable fruit quality by using these joint integration methodologies, which provide useful information on the precise regulation of fruit texture (Di Guardo et al. 2017).

Figure 9.1. (a) Sweet (*Citrus sinensis*) orange tree. (b) Mandarin orange (*Citrus reticulata*). (c) Sweet oranges. (d) Lemon orange (*Citrus limon* L.). (e) Sweet orange seeds. (f) Tangerin orange seeds. (g) Lemon orange seeds.

Figure 9.2. (a) Grape (*Vitis vinifera* spp. *sativa*) nursery plants. (b) Grape plantation. (c) Grape fruits. (d) Grape seeds.

Figure 9.3. Pomegranate (*Punica granatum*) (a) The plant. (b) The flower. (c) The fruits. (d) The fresh seeds.

Figure 9.4. (a) Banana (*Musa* spp.) garden plantation. (b) Banana home garden (c) Banana fruit.

Figure 9.5. (a) Ripe avacado (*Persea americana* Mill.) pear. (b) Avacado flesh and nut. (c) Fresh avocado kernel. (d) Semi-dried avocado kernel.

Figure 9.6. (a) Mango (*Mangifera indica* L.) crop. (b) and (c) Mango (*Mangifera indica* L.) fruit varieties.

Figure 9.7. (a) Cashew (*Anacardium occidentale* L.) nursery. (b) Cashew (*Anacardium occidentale* L.) tree. (c) Cashew (*Anacardium occidentale* L.) fruit early growth. (d) Cashew (*Anacardium occidentale* L.) ripe/mature fruits.

Figure 9.8. (a) Apple (*Malus domestica* Borkh) plant. (b) Apple (*Malus domestica* Borkh) fruit green yellow variety. (c) Apple (*Malus domestica* Borkh) fruit purple variety. (d) Fresh apple (*Malus domestica* Borkh) seeds. (e) Dried apple (*Malus domestica* Borkh) seeds. (f) Apple (*Malus domestica* Borkh) seed oil.

Cucurbit Crops

10.1 Cucumber (*Cucumis sativus* L.)

The term "Cucurbit" generally designates all species belonging to the Cucurbitaceae family, which comprises approximately 800 species in 130 genera. Cucurbits are mainly annual, herbaceous, tendril-bearing and frost-sensitive vines and are among the economically most significant vegetable crops globally (Weng and Sun 2011). For effective genetic improvement and conservation, it is important to know the geographical origin of economically important plants, However, this has been slowed down by erroneous geographic sampling, where relatives are more likely to show up in far-off places. It was observed that using herbarium collections as DNA sources can result in less biased species sampling (Schaefer et al. 2009). The plastid rbcL, matK, ndhF, atpB, trnL, trnL-trnF, rpl20-rps12, trnS-trnG, and trnH-psbA genes, spacers, and introns were used to analyze the phylogenetic connections in the order Cucurbitales utilizing 14 DNA sections from the three plant genomes. The collection includes 664 ingroup species, which accounted for more than 25% of the nearly 2600 species in the order and spanned all genres but two. For the datasets for the three genomes, Topologies generated by maximum likelihood analyses were mostly consistent. The eight families of Cucurbitales were related to one another in the following ways: (Anisophylleaceae, Apodanthaceae, Cucurbitaceae, Corynocarpaceae, Coriariaceae), (Tetramelaceae, Begoniaceae, Datiscaceae). Based on this DNA data and morphological information from the literature, the authors rearranged tribes and genera within Cucurbitaceae and offered a more natural categorization for this family. Actinostemmateae, Indofevilleeae, Thladiantha, Momordiceae, and Siraitieae are five of the 95 genera that make up their 95-genus revised classification, which also contains 15 tribes. Formal naming requires two new names and 44 new combinations for the Cucurbitaceae family (Schaefer and Renner 2011). Zhang et al. (2006) presented the phylogeny of the Cucurbitales based on DNA sequences from nine different locations across three genomes, focusing on the implications for the evolution of morphology and sexuality.

10.1.1 Genetic diversity and improvement

The cucumber (Fig. 10.1) is representative of other Cucurbitaceae and the fourth most important vegetable crop in the world. Results are based on a core collection of 115 accessions that contains over 77% of the SSR alleles, and they suggest that the knowledge of cucumber's genetic makeup can help in creating appropriate conservation strategies and lay the groundwork for population-level genome sequencing in cucumber (Lv et al. 2012). Gene action studies showed that the predominant function of non-additive gene action is for the control of all the traits under study; hence heterosis breeding can be used for the genetic improvement of seed vigor and yield traits in cucumbers (Thakur et al. 2017). Characterization of the cucumber genotypes based on their performance and recommendation of genotype selection for cultivation in the Derived Savannah, Southeast Nigeria agro-ecological zone was reported (Ene et al. 2016).

10.1.2 Medicinal and environmental potential

The powder form of C. sativus extracts contains phytoconstituents that have successfully been produced using the spray dry method to exhibit potential cytotoxic effects on specific human cancer cell lines. The biological properties exhibited by the extracts were further supported by the presence of the phytoconstituents like alkaloids, saponins, flavonoids, and steroids (Foong et al. 2015). Report proposed *Cucumis sativus* as a promising candidate plant for phytoextraction of arsenic (As) from soils and water. *Cucumis sativus* (cucumber) exhibited the highest tolerance among the 4 plants tested (cucumber, sorghum, corn and wheat) (Hong et al. 2011). Cucumber plants (*Cucumis sativus* L.) are capable of accumulating high levels of Persistent Organic Pollutants (POPs) such as polychlorinated dibenzo-*p*-dioxins (PCDD), polychlorinated dibenzofurans (PCDF) and polychlorinated biphenyls (PCB), in their tissues (Warwick et al. 2019).

10.2 Sweet melon (*Cucumis melo*)

One of the most notable and well-known farmed cucurbits is *Cucumis melo*. C. melo is produced largely for its fruit, which is characterized by a sweet aromatic flavor, with considerable variation and size (50 g to 15 kg), the color is flesh (white, orange, pink, and green), rind color (orange, red, green, gray yellow, and white), form (flat, round and elongated), and dimension (4 to 200 cm) (Nũnez-Palenius et al. 2008). Melon (*Cucumis melo* L.) (Fig. 10.2a-d) is an extremely varied species globally cultivated. The hunt for nucleotide diversity in this species is currently made possible by advances in extremely parallel sequencing. The largest melon SNP dataset to date was produced by thorough data resequencing for wild, exotic, and domesticated (landraces and commercial) melon transcriptomes, providing an amazing representation of the species diversity. The data is a significant resource for creating a library of allelic variants of melon genes that can help in future in-depth research of population genetics, marker-assisted breeding, and gene identification targeted at developing superior varieties (Blanca et al. 2012).

10.2.1 The genome and genetic variation

In future, investigations into the utilization of the genome sequence will hasten our understanding of cucurbit evolution and advance our breeding tactics. Melon is a significant crop in the world of horticulture, and its genomic sequence was reported (Garcia-Mas et al. 2012). The relatively high level of genetic variation of *Cucumis melo* L. revealed that further investigation will provide information leading to more effective strategies and goals for hybridization and long-term germplasm management (Wolukau et al. 2018). Crop advances via biotechnological protocols for the current changing climatic conditions is significant for development and commercialization (Sultana and Rahman 2014).

10.2.2 Bio-products

Pectin from the waste of cucumis *melo* is a new ingredient used for food in Malaysia and can be used as a good substitute for gelatin in food industry uses and other non-food potential interventions. Optimizing new pectin acid extraction from honeydew (*Cucumis melo* L. var. inodorous) peels as a thickener for halal food was reported (Omar et al. 2020). Melon residues are found to be an excellent source of natural phytoconstituents of various uses, such as for nutraceutical, cosmetic, or pharmaceutical industrial ingredients, for making novel foods with useful components, as well as fertilizers and animal feed (Vella et al. 2019). The tocopherol content of the seed oil was discovered to be high, with a preference for the + tocopherol fraction. The melon seeds are used as an alternative source of plant oil, a useful raw material for food applications (Mallek-Ayad et al. 2018). The many uses of the melon and its relevance for value aggregation are encouraged, based on the fruit's nutritional qualities, antioxidant capacity, antiproliferative effect, and potential as a substrate for solid-state fermentation. The current state of knowledge regarding melon residues (*Cucumis melo* L.) and their function in health promotion and biotechnology applications were reviewed. Due to their bioactive contents, melon waste such as seeds and peel are of interest because they can be used as a good substitute for the valorization of fruit processing by-products (Rolim et al. 2019).

10.3 Canary melon (*Cucumis melo*)

Canary melon is a bright-yellow cucurbit having white color inner flesh. *C. melo* belongs to the general family of melons, the cucurbit family (Cucurbitaceae). Hence, the general scientific name is *Cucumis melo* (abbreviated; *C. melo*).

10.3.1 Seed oil quality and industrial applications

The first literature report on canary melon seed oil aimed at quality assessment (Table 10.1a), fatty acids elucidation using GC-MS (Table 10.1b) hexane extract from the seed oil was reported to be coldly saponified for use in the cosmetic and medicinal sectors (Warra et al. 2015). Seeds recovered from industry by-products of cucumis melo showed a high potential utility as a source of unconventional oil for biodiesel, cosmetic and pharmaceutical sectors (Fig. 10.3) (Górnaś and Rudzińska 2016).

Table 10.1a. Physicochemical properties of Canary melon seed oil (Warra et al. 2015)

Parameters	Values*
Oil yield (%)	50.42±0.01
Color	Light cream
Acid value mgKOH/g	0.35±0.01
Iodine value gI_2/100 g	135.6±0.07
Peroxide value meq H_2O_2	1.80± 0.00
Saponification value mgKOH/g	233.62±0.01
Relative density (g/cm³)	0.82±0.01
Refractive index	1.44±0.00

* Values are expressed as mean and ± standard deviation of triplicate determinations.

Table 10.1b. Major fatty acids derived from hexane extract of Canary melon L. seed oil (Warra et al. 2015)

S.N.	Name of fatty acid	MF	MW	RI	SI% to T.C.
1.	Palmitic acid	$C_{17}H_{34}O_2$	270	1878	91
2.	Stearic acid	$C_{18}H_{36}O_2$	284	2167	90
3.	11-Octadecenoic acid	$C_{19}H_{36}O_2$	296	2085	94
4.	Octadecenoic acid	$C_{19}H_{36}O_2$	296	2085	90
5.	Oleic acid	$C_{18}H_{34}O_2$	282	2175	90
6.	Octadecenoic acid	$C_{18}H_{34}O_2$	296	2085	90
7.	n-Hexadecanoic acid	$C_{16}H_{32}O_2$	256	1968	89
8.	Ricinoleic acid	$C_{18}H_{34}O_3$	298	2337	80
9.	Docosanoic acid	$C_{23}H_{46}O_2$	354	2475	90

Note: S/N = Serial number, M.F. = Molecular formula, M.W. = Molecular weight, RI = Retention index, SI% = Similarity index, T.C. = Target compound.

10.3.2 Other potential products

Assessment of the crop waste resources of the melon yellow canary as animal feed was reported (Fodil and Arbouche 2007). Canary melon-seed oil, a high-quality product made from the leftovers from the agro-industrial processing of melons, has a high concentration of vitamin E and a high quantity of polyunsaturated fatty acids, particularly linoleic acid (Rabadán et al. 2020).

10.4 Bitter gourd (*Momordica charantia*)

Bitter gourd (*Momordica charantia*) is rich in health-promoting substances that make it an economically important crop species for plant improvement. Nutritional quality development in this species will provide a sustainable, inexpensive complement to medical programs aimed at alleviating certain human diseases (Behera et al. 2007). Due to oxidative stress, which causes cell membrane degradation and many clinical disorders, antioxidants and secondary metabolites have recently received a lot of attention for their potential to prevent disease. Momordica charantia, also known as "Karela", is a promising source of both macronutrients (ash) and micronutrients, according to studies (Fig. 10.4), (carbohydrate, protein, fiber contents) as well as a potent modulator of antioxidant and liver function indices (Orji et al. 2018, Saeed et al. 2010).

10.4.1 Genetic diversity and breeding

Strategies for the collection and conservation of this important cucurbit vegetable crop reportedly showed that sample variability present in the collected germplasm could produce valuable breeding lines upon using appropriate recombinant breeding methods among the diversified landraces (Raja et al. 2019). It is significant to find out accessions with desirable characteristics for breeding programs by classification (Pramote et al. 2011).

10.4.2 Bioresource potential of Momordica species

Triterpenoids, carotenoids, and phenolics, which have the potential to be employed as antioxidants in the nutraceutical, cosmetic, and pharmaceutical industries, were abundant in the green fruits, leaves, and stems of Momordica species (Nagarani et al. 2014).

10.5 Watermelon (*Citrullus lanatus*)

Watermelon (*Citrullus lanatus* (Thunb.) Matsumand Nakai) is a useful crop (Fig. 10.5) for genetic research due to its small genome size, and the many available gene mutants (Guner and Wehner 2004). An increase in breeding is being provided by private seed firms thanks to the development of new technologies and watermelon genomic information. These new discoveries and understanding are also making watermelon more desirable for public-domain research (Sarria 2017).

10.5.1 Uses of watermelon products

Plant oil and proteins extracted from watermelon fruit waste are regarded as a good source of Omega-6 fatty acid and proteins (Fakir and Waghmare 2017). Watermelon seed has viable characteristics for making biosorbents. An investigation of the efficacy of the ground seed of watermelon as substitute low-cost biosorbents for the Pb (II) ions removal from aqueous solutions as reported (Adeoye et al. 2020). The jam preparation from watermelon waste with its important acceptable characteristics capable of being commercialized for industrial use was reported (Souad et al. 2012). The utilization of watermelon seed oil as feedstock for oleochemicals that contribute significantly to the economy with little or no environmental impairment was reported (Akinsanoye and Omotoso 2018).

10.6 Sponge gourd (*Luffa cylindrica*)

The manufacturing of loofa sponge (Luffa cylindrica), which has been effectively used in a variety of applications since it was first identified as a matrix for the immobilization of microbiological cells in 1993, has undergone biotechnological advancement. For the production of ethanol, organic acids, enzymes, and secondary metabolites, the cells immobilized in loofa sponge have performed well and more effectively than free suspended cells and those immobilized in conventionally utilized natural and synthetic polymeric materials. The technology has been used to build biofilms for the remediation of home and industrial wastewater rich in inorganic and organic debris as well as for the treatment of wastewater containing hazardous metals, dyes, and chlorinated chemicals. Moreover, the construction of a bioartificial liver device using three-dimensional loofa sponge scaffolds for hepatocyte cultivation has been suggested (Saeed and Iqbal 2013).

10.6.1 Bio-products

The application of fibers from *Luffa cylindrica* in a biomass-packed bed for the treatment of paint industry effluent before releasing it into the environment was investigated. It demonstrated that Luffa cylindrical fibers can be used as crucial packing material in a biomass filter to treat paint industry effluent before releasing it into the environment in accordance with WHO limits, indicating that the release would not result in an obvious environmental degradation and the extinction of the flora and fauna in the ecosystem (Ighalo et al. 2020). "*Luffa Cylindrica*" natural fibers can also be processed into industrial products such as insulation (Daniel 2016). The usage of LUFFACHITIN derived from the residue of the sponge-like dried fruit of *Luffa aegyptiaca* as a weavable skin substitute was reported (Jiang et al. 2014).

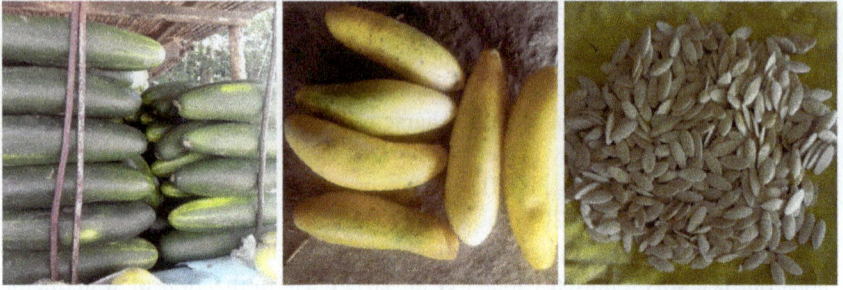

Figure 10.1. (a) Fresh cucumber (*Cucumis sativus* L.). (b) Dried cucumber (*Cucumis sativus* L.). (c) Cucumber (*Cucumis sativus* L.) seeds.

Figure 10.2. (a) Ripe *Cucumis melo.* (b) Cross-sectional view of *C. melo.* (c) *Cucumis melo* seeds. (d) *Cucums melo* seed oil.

Figure 10.3. Canary melon: (a) Fruits. (b) Seeds. (c) Seed oil. Seeds are the source of oil for biodiesel, cosmetic and pharmaceutical sectors.

Figure 10.4. Bitter gourd (*Momordica charantia*): (a) Plant, (b) Fruit, (c) Seeds, (d) Seed powder and (e) Seed oil.

Figure 10.5. (a) Water melons (*Citrullus lanatus*). (b) Cross-section of water melon. (c) and (d) Water melon seed varieties. (e) Water melon seed oil.

Herb and Spice Crops

Herbs are plants valued for their aromatic and medicinal properties, they grow in the wild while spices are seeds, fruits, roots, barks, or other plant parts mainly known for their utility as food, coloring or preservative. Spice and herbs are rich store houses of different bioactive compounds and are well known for their beneficial effect on health, they are grown both as wild or cultivated medicinal plant species (Hiltan Mathew 2000).

11.1 Black cumin (*Nigella sativa*) crop

In the history of spices, black cumin (*Nigella sativa* L.) in the family Ranunculaceae is one of the most important earliest cultivated plants with worldwide distribution. It serves as a reservoir of bioactive compounds of substantial medicinal importance and contains essential macro and micronutrients which function as structural and important components of metalloproteins and enzymes in living cells; it has been used in the treatment of various ailments (Gashaw 2020, Yessuf 2015). *Nigella sativa*, also known as *black cumin*, is mentioned in the earliest Islamic medical texts. It belongs to the Ranunculaceae family and is an annual herbaceous or peppery plant. Cumin seeds (black) (Fig. 11.1) are referred to as thymoquinone and are known to cure some diseases and ailments in humans and were reported to have contained fatty acids such as palmitic acid, stearic acid, oleic acid, and linolenic acid. The use of seed oil as an antibacterial agent against multi-drug resistant pathogenic strains in hospitals was reported by Hussain et al. 2016, Hussein et al. 2016, Hasan et al. 2013, Inayatullah Bhatti et al. 2013). *Nigella sativa* seed and its products are one of the most important medicinal products, with a long history of usage in therapeutics. The literature viewed from 1960–2015 revealed its pharmacological and medicinal effects (Hussein et al. 2016). It has been utilized for thousands of years ago as a substance for curing many disorders. The *N. sativa* seed has been found to have properties encompassing a wide spectrum of pharmacological activities include immunopotentiation, antihistaminic, anti-diabetic, anti-hypertensive, anti-inflammatory, and antimicrobial activity. In addition to other compounds, it is known to be a source of thymoquinone, thymohydroquinone, dithymoquinone, p-cymene, carvacrol, 4-terpineol, t-anethol, sesquiterpene longifolene, nigellicimine and nigellicimine-N-oxide, -pinene, and thymol (Sultana et al. 2015).

11.1.1 Toxicity removal

Complexation of thymol (Fig. 11.1.1) using a pH meter, the effects of several metals, including Fe(III), Cr(VI), Cu(II), V(IV), and Co(II), were investigated. The research proved that thymol interacts strongly with all metals to create complexes. Out of the aforementioned metals, it was found that Fe(III) and V(IV) form relatively stable complexes with thymol. Moreover, compared to other metals, these metals formed complexes at lower pH levels. The pKa value of thymol was also determined. The human body needs each of the aforementioned metals in trace amounts, and when they are present in excess, they can cause heavy metal toxicity, which can have a number of negative effects. Thymol may be useful in removing any harmful metals from the body which were caused by any of the aforementioned metals, according to reports. By building complexes with these metals or by reducing the metals to their reduced state, thymol may be able to eliminate these metals from the body (Kishwar et al. 2013).

11.1.2 Potential for energy, industry and environment

Biodiesels for utilization in an unmodified CI engine can be produced from the seeds of black cumin as one of the identified biomasses; sustainable biofuel for many energy applications is developed from *Nigella sativa* methyl ester (Muniappan et al. 2020). The seed waste of *Nigella sativa* left after the oil extraction is found to be useful as a bio-adsorbent for reactive dye removal from dye effluents (Raja and Kiruba 2020). The waste biomass of *Nigella sativa* seeds was reported as an important environment-friendly and low-cost sorbent for the absorption of lead from wastewater (Addala et al. 2018). Thymoquinone, saponins, tannins, flavonoids, triterpenoids, and glycoside are reported as active compounds from oil products and waste of *Nigella sativa* potentially useful as a material for the pharmaceutical industry (Barkah et al. 2021 Hannan et al. 2021, Ermumcu and Şanlıer 2017, Khader and Eckl 2014).

11.2 Fenugreek crop

Fenugreek (*Trigonella foenum-graecum* L.) plant belongs to the family Fabacecaeis and is cultivated globally as a dye, condiment, medicine and food. It is one of the good medicinal herbs with nutritional value found on the continents of Asia, Europe, Africa and Australia. Even though fenugreek plant biotechnology is still in its infancy, groundbreaking biotechnological research was conducted in 1945 and provided the first report on the effects of diniconazole, a fungicide of the triazole class, on fenugreek cell suspension cultures. The increased production of protein and commercially significant metabolites like trigonelline, sapogenin, isoflavonoid pterocarpans, diosgenin, gitogenin, and tigogenin from callus, leaves, stems, and roots explants were highlighted in numerous studies using cell suspension and callus culture in the studies that came after it. Callus and cotyledon utilization have been emphasized in studies on plant tissue culture.

It was reported against this background that persistent efforts are required to improve fenugreek via biotechnological approaches in addition to conventional plant breeding tools (Aasim et al. 2014, Aher et al. 2016). It contains phytoconstituents like saponins, steroids, flavonoids, alkaloids, etc. It is well known that fenugreek has hypoglycemic, hypocholesterolemic, and anti-inflammatory properties. Fenugreek is now known to be a helpful medicinal plant with the potential to treat ailments as well as a source for creating the raw chemicals used in the pharmaceutical industry, such as steroidal hormones. Due to the fact that fenugreek is a self-pollinated plant, a mutation breeding strategy can be used to produce mutants with a predictable growth pattern. Point mutations in fenugreek can be created by radiation and chemical mutagens (Moradi Korand Moradi 2013). For plant growth and development, significant development of raw materials for genetic advancement in the genetic variability and quantitative traits was induced and the promising mutants associated with an increase in yield potential of the crop were isolated (Rajoriya et al. 2016).

11.2.1   Potential for health, industry, and environment

Diosgenin, 4-hydroxy isoleucine, and the fiber component of fenugreek are the three most thoroughly investigated bioactive ingredients. These substances are known to have a positive impact on a number of physiologic indicators, such as glucose tolerance, inflammation, insulin action, liver function, blood lipid levels, and cardiovascular health. Knowledge of the molecular processes behind the beneficial effects of fenugreek has been gained (Fuller and Stephens 2015). A relationship was investigated between diosgenin (Fig. 11.2c), a steroidal saponin (see section 11.2.2) for medicinal uses and fenugreek as a source of raw material for the production of steroidal hormones (Chaudhary et al. 2018). Fenugreek seeds (Fig. 11.2a) are rich in fiber, phospholipids, glycolipids, oleic, linolenic, and linoleic acids, choline, and vitamins A, B1, B2, C, nicotinic acid, and niacin, among many other essential nutrients. It can thrive in a broad variety of environments, is tolerably tolerant of salinity and drought, and can even be successfully produced in marginal areas. These characteristics and the potential to remove heavy metals from soil make fenugreek suitable for a variety of farming systems (Ahmad et al. 2016). High chemical compositions of fat, protein, fiber, ash and carbohydrate were reported in fenugreek seed oil (Fig. 11.2b) (Al-Sebaeai et al. 2017).

11.2.2 Diosgenin

Diosgenin belongs to the triterpene group and is the typical bioactive compound of the Dioscorea family. It is a saponin that is spirostan (A spirostan found in DIOSCOREA and other plants). Yamogenin is the name given to the 25S isomer, which has a double bond at positions 5 and 6, a hydroxyl group replaced at position 3 betas, and an R-configuration at position 25. It is a naturally occurring substance found in the Dioscorea (wild yam) species that is used as the precursor for the synthetic production of several steroids, including cortisone, pregnenolone, and progesterone. It functions as a metabolite, antineoplastic agent, antiviral agent, and apoptosis inducer. It is a spiroketal, a sapogenin, a hexacyclic triterpenoid, and a 3beta-sterol. It comes from a spirostan hydride (National Center for Biotechnology Information 2019).

11.3 *Artemisia annua* L.

Artemisia annua L. (Asteraceae) is a popular source of the unique sesquiterpene endoperoxide lactone artemisinin and is utilized in the treatment of chloroquine-resistant and cerebral malaria. Artemisinin is a known strong antimalarial drug which is naturally occurring from the plant artemisinin and is the most effective anti-malarial drug. It is considered to be a safe drug even for pregnant women, with no adverse reactions or obvious side effects. A combination of derivatives of artemisinin, an artemisia plant extract (Artemisia annua L. or A. annua), with other drugs was found to be a very effective treatment of malaria. They are known as artemisinin-based combination therapies or ACTs. The antimalarial active compound in Artemisia annua is artemisinin (Fig. 11.3a-b) other compounds present include; flavonoids, and aromatic oils (Namuli et al. 2018, Dalrymple 2013). Artemisinin is the first choice of treatment for malaria. However, malaria parasites become drug-resistant when Artemisinin is administered as a single compound. To prevent artemisinin resistance, WHO has advised ACTs in combination with the medications mefloquine and chlorproguanil. It was realized that the supply of artemisinin is not stable or sufficient and as a result, treatment remains expensive. Therefore, expanding prevention and control interventions and putting all of their recommendations into practice requires a sizable budget, sustained political commitment, and robust cross-border cooperation. The removal of oral artemisinin-based monotherapies from the market and antimalarials that do not meet international quality standards were both advised that governments of endemic countries take (Malaviya et al. 2019). The essential oil from the plant has medicinal properties and is rich in mono- and sesquiterpenes aside from important variations in its percentage and composition have been reported (major constituents can be camphor (up to 48%), germacrene D (up to 18.9%), artemisia ketone (up to 68%), and 1, 8 cineole (up to 51.5%)), studies have shown that the oil has antibacterial and antifungal activities (Bilia et al. 2014, Malaviya et al. 2019).

11.3.1 Genetic adaptability, regeneration and genomics

The adaptability of some artemisia mutant genotypes, which were planted in different locations, was identified as a result of gamma irradiation (Syukur et al. 2011). The Artemisia annua L. plant was used to test an effective in vitro method for multiple shoot bud induction and regeneration using leaf and stem explants in different concentrations and combinations of plant growth regulators. *A. annua* L. plants with high artemisinin yields could potentially be developed through the use of this system, which demonstrated the potential for rapid propagation of shoots from leaf and stem explants (Dangash et al. 2015). It was reported that *A. annua* L. culture protocol was optimized to produce yields of biomass, dry matter, and artemisinin—key components of cost price—in order to improve farmers' access to the new product and to sustainable development technology (Pop et al. 2017). A genuine *A. annua* plant genome size is a special, important, and fundamental parameter that can provide additional insight into genomic studies of artemisinin biosynthesis and development. Research reported the facilitation of the schedule of the *A. annua* whole genome sequencing project, optimization of assembly techniques and insight into its subsequent genetics and evolution (Liu et al. 2018).

11.3.2 Commercialization

For large-scale commercial cultivation of transgenic *A. annua*, the use of physical and biological isolation is required to avoid the spread of pollen. Findings such as the one that uses plant characterization and comparison in an environmental release trial may give a chance for risk evaluation in the future commercialization of transgenic *A. annua* varieties (Liu et al. 2016). Regeneration of Artemisia via leaf primordial provides biomass of leaf material required for Artemisinin development and also as a feasible approach to the supply of the raw materials needed for the development of antimalarial drugs for the fight against malaria fever (Tahir et al. 2013). The use of controlled environments can overcome cultivation difficulties and could be a means to manipulate phenotypic variation in bioactive compounds and toxins, and offers the opportunity to optimize yield and achieve a uniform, high-quality product for commercialization (Jelodar et al. 2014). Aspects that may be beneficial in the successful large-scale cultivation of *A. annua* are agricultural, environmental and genetic and for producing the antimalarial artemisinin was addressed which includes geographic aspects (latitude and altitude), which will help in making decisions about crop establishment in tropical countries, and includes a list of Good Agricultural and Collection Practices for *A. annua* (Ferreira et al. 2005). Suitable soils for the commercial cultivation of this medicinal plant are a necessary requirement (Abubakar et al. 2018).

11.4 Garlic (*Allium sativum* L.)

Garlic and its oil (Fig. 11.4a-b) are regarded as a worldwide popular condiment with medicinal and nutraceutical properties. The garlic plant appears to not be a major concern of either classical breeding or genetic studies because of not producing seeds. However, recent achievements have enabled genetic studies and breeding because of the restoration of fertility in a number of genotypes which resulted in flowering and seed development. For the study and breeding of this significant crop, as well as for the creation of efficient molecular markers for desirable features including fertility and seed production, pest resistance, and nutraceutical qualities, the de novo transcriptome of fertile garlic represents a fresh resource (Kamenetsky et al. 2015). The review has examined the usefulness and nutritional value of garlic in combating a variety of dangers, such as hyperglycemia and dyslipidemia, cardiovascular diseases, antioxidant capacity, and carcinogenic aspects (Suleria et al. 2015).

11.4.1   Disease tolerance

A claimed novel report on fungal resistance gene's introduction in garlic showed the successful transformation of garlic that is disease tolerant. A. tumefaciens-mediated transformation system was produced which enabled the incorporation of genes of tobacco chitinase and glucanase into garlic cultivar 'ABEN'. The agrobacterium-mediated transformation technique reported could be a successful technique for increasing white root disease tolerance and reducing crop losses (Lagunes-Fortiz et al. 2013).

11.5   Ginger (Zingiber officinale)

Ginger (*Zingiber officinale* Roscoe; from the Family: Zingiberaceae) is a spice-dried underground stem of the herbaceous aromatic tropical plant. Fresh ginger (Fig. 11.5), dry ginger powder, oleoresin, and oil are utilized in food processing. It is utilized in the production of curry powders, gingerbread, ginger ale, confectionary, ginger cocktail, certain curried meats, table sauces, pickling, and in the production of carbonated drinks, certain cordials, liquors etc. It is utilized as a carminative and stimulant in indigenous medicine (Satyagopal et al. 2014).

11.5.1   Genetic Engineering

The cell- and tissue-culture methods have a great advantage in vegetatively propagated horticultural crops. Ginger is no exception. Techniques for micropropagation, plant regeneration, in vitro pollination, protoplast culture, the production of synseeds, and cryopreservation are available. These can be utilized successfully in ginger crop production programs (Babu et al. 2015). Callus production, regeneration, and molecular characterization of ginger were reported. The biochemical markers showed a strong similarity between them (Taha et al. 2013).

Figure 11.1. (a) Black cumin (*Nigella sativa*) seeds. (b) Black cumin seed oil.

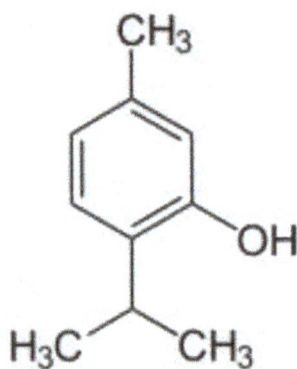

Figure 11.1.1.

Figure 11.2. Fenugreek (*Trigonella foenum-graecum*) seeds. (b) Fenugreek seed oil. (c) Chemical structure of Diosgenin.

Figures 11.3. (a) *Artemisia annua* cultivation. (b) Seeds of *Artemisia annua*.

Figure 11.4. (a) Garlic (*Allium sativum* L.). (b) Garlic oil.

Figure 11.5. Fresh ginger (*Zingiber officinale*).

Timber Crops

Timber crops are trees that are usually cut to extract wood which is useful for construction and papermaking.

12.1 Teak (*Tectona grandis*)

Teak (*Tectona grandis*) is considered among the greatest tropical timber species both in natural forests and in plantations (Graudal and Moestrup 2017). Domestication through plantations (Fig. 12.1) over the last one and a half centuries has made teak one of the most widely planted and researched tropical hardwood species domesticated through plantations (Katwal 2003). Sustainable yield from plantations of teak exceeds its high demand and that creates opportunities for enterprising farmers (Roshetko et al. 2013).

12.1.1 Genetic resources and diversity

Deforestation and habitat fragmentation are some factors affecting teak (*Tectona grandis*), one of the most valuable tropical timber species. Suitable gene conservation strategies and recommended seed zones are recommended for safeguarding genetic resources (Thwe-Thwe-Win 2015). A genotypically unique teak population was reported in India. The data collected via microsatellite marker studies from 550 trees of the natural teak populations in different geographic areas indicated that it has varied genetic structures forming separate genetic clusters. Molecular markers also supported that gene diversity within populations of teak in India is in a higher percentage than the gene diversity among populations elsewhere (Nair et al. 2015). Genetic knowledge is said to be essential for programmes to develop teak varieties and to improve the quality of plantations (Verhaegen et al. 2010).

12.1.2 Biomass for bioenergy

Distribution, growth and above-ground biomass of teak (*Tectona grandis* L.) in a site having suitable conditions for the growth and development of a teak plantation and regeneration was reported (Satheesan et al. 2016). A study to find the efficient genotypes suited for large clonal plantations considering the rate of biomass accumulation and carbon stocks of various clones of teak was reported (Behera and Mohapatra 2015). Fallen teak leaves are reported to be very suitable as a substrate for biogas production. Biogas is a renewable gaseous fuel. This biomass for energy generation is a resource for sustainable energy (Wannapokin 2016).

12.1.3 Carbon sequestration

Carbon emission in the atmosphere is said to have increased tremendously due to industrialization and human inference, the resultant green gas damaging the ozone layer of the earth which is a matter of concern. Forests form a carbon sink that absorbs a large quantity of carbon dioxide and converts it into fruitful bio-products like fiber, timber, fruits and food for human consumption. The teak's great potential for carbon sequestration and significance in climate change mitigation has been reported, even though the mixing of other tree species along with teak plantation was suggested to have important uses for carbon sequestration and climate change management (Pichhode and Nikhil 2017).

12.1.4 Pharmacological, antimicrobial and disease control

Teak has high pharmacological and timber value. Some of the biological activities on various parts of teak notably, leaf, bark, and wood have been highlighted and more studies on the isolation of active fractions responsible are reportedly under process and are suggested to be a boon to mankind concerning the present scenario of emerging diseases (Krishna and Jayakumaran 2010). Drug resistance by pathogenic species necessitated the search for new molecules against it. A detailed study on the antibacterial properties of Anthraquinones from leaves of *Tectona grandis* and other biological activity was reported (Krishna and Jayakumaran 2011). It has been claimed that teak leaves, an agricultural bio waste from the world's largest hardwood timber sector, can be used effectively for the "green" production of silver nanoparticles (AgNPs). The biosynthesized AgNps are advantageous from an environmental, energy consumption, and economic standpoint. They can be employed as water purifiers, antibacterial fabrics, sports apparel, and cosmetics with the potential to be an antibacterial agent (Devadiga et al. 2015). Production of Transgenic Teak (Tectona grandis) expressing a cry1Ab gene for control of the Skeletonizer, *Paliga damastesalis*, which causes minor to considerable damage to teak trees in the plantations as reported (Norwati et al. 2011). Research on clonal resistance in teak to its pernicious insect pests, *Hyblaeapuera cramer* (Lepidoptera: Hyblaedae) commonly known as teak defoliator and *Eutectona machaeralis* (Walker) (Lepidopera Pyralidae), teak lafskeletonier was reviewed. Breeding for insect resistance in teak and the role of biotechnology in the development of resistance was also highlighted (Roychoudhury 2012).

12.2 Beach hibiscus (*Hibiscus tiliaceus*)

Hibiscus tiliaceus L. (Malvaceae) is a flowering tropical plant characterized by leathery leaves that are heart-shaped, hairy beneath, and having 1–3 nectary glands at the base of the underside mid-rib. Flowers appear singly and bell-shaped, each with a maroon heart and stigma (Fig. 12.2). They appear yellow in the morning and turn to orange-red in the evening. The floral color change is a characteristic of the flowers of Hibiscus (Wong et al. 2010).

12.2.1 Genetic diversity and resources

The retrotransposon sequence-specific amplified polymorphism (SSAP) was reported as a technique utilized in understanding the genetic variation between four population pairs of H. tiliaceus, retrotransposon-based SSAP marker revealed habitat differentiation between estuarine and inland *Hibiscus tiliaceus* L. (Tang et al. 2011).

12.2.2 Commonly used Beach hibiscus products

The bark is used in the making of rope and cordage. It serves as a source of firewood. There are varieties of *Hibiscus tiliaceus*. However, Beach hibiscus is a tropical ornamental, prized for its showing flowers and leaves. It is utilized as a hedge or privacy screen in urban environments. It is commercially useful in the landscaping industry. The wood is utilizable as craft wood. For light and transient construction, wood is used as timber due to its softness and durability (Elevitch and Thomson 2006). According to preliminary research, *H. tiliaceus* leaf and bark extracts both contain medicinally significant bioactive components and are therefore utilized in traditional medicine. It's possible that the extracts' combined cytotoxic, analgesic, and neuropharmacological effects result from the existence of various active secondary metabolites (Abdul Awal et al. 2016). For animal nutrients production, *Hibiscus tiliaceus* leaf powder extract carrier as an additive in the diets for fattening of local cattle (in vitro) was reported (Bata and Rahayu 2016).

12.3 Mahogany (*Swietenia macrophylla*)

12.3.1 Genetic resource management

For the protection of the rich diversity of mahogany, conscientious efforts are needed in policymaking and the enforcement of existing laws about to the management and conservation of mahogany species. In addition, it is also important that mahogany is utilized as an integral component in agroforestry systems and reforestation efforts, likewise in the reclamation of degraded forest ecosystems so that the genetic resources of the species can be protected (Danquah et al. 2019). The foundation for the creation of a strategy for the sustainable management and conservation of mahogany is a thorough understanding of the degree of genetic heterogeneity within and across populations as well as an understanding of the mechanisms sustaining this variation. According to reports, many methods have recently been offered that could improve the sustainable management of mahogany's genetic resources (Newton et al. 2000).

12.3.2 Bioresource to biomass and biofuel

The elucidation of the carbon storage potential of the Mahogany (*Swietenia macrophylla*) sapling highlights the contribution of this sapling to greenhouse reduction. Research has evaluated the above ground standing biomass of the species, CO_2 capture, and carbon storage potential of the leaves, bark, and wood of this species (Superales 2016). Physico-chemical analysis of pyrolyze bio-oil of mahogany has demonstrated that its wood and fallen leaves form a useful biomass with the potential as a substitute for fossil fuel, and find many applications in the biofuel industry likewise its falling leaves (Chukwuneke et al. 2019, Ibrahim et al. 2021).

12.3.3 Medicinal, economic and environmental potential

African mahoganies' timber is usually used in traditional medicine. The decoctions of the bark are utilized in treating ailments such as ulcers, fever, skin diseases, and gastric pain after birth. There is an increasing demand for mahogany products in international trade because of the durability of the wood, considerable physical properties, and its pinkish to dark brown color after polishing make it recommendable for furniture, decorative veneer and panelling. They are also good for water conservation (Opuni-Frimpong et al. 2016). The recovery of underutilized mahogany (*Swietenia macrophylla*) can be intensified through silviculture (Verwer et al. 2008).

12.4   Melina (*Gmelina arborea* Roxb.)

Melina (*Gmelina arborea* Roxb.) (Fig. 12.3a-b) is a plant belonging to the family Verbenaceae. Since ancient times, it has been known to possess various medicinal properties and biological activities. Bio-prospecting this plant is of industrial and environmental importance. It was reported to contain chemical constituents which include flavonoids, sterols, lignans, iridoid glycoside flavons, flavones, and glycosides (Arora and Tamrakar 2017).

12.4.1 Potential uses

Apart from its use in forestry and large-scale afforestation/reforestation initiatives, melina (*G. arborea*) is useful for forest-based industries in the production of timber, furniture wood, pulp and paper making. It is a most important multipurpose tree which is significant in the development of wood, fodder, fuel and medicinal products, it has carbon sequestration potential, therefore, *G. arborea* is desirable for cultivation in home gardens and agroforestry systems (Verma et al. 2017). The chemical composition and fiber morphology of Nigerian-grown G. arborea and B. vulgaris showed that the biomass blends are suited to chemical materials for c pulping and paper production (Azeez et al. 2016). For the production of briquettes, sawdust particles of *Gmelina arborea* are considered suitable sources of material (Tembe et al. 2017, Isaac 2012). Studies showed insecticidal properties of extracts of *Gmelina arborea* products for pest control (Oparaeke 2005). Medicinally it could be used as a naturally occurring antioxidant and anticancer agent (Ghareeb et al. 2014).

12.4.2 Biomass and carbon stock

Gmelina was reported to have a high biomass yield per hectare, even at a young age. The high proportion of Gmelina wood which is merchantable for timber is associated with a high stem biomass. Trees are regarded as a terrestrial carbon sink. Managing Gmelina forests can therefore sequester carbon both in situ (biomass and soil) and ex-situ (products). This is an indication that *Gmelina arborea* has great potential in promoting carbon sequestration (Ige 2018). For the accurate and reliable estimations of the biomass and the amount of carbon being sequestered in yemane (*Gmelina arborea* Roxb.) trees were provided as basic information (Tandug 2008).

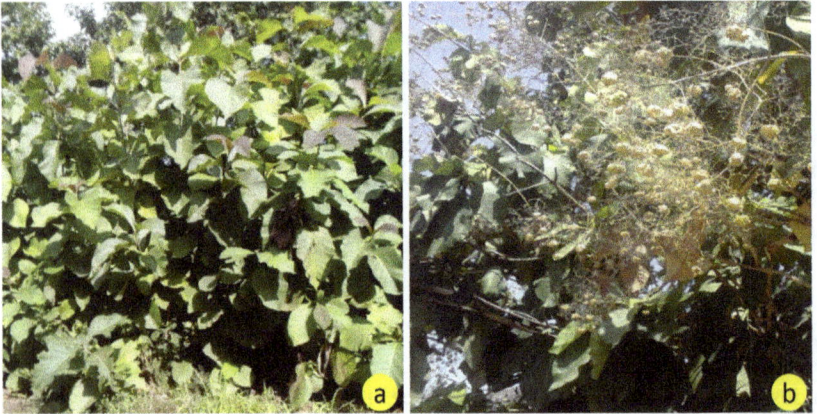

Figure 12.1. (a) Teak (*Tectona grandis*) plantations. (b) Teak seeds.

Figure 12.2. (a-c) Beach hibiscus (*Hibiscus tiliaceus*) and flowers occurring singly are bell-shaped, each with a maroon heart and stigma.

Figure 12.3. Melina (*Gmelina arborea* Roxb.) (a) The plant. (b) The fresh fruits.

Flowering and Ornamental Crops

The world is faced with environmental erosion such as deforestation, water, soil, and air pollution, due to dumping of toxic waste, industrial pollution, etc. Further, the globe is also witnessing periodic instances of floods, drying up of rivers and lakes, drop in agricultural yield, and loss of biodiversity. All these problems have contributed to rapid negative effects on climate, sanitation, and health. Ornamental crops play a great role in environmental remediation and resource recovery. These plants are used as bio-fences to protect crop fields from grazing animals and to prevent erosion. Ornamentals plants have gained recognition in the development of mega/smart cities around the world. They are used for decorative purposes in landscaping; they beautify the home and places such as streets in general public places and institutions. They contribute to environmental and socio-economic development. Cultivation of ornamental crops has found wide application for economic sustainability and the rehabilitation of environmentally constrained lands. In some ancient and modern communities, respondents have claimed ornamental plants supply oxygen, arrest dust, and help as windbreaks. These plants produce value-added non-edible products and are traditionally used as cosmetics for e.g., the manufacturing of soap. Researchers suggest that the detoxification or complete removal of the toxicity is essential before their use in industry or medicinal applications. Products from some ornamental crops and their environmental and economic applications are presented in some available literature. The scope of the market for these crops in agri-bio business for value chain and value additions needs to be widened in this regard. Although the majority of biotechnological applications for aesthetic crops are proving to be quite challenging, there have been some advances made with genetic engineering of a wide range of genetic features in turfgrass and floriculture crops. Cloning crucial genes that are demonstrated to be engaged in biological processes that researchers plan to alter in ornamental crops in the coming years has also seen significant advancements. A status report was provided on the achievement being made in biotechnology applications to plants developed for ornamental qualities and to projects where future research works will be concentrated. The decorative plant industry is technically varied and is distinguished by the usage of literally hundreds of different plant species. Often, there are numerous cultivars of a single species that can be grown, and plants can be propagated sexually or vegetatively. As a result of the huge quantity of genetic variation used by this business, there are often quite complicated challenges that occur throughout crop production and post-harvest management throughout the wholesale and retail marketplaces. From a broader perspective, however, it seems that the use of biotechnology in ornamental crops has a bright future. There have been numerous technological developments that have demonstrated their commercial utility in decorative crops (Clark 2004).

13.1 Biotechnological improvement of floricultural crops

Across the world, the flower industry thrives mostly on novelty. Recombinant DNA technology is offering a useful way to increase the gene pool in floriculture, which is encouraging the development of new commercial types. Engineered features have value for both producers and consumers. Currently, only consumer qualities appear to provide a return sufficient to sustain what still seems to be an expensive molecular breeding tool. The purpose of genetic engineering is to improve the characteristics of flowers, including their shape, aroma, color, vase life, disease resistance, productivity, and timing and synchrony of flowering (Yadav and Singh 2017). Because of low maintenance costs, including low labor costs, the floriculture business has made significant progress worldwide, particularly in developing nations. The industry must satisfy consumer demand by offering novel value additions for flowers that are affordable and distinctive in their characteristics, such as flower color, flower shape, and long shelf life. Hence, one of the main objectives of floriculture is to create unique aesthetic cultivars with enhanced floral characteristics. Many types of ornamental plants have been developed using biotechnological methods, including tissue culture and micropropagation techniques, polyploidy induction, mutation breeding, and genetic engineering. Biotechnology can be used to create plants with novel traits like floral architecture, color, aroma, tolerance to abiotic stress, and post-harvest life (Miri and Roughani 2018).

13.2 Achiote (*Bixa orellana* Linn)

Bixa orellana L. popularly familiar as the annatto plant (Fig. 13.1a-g) belonging to Bixaceae is historically known as a natural dye-yielding plant. The reddish-orange dye is produced in the aril portion of its seeds, it is one of the significant food-grade natural colorants used largely in bakery, dairy industry and confectionary products. Also used for various non-edible purposes like in cosmeceutical and dyeing leather. Different parts of annatto have mostly been utilized in traditional medicine for the control and treatment of quite a number of health problems.

Phytochemicals are reportedly isolated from all parts of the plant which includes terpenoids, carotenoids, apocarotenoids, sterols, terpenes, and aliphatic compounds possessing a wide range of pharmacological activities. Annatto was found to be significantly used and exploited as a source of drugs and a potential natural dye (Shahid-ul-Islam et al. 2016, Venugopalan et al. 2011, Deshmukh et al. 2013). Annatto extract was reported to have the highest total carotenoid content as a secondary metabolite included in the class of tetraterpenoids and the extract also showed potential antimicrobial activity (Natividad and Rafael 2014). Well-characterized pharmacological actions of *Bixa orellana* L. "urucum", that may be regarded as suitable for the future production of an innovative therapeutic agent were reported (Vilaret al. 2014). Control of carotenoid gene expression in *Bixa orellana* L. leaves treated with norflurazon was reported. The concentrations of herbicide norflurazon (NF), potentially reflected the roles of carotenoid genes in carotenoid synthesis in B. orellana (Rivera-Madrid et al. 2013).

13.2.1 Biotechnological transformations

In a recent article, the New York Times reported that the U.S. Department of Agriculture (USDA) is considering expanding its program to include agrochemicals. Moreover, the two wild populations of Annatto had higher apparent outcrossing rates than the two cultivated populations (Bixa orellana). All of the markers mentioned here could be used in future studies to examine the genetic diversity, population dynamics, domestication, breeding, and conservation genetics of annatto (Dequigiovanni et al. 2018). Achiote can develop bixin or norbixin more fully thanks to genetic engineering. The development of focused transgenic plants will be made possible by understanding the genes that regulate the manufacture of these significant secondary metabolites. The achiote (Bixa orellana L.) seeds, or arils, are the source of annatto, a natural dye or coloring ingredient also known as E160b in the marketplace. The main active component of annatto dye is water-insoluble bixin, although water-soluble norbixin also has industrial applications. Bixin is safe to consume because, unlike other antioxidants, it is light- and temperature-stable. Bixin is thus frequently utilized in the food, pharmaceutical, and cosmetic industries as a color and as an antioxidant. Although the yield is lower than from seeds, bixin has also been extracted from leaves and bark. Further research on this industrial and medical plant using genetic engineering is required. It would be useful to transform genes that could increase the production of bixin or norbixin based on preliminary genetic transformation experiments. Targeted amplification in transgenic plants will be possible with the development of reliable methods for the extraction of bixin and norbixin as well as an improved understanding of the genes that govern the manufacture of these significant secondary metabolites (Teixeira da Silva et al. 2018). Molecular method intervention was recommended to provide an opportunity to not only screen the germplasm to identify the elite variety via

Random amplification of polymorphic DNA (RAPD) markers and other marker registered selection but also to regulate the bixin biosynthetic pathway in order to increase the yield of pigment of the plant. Since heterologous expression of the genes involved in the bixin biosynthetic pathway showed that bixin could be produced in large quantities, it may now be viable to use recombinant DNA technology to increase the production of the annatto pigment. To identify the critical gene and its subsequent overexpression in this case, it would be helpful to examine the differential expression of genes involved in the bixin biosynthesis pathway during the development of flowers and fruit (Venugopalan et al. 2011).

13.2.2 Industrial resource and applications

The application of *Bixa orellana* seed (water extract) as a natural dye for the dyeing of cotton knitted fabric was reported to satisfy the future demand for an eco-friendly as well as sustainable dyeing of cotton fabric performing numeric exploration (Nasim-Uz-Zaman et al. 2018). Dye from Bixa (Annatto) seeds can be used for dyeing fibers like cotton, wool, and silk as well as in the food industry. Bixin used for this purpose is non-carcinogenic in nature and therefore healthy and eco-friendly for the human body and environment (Saha and Sinha 2012). The role of dyes in the photoinitiation processes of polymerization reactions was reported (Fouassier et al. 2010). Preparation of dye-sensitized solar cells (DSSC) using natural dye extracted from Annatto (Bixa Orellana) seeds as the sensitizer is prominent in recent literature reports. The fabrication of dye-sensitized solar cells (DSSC) using Annatto seeds has been conducted by Haryanto et al. (2014). The main pigments are bixin and norbixin, which were obtained by separation and purification from the dark-red extract (annatto). The dyes were characterized using 1H-NMR, FTIR spectroscopy, and UV–Vis spectrophotometry (Gómez-Ortíz et al. 2010).

13.3 Thevetia peruviana

Thevetia peruviana (Yellow oleander) is a large flowering tree cultivated as an ornamental plant (Kumar et al. 2017). Knowledge of its chemical composition is useful in developing complex biochemical substances and assists in the formation of a scientific basis of the therapeutic effect of this plant biologics. Leaf extract of yellow oleander (*Thevetia peruviana*) showed the presence of phytoconstituents of phytomedicinal importance such as tannins, flavonoids, alkaloids, terpenoids, phenols, cardiac glycosides, steroid glycosides and anthraquinones (Edo 2022).

13.3.1 Future genetic improvement

Survival of polyembryony seedlings was recorded for the first-time survival polyembryony seedlings in *T. Peruviana* which will serve as a utility in various fields such as forestry, environment, genetics, and embryology (Rai 2014).

13.3.2 Energy and non-energy uses

Plants bearing non-edible seeds such as *Thevetia peruviana* can be used in the revegetation of wasteland and are not a competition with food crops, and using up of these non-traditional and non-edible raw materials can be sustainable for the development of biodiesel (Kumar et al. 2017). The plant seed oil (Fig. 13.2a-e) is utilizable in producing oleo chemicals such as alkyd resin, and fatty acid methyl ester. The products are expectedly comparable to products from other lesser-known oil seeds (Warra 2017, Usman et al. 2009).

13.4 Sunflower (*Helianthus annuus* L.)

Early on, the sunflower was a common ornamental. However, its use as an environmental crop has recently gained significant credibility. Dehulled seeds are used as chicken feed. The growth and productivity have been effectively boosted through agronomic field study at a farm research site in India using recycled organic manure from an integrated farming system (goats, cows, poultry, etc.). The performance and productivity of brackish water irrigated sunflower crops are satisfactory according to agronomic trials in the typical Mediterranean climate, where winter precipitation amounts to about 500 mm on average. This helps to support agricultural sustainability and also provides an additional drought remedy. Sunflower offers potential in phytotechnologies for the removal of pollutants and toxins, both organic and inorganic. For the success of this strategy, there is a need for high bio-productivity and biomass yield (Prasad 2007). The major products derived from sunflowers are seeds and oil (Fig. 13.3a-c). Value addition products include soap and skin petroleum jelly (Fig. 13.4a-c). Sunflower is also an alternative crop for the development of biodiesel via cultivation on metal-contaminated soils (Prasad 2007).

Figure 13.1. (a) Bixa nursery. (b) *Bixa orellana* tree crop. (c) *Bixa orellana* flowers. (d) Bixa nuts early growth. (e) Bixa nuts matured and (f-g) Bixa seeds.

Figures 13.2. (a) *Thevetia peruviana* plant. (b) *Thevetia peruviana* flowers. (c) *Thevetia peruviana* nuts. (d) Deshelled *Thevetia peruviana* nuts. (e) *Thevetia peruviana* oil.

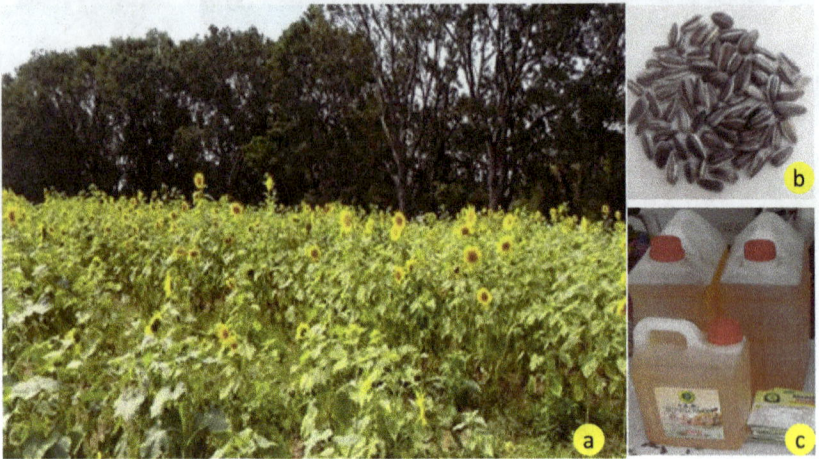

Figure 13.3 (a) Sunflower (*Helianthus annuus* L.) farm. (b) Sunflower seeds.
(c) Sunflower oil.

Figure 13.4. Sunflower (*Helianthus annuus* L.) oil value addition products: (a) Soap,
(b) Skin jelly, (c) Body cream.

Tree Crops

Trees are an important part of biodiversity. Some trees are renewable resources for fuel, timber, wood, fodder, food, and other useful non-timber products. As a result of the rapid growth of the human population and the human desire to progress, a drastic reduction in forest cover from the earth's surface has been experienced. For sustaining forest vegetation, previously, traditional approaches were used for propagation and improvement. However, such efforts faced many inherent bottlenecks. Biotechnological interventions for in vitro regeneration, mass micropropagation, and gene transfer techniques in forest tree species are now successfully practiced (Giri et al. 2004).

14.1 Neem tree (*Azadirachta indica*)

Neem belongs to the family of Mahogany. Its attributes resemble that of *Melia azederach*. *Azadirachta* is the word derived from the Persian azaddhirakt (meaning 'noble tree') (Ogbuewu et al. 2011). The *Azadirachta indica* tree is a member of the Meliaceae family (Csurhes 2008). The Neem tree is planted usually along roadsides in the northern part of Nigeria (Fig. 14.1). Mechanical pressing is the most common method used for oil extraction from its seed (Liauw et al. 2008). The main composition of oil extracted from its seeds is triacylglycerols of linoleic, oleic, palmitic, and stearic acids. The oil yield from the seeds is 40%, deep yellow in color, and is well-known as 'Margosa oil' (Girish and Shankara Bhat 2008). During the last five decades, aside from the chemical nature of the neem compounds, substantial achievements have been recorded regarding the biological activity and medicinal potential of neem. It is now considered a valuable source of unique natural products for the production of medicines against various diseases and also for the production of industrial products (Biswas et al. 2002). Other industrial uses of neem seed oil, are as a lubricant and in pharmaceutical preparations like ointments, liniments, emulsions and poultices and as in soap ingredients (National Research Council 1992). It is also useful in the treatment of dandruff, dry and itchy scalps and also in the restoration of dry and damaged hair (Drabu et al. 2012).

14.1.1 Micropropagation

Azadirachta indica A. Juss (Neem) is an unexploited natural resource as far as its genetic potential is concerned. A technique that is reproducible was reported for the regeneration of complete plants from nodal segments containing axillary buds. The seed oil containing Azadirachtin is the major industrial utility. Neem trees exhibit important variability in the azadirachtin content in seeds irrespective of the habitat (Gehlot et al. 2014). The current status and future challenges of plant tissue culture for azadirachtin production were reviewed by researchers (Prakash et al. 2002).

14.1.2 Neem oil limonoids as bioactive cosmetic component

Azadirachta Indica A. Juss. (Neem) is a source of several bioactive triterpenoids; however, the component that is commercially utilized is azadirachtin. Neem (*Azadirachta indica*) fruit and seeds were found to have substantial triterpenoids and oil (Fig. 14.1a-c) (Johnson et al. 1996). The other two key bioactive substances that have the potential to be used in the development of new products, as well as markers for neem improvement, are nimbin and salannin (Fig. 14.2b-c) (Sidhu et al. 2003). Using supercritical carbon dioxide, nimbin extraction from neem seeds was researched (Mongkholkhajornsilp et al. 2005). Nimbidin (Fig. 14.2d) is effective in treating skin conditions such as furunculosis, eczema, arsenical and seborrheic dermatitis, and scabies (Upma et al. 2007). Hemp oil natural extract called Limonoids is made from cold-pressed neem seed oil and is standardized to include at least 50% of total Limonoids and 1000 ppm of Azadirachtin. Preparations for topical use in skin and hair care that are antifungal, antiparasitic, antibacterial, insect repellent, and anti-pediculosis have some cosmeceutical promise. Cold-pressed Neem seed oil is used to create Neemoids, a free-flowing pale brown to yellowish brown powder that is used in the creation of hand and body washes, lotions, shampoos, creams, and other products (Majeed et al. 2007). Neem oil limonoids containing Azadirachtin (Fig. 14.2a) have promising potential for hair care preparations owing to their anti-head lice, anti-dandruff, and antifungal properties (Majeed et al. 2007). A novel shampoo based on neem (*Azadirachta indica)* which is highly effective against head lice in vitro was reported (Heukelbach et al. 2006). It has been reported to use Azadirachta indica (neem) seed oil as an adjuvant for antibacterial action (Chindo et al. 2011). Neem (Azadirachta indica) seed oil cream stability experiments were published (Aremu and Femi-Oyewo 2009). Chemical characteristics of toilet soap made from neem (Azadirachta indica A. Juss) seed oil were reported (Mak-Mensah and Firempong 2011). The chemical characteristics of the soap were reported as total fatty matter, total alkali, free caustic alkali, percentage chloride (% Cl⁻), percentage moisture, and pH, respectively, at 63.75, 0.24, 0.06, 1.15, 12.6, and 10.4%. Neem oil contains phytochemicals, and the soap has important chemical properties that lead to the conclusion that it can be used as toilet soap for both medical and cosmetic purposes. Such neem soap might be helpful for skin protection (Mak-Mensah and Firempong 2011). It was also claimed that the Nigerian species of neem tree seed oil may be used to make soap (Warra et al. 2011). The soap solution's characteristics demonstrated that it is suitable for commercial preparation (Warra et al. 2011). Azadirachta Indica (neem) seed oil was used to create a biodegradable detergent, which was described and compared favorably with detergents currently on the market (Ameh et al. 2010). Nowadays, natural product chemistry continues to be one of the major areas that contributed tremendously to medicinal and cosmetic delivery. Previous research works reported that neem seed oil, a natural product from the neem tree, is resourceful in this direction.

14.2 *Melia azedarach* L.

Melia azedarach L. (Family: Meliaceae) is a deciduous tree that goes by different names, commonly as Persian lilac, Tulip cedar, white cedar, and Chinaberry (Mishra et al. 2013). It is an important medicinal plant. Different parts such as the leaf, flower, seed, fruit (Fig. 14.3a-b), and young branches have been utilized for the treatment of malaria, as a purgative, for cough, diabetes, skin disease, and so on. Experimental and clinical studies prove that it has antimicrobial, anticancer, antioxidant, antipyretic, anti-inflammatory, cardioprotective, analgesic, antiulcer, antiplasmodial, and male contraceptive properties (Azam et al. 2013).

14.2.1 Genetic modification and transformation

A study observed the possibility of introducing specific foreign genes into *Melia azedarach* L. using mediated transformation which is part of the "Development of shoot organogenesis system of *Melia azedarach* L. and its applications for genetic transformation" (Nirsatmanto and Gyokusen 2007). Research reports the development of limonoids via callus and cell suspension cultures of chinaberry (*Melia azedarach* L.) Plant callus cell suspension cultures via an optimized medium may be regarded as a significant source for the development of bio-products and its purified form could be utilized for medicinal purposes (Rind et al. 2021).

The development of an efficient somatic embryogenesis system and subsequent plant regeneration of the chinaberry tree (*Melia azedarach* L.) was reported. This is an alternative pathway for the plant in vitro regeneration of the chinaberry plant (*Melia azedarach* L.) derived from the induction of somatic embryogenesis. The established technique developed provides a great opportunity for the production of plantlets from cotyledon explants via somatic embryogenesis. It also serves as a very effective protocol for mass clonal propagation and conservation of *Melia azedarach* L. (Sharry et al. 2006).

14.2.2 Useful applications

It has long been known that the plant family Meliaceae is a significant source of chemicals with insect-controlling properties. For example, Melia azedarach L., a member of the Melia and Azadirachta genera, was regarded as a tree having biopesticide properties, particularly for its insecticidal, larvicidal, and insect repellent properties (Hammad 2008, Sabiha et al. 2017). Extracts of *M. azedarach's* fruits, seeds, and leaves have demonstrated a variety of therapeutic and pesticidal activity against various pathogenic and pest species, respectively (AL-Rubae 2009). There have been reports of a potent limonoid insect antifeudant from Melia azedarach. It was made able to isolate the limonoid antifeedant meliartenin by carefully fractionating the fruit extract (Carpinella et al. 2002).

14.3 Shea (*Vitellaria paradoxa*) tree

Shea-tree (*Vitellaria paradoxa*) (Fig. 14.4a) thrives in the West African savannah zone; its fruit contains one or two nuts, which are brown and shiny. Although the fruit pulp (Fig. 14.4b-c) is edible, the nuts are the major commercial products (Fig. 14.4d), which contain a kernel (Fig. 14.4e) with an oil content ranging from 45–60%. The shea butter which is its (Fig. 14.4f) fat is used for cooking and the production of butter substitutes, candle, soap, cosmetics and pharmaceutical products (Oluwole et al. 2004).

14.3.1 Genetic improvement

The knowledge and understanding of the extent of genetic variation of shea genotypes are important for the conservation and improvement of the crop (Enaberue1 et al. 2014). There is a need to utilize molecular markers to unravel the underlying variation for use in the selection and genetic improvement of the shea tree which can be achieved through the quantitative morphological characterization of leaf, fruit and nut (Djekota et al. 2014).

14.3.2 Chemical constituents of shea by-products of industrial potential

Shea nut fat extract from the nuts contains several chemical components of personal care applications (Warra 2017). Evaluation of the constituents of phenolics of shea (Vitellaria paradoxa) kernels by LC-MS showed the presence of eight catechin compounds-gallic acid, catechin, epicatechin, epicatechin gallate, gallocatechin, epigallocatechin, gallocatechin gallate, and epigallocatechin gallate as well as quercetin and trans-cinnamic acid (Maranz et al. 2003). The nutritional value of several regional by-products, such as Shea-Nut (Vitellaria paradoxa) cake, was assessed by Pousga et al. in 2007. It draws interest from all around the world due to its application as an alternative to cocoa butter in the pharmaceutical and cosmetics industries (Elias et al. 2006). Shea butter is mostly used in chocolate where the similarity in composition and crystallization properties between shea butter and cocoa butter is utilized. On the other hand, raw, unrefined shea butter can be used to treat eczema, dry skin, scars, frostbite, itchy skin, blemishes, skin rashes, diaper rash, muscle soreness, stretch marks, skin peeling after tanning, sunburn, cracked heels and skin, chapped lips, small cuts or scrapes, burns, athlete's foot, insect bites and stings, arthritis, and as a hair product (Lamien et al. 2006).

14.3.3 Industrial potential of shea butter fatty acids

Here is a description of the fatty acids found in the GC-MS fragments: Oleic acid, often referred to as cis-9-octadecenoic acid, cis-oleic acid, oleate, elaidoic acid, metaupon, and delsauere. Delsauere is the most prevalent and common unsaturated fatty acid in nature. Commercially, it is utilized as a medicinal solvent and to make oleates and lotions. Oleic acid has a high lipid content, which makes it a superb moisturizer. Several cosmetic businesses add oleic acid to lotions and soaps to increase the skin-nourishing properties of these products. The majority of products made with palmitic acid are soaps and cosmetics. Several applications make use of sodium palmitate. Cetyl alcohol, produced by hydrogenating palmitic acid, is used to make detergents and cosmetics. Stearic acid, also known as octadecanoic acid in the IUPAC, is a saturated fatty acid having an 18-carbon chain. Shampoos and other cosmetic goods like shaving cream and soap are made mostly using stearic acid. Stearic acid is not used to make soap directly, but rather indirectly through the saponification of triglycerides that include stearic acid esters. In shampoos, soaps, and other cosmetic products, stearic acid esters with ethylene glycol, glycol stearate, and glycol distearate are used to create a pearly look. They are added to the product in molten form and allowed to crystallize under regulated circumstances. Amido and quaternary alkylammonium derivatives of stearic acid are used to make detergents. Surfactants, cosmetics and personal hygiene products are in fact prospects of stearic acid. Undecylenic acid, also known as 10-undecenoic acid, is another fatty acid that can be produced. It is a monomer that is created from ricinoleic acid and is used to make anti-dandruff shampoos and antibacterial powders. Ninety percent of the fatty acids in castor oil are naturally occurring 12-hydroxy fatty acids called methyl ricinoleate or ricinoleic acid methyl ester. Another fatty acid fragment known as behenic acid It has a melting point of 80°C and a boiling point of 306°C, and it appears as white to cream-colored crystals or powder. It is dispersible in ether and ethanol. It is a key ingredient of Ben oil, which is made from Moringa oleifera tree seeds. Hair conditioners and moisturizers frequently contain behenic acid to impart their smoothing effects. Also utilized as an antifoam in the production of detergents. Arachidic acid (eicosanoic acid) was also identified, it is used in the production of detergents, and due to its surfactant-like properties, it is used in the manufacture of soaps and cosmetics. Tridecene was also obtained from GC-MS analysis (other names: alphaTridecene, Tridec-1-ene, Tridecylene, Undecylethylene). Chemical intermediates for detergent alcohols, linear alkylbenzenes, internal olefin mixtures, alpha-olefin sulfonates, and alpha-olefin epoxides are among the main application areas (as alpha-olefin mixt). Erucic acid, an omega-9 monounsaturated fatty acid, is readily available. Due to the fact that its fatty acid chain contains just one double-bond carbon atom, it is known as monounsaturated. A group of fatty acids known as omega-9, which includes erucic and oleic acid and has its double carbon bond located at the ninth position from the end of its acid chains, is referred to as this. The n-9 position is what it is called. Nonadecyclic acid, also known as nonadecanoic acid, is a long-chain saturated fatty acid with a carbon chain length of 19, and its chemical formula is CH_3 $(CH_2)17COOH$. It creates salts termed nonsalicylates. You can find nonadecylic acid in fats and vegetable oils. Moreover, insects use it as pheromones. It is an intermediate in the biodegradation of n-eicosane which showed it to inhibit cancer growth. Another fatty acid that was discovered was myristic acid, also called tetradecanoic acid and found in nutmeg, palm oil, coconut oil, and butterfat. It can be used as a surfactant, emulsifier, opacifying agent, fragrance ingredient, and more in the cosmetics sector. Because of the fast rate of skin absorption, one of its main characteristics is as a lubricant. In cosmetic and topical medicinal preparations where good skin absorption is sought, the ester isopropyl myristate is employed. It is also employed as a pesticide-free treatment for head lice that eliminates the wax that protects the lice's shell, killing them by dehydration; it is used as a solvent in perfume materials. Lastly, Methyl ricinoleate or Ricinoleic acid which is a methyl ester and a naturally occurring 12 hydroxy fatty acid constituting about 90% of the fatty acids in castor oil was also obtained. Shea Nut fats have potential in the production of pharmaceuticals, cosmetics, and perfumery (Warra 2017).

14.4 *Adansonia digitata* tree

Baobab (*Adansonia digitata* L., Malvaceae) is a tree species that flourishes in Africa (De Caluwé et al. 2010). The name honors French botanist Michel Adanson (1727–1806), who spent six years living in Senegal and penning a book on its natural history. He was honored with the genus and species name, "digitata", which refers to the leaf's form and means "hand-shaped" in Latin. It is called different names in many languages; Hausa (kuka); Hindi (gorakh-imli, goraklichora) (Orwa et al. 2009). Baobab seeds have a pleasant flavor and are a good coffee alternative when roasted and ground. The seeds have a very high oil content (Fig. 14.5a-c) (De Caluwé et al. 2005).

14.4.1 Potential food and non-food products

The Baobab tree has various uses, food and non-food products can be derived from the tree such as timber, medicines, fodder, fuel, and dye. Every baobab tree part is useful. The flowers, fruit pulp, seeds, leaves, roots, and bark of baobab are edible. Traditional recipes call for creating soup with baobab leaves. The green bark can be used as a decorative dye. Vitamin C content is especially high in baobab pulp. Seeds are used as a thickening agent in soups, as well as a flavoring agent when fermented and as a snack when roasted. The pulp can be sucked or used in the creation of drinks. Baobab pieces are used as a source of nutrients because they include antinutrients such as tannins, phytates, and protease inhibitors, however, processing techniques may eliminate or greatly reduce these antinutrients. Baobab bark, pulp, leaves, roots, and seeds are potential medicinal products in many parts of Africa processing medicinal properties including analgesic, anti-dysentery activity anti-diarrhoea, anti-inflammatory, antioxidant, prebiotic-like activity, antipyretic activity, and excipient (Rahul et al. 2015, Kaboré et al. 2011, Chadare et al. 2008). Some statutory bodies approved its use in certain food products (Zahra'u et al. 2014). Production of Baobab yogurt using Lactobacillus bulgaricus was reported (Aisha et al. 2017). The baobab tree is a treasure trove of possible compounds with therapeutic application value and powerful antimicrobial potential against particular microbes. Actually, due to this species' exceptional therapeutic qualities and abundance of chemical components, interest in its rehabilitation has grown (Sugandha and Shashi 2017).

14.4.2 Mucilage

Mucilage produced from baobab leaves serves as a constituent of foodstuffs in soups and stews, particularly in West Africa. The key aspect of interest is the high protein and mineral content in both the raw and purified mucilage. Rhamnose and galactose, two neutral sugars, are present in extremely minute quantities in the mucilage. Uronic acid is present as a combination of galacturonic and glucuronic acids. The mucilage is categorized as an acidic galactomannan polysaccharide because of the comparatively high concentration of uronic acids in it (Sidibe and Williams 2002).

14.4.3 Fiber

The strong fiber found in the inner bark is used to make rope, snares, basket nets, fishing lines, and weaving, among other things. Also detected are fibers with a high protein content. The fiber in the wood fragments is very abundant and has been used for packing. Other fibers for creating rope can be gotten from the root bark (Sidibe and Williams 2002).

14.4.4 Fatty acids for cosmetic utilization

Various fatty acids useful for cosmetic industries were found in baobab seed oil. The result of the GC-MS analysis showed the presence of the following fatty acids; linoleic, palmitic acid, behenic and stearic acids which were reported previously as having potential in pharmaceutical industries, and cosmetic (Warra 2015a, Warra 2015b). Furthermore, it was discovered that they were myristic and valeric acids; the former is a saturated fatty acid called myristic acid (tetradecanoic acid). An ester or salt of myristic acid is a myristate. The nutmeg Myristica fragrans is where myristic acid gets its name. When producing cosmetics and topical medications, ester isopropyl myristate is used since effective skin absorption is crucial. The safety of the inorganic salts and esters of several myristic acid fatty alcohols has been documented in the literature. Most esters were recognized to work well as skin conditioners in a variety of cosmetic products at various concentrations. The latter, also known as valeric acid or pentanoic acid, is an alkyl carboxylic acid with a straight chain. Valeric acid's volatile esters are employed in cosmetics and fragrances because of its generally pleasant scents. Most esters were recognized to work well as skin conditioners in a variety of cosmetic products at various concentrations. The latter, also known as valeric acid or pentanoic acid, is an alkyl carboxylic acid with a straight chain. Valeric acid's volatile esters are employed in cosmetics and fragrances because of its generally pleasant scents. Pharmaceutical soap and other cosmetic companies could use baobab seed oil. It was stated that the hexane extract of the seed oil might be used to make a light cream soap (Warra et al. 2015).

14.4.5 Potential for traditional medicine

The medicinal potential of the tree parts due to the medicinal compounds contained are described in the literature (Zahra'u et al. 2014). The leaves and fruit pulp provide antipyretic or febrifuge effects in overcoming fevers. Powdered seeds and fruit pulp are utilized in the treatment of dysentery and in the promotion of perspiration (i.e., a diaphoretic). Powdered leaves possess anti-asthmatic, anti-tension and antihistamine properties. They are also utilized in the treatment of fatigue, as a tonic and for guinea worm, insect bites, and internal pains, in the treatment of dysentery, antiviral, and anaemia and asthma. Leaves are utilized in curing ailments like opthalmia, urinary tract, and otitis. Seeds are also used as a remedy against hiccoughs and diarrhea. The seeds oil extract is a remedy for diseased teeth and inflamed gums (Rana et al. 2022, Sidibe and Williams 2002).

14.4.6 Development of recombinant DNA technology

Co-dominant markers suited for molecular ecology investigations in the genus Adansonia are greatly desired in order to answer interesting research questions about the unique life cycle features, gene flow, and distribution dynamics of the Adansonia species. The Adansonia genus was evaluated for cross-amplification using a set of 18 SSR primers created for Adansonia digitata (Larsen et al. 2009). Populations collected in the different climatic zones were reported as per some degree of physical isolation and a substantial amount of genetic structuring between the analyzed populations of baobab was suggested. For in-situ or ex-situ conservation, many natural population sampling options are recommended (Assogbadjo et al. 2006). To raise the nutritional status of the trees, selection and hybridization are required among populations with superior features. It is possible to cross the genetically distinct populations of baobab to create genotypes that have a higher nutritional value in the leaves and can tolerate both biotic and abiotic environmental stress (Ibrahim et al. 2014).

14.5   Cocoa (*Theobroma cacao* L.) crop

The principal raw ingredient used to make chocolate, confections, and various cosmetic products is derived from the cacao tree. Cocoa serves the livelihood of over 6.5 million smallholder families, is a potential source of revenue for a small number of tropical countries, and is a major ingredient for sustaining the chocolate industry (Pye-Smit 2011). Future global cocoa economies depend heavily on our capacity to create superior, sustainable varieties using this varied genetic base. An irrevocable loss of the cacao genetic variety so important to farmers, breeders, and consumers has been observed as a result of structural changes in land use, unexpected climate changes, the expansion of pests and diseases, harsh weather, and natural disasters. International cooperation is crucial for the efficient management of cacao genetic resources because the majority of the countries involved in cacao production and its improvement heavily rely on the genes and varieties characterized and conserved in other countries and regions (Laliberté et al. 2012). Most significantly, the magnitude and diversification of coca landscapes could greatly increase productivity, incomes and biodiversity and help conserve forests (Alemagi et al. 2015)

14.5.1   Genetic diversity for crop improvement

The diversity reported to exist in cocoa plantations is a result of analysis of genetic similarities and differentiation for the cocoa trees which can be exploited in crop improvement research (Thondaiman et al. 2013). Variability among the core of cocoa clones qualifies them as vital materials for parent selection so that hybrid progeny can be obtained (Lins et al. 2016). The black pod is a serious fungal disease of cocoa (Theobroma cacao L.) that has a significant economic impact. The disease primarily affects the roots, stems, beans, cushions, pods, leaves, and flowers. Assessment of the efficacy of the resistance screening technique used in breeding black pod disease resistance in cocoa was reported. In five laboratory studies and one field observation, the resistance of cocoa leaves and pods to Phytophthora palmivora was tested in 25 international genotypes of cocoa at the Cocoa Research Institute of Ghana. At the penetration and post-penetration stages of infection, a significant clonal variation for resistance to leaf and pod infection was seen. At both the penetration and post-penetration stages of infection, the connection between leaves and pods resistance was considered positive. When these traits of the cocoa leaf appear, consider the significance of using the leaves of cocoa seedlings to anticipate the resistance of the pods to the black pod disease (Nyadanu et al. 2009). In order to circumvent the situation in the 1930s when, as a result of a shortage of genetic variability in cacao collections, Microsatellite markers were used to map the genetic makeup of cocoa collections in West Africa after the swollen shoot virus epidemic nearly destroyed the sector. This will serve as one of the bridges for securing the future of the world's cocoa economy (Aikpokpodion 2012). Identification of Phytophthora spp. isolates in the cocoa orchard by the investigation of their molecular structure and evaluation of their pathogenicity on cocoa tree leaf disks having different sensitivity was reported as an indication of the highly heterogeneous nature of the aggressiveness level of isolates in this population. This heterogeneity is observed in isolates belonging either to the same plot of land or plots located in different areas. The results reveal that there exists the possibility of utilizing this criterion to evaluate the aggressiveness of P. palmivora isolates and also to quantify the intensity of resistance as part of an early breeding program of the cocoa tree to the disease (Coulibaly et al. 2018). It is interesting to note that disease resistance and defense gene analog (RGA/ DGA) sequences were found in cocoa utilizing a PCR method and degenerate primers made from conserved domains of plant resistance and defense genes. Isolated candidate genes may have a greater promise for managing cocoa disease (Lanaud et al. 2004).

14.5.2 Prospects for major cocoa products

The major intermediate cocoa products are utilized in the industrial production of food and non-food products.

14.6 Bitter kola (*Garcinia kola*)

14.6.1 Genetic propagation

Studies on the grafting and budding of *Garcinia kola* have indicated that, there is a probability of utilizing identified significant traits from a mother tree, such as a time reduction in bearing, increase in fruit size, higher yield, tree height reduction, resistance to pests, and diseases for establishing domestication and plantation with which conservation of G. kola and its attraction in Agroforestry system could be strengthened (Yakubu et al. 2014).

14.6.2 Medicinal potential

Every part of *Garcinia kola* exhibits medicinal properties, the seed (Fig. 14.6) is utilized as an antipyretic agent in traditional medicine. Pharmacologic investigation on the root, leaf and seed of this plant has shown antimicrobial, adaptogenic, antihepatotoxic, antiulcer, anti-inflammatory, antihypertensive, aphrodisiac, antiasthma, antidiabetic, and antiviral activities (Buba et al. 2016). The triterpenoid (and likely also glycoside) components of the extract di-ethyl ether Garcinia kola have been linked to anti-oxidant and inhibitory activities, and they have also been found to have strong anti-bacterial and weakly antifungal (inhibitory) properties (Lovet et al. 2014).

14.6.3 Bioeconomy and product development

Different improved marketing and trade networks of Bitter kola, starting from the point of production in order to enhance the rural economy and product development were reported. A recommendation was made for urgent attention to the establishment of Garcinia plantations, the encouragement of diversification of the products to create more market opportunities (Babalola and Agbeja 2010). Methods that could optimize ethanol production from bitter kola pulp wastes were reported (Nzelibe and Okafoagu 2007).

14.7 Ackee (*Blighia sapida* K.D. Koenig)

14.7.1 Industrial utilization

Properties of Blighia sapida seed starch showed the tendency for it to be useful in food and non-food industries. The potential of under-utilized *Blighia sapida* seed starch for industrial use was reported. It is among the wasted and under-utilized crops. The properties of the starch of *Blighia sapida* seed were comparable to other crops in granule sizes. High pasting properties observed in the starch indicated its potential use in highly viscous products. The use of this wasted seed in starch production could also be a means of generating another industrial resource for textiles, adhesives and food (Abiodun et al. 2015).

14.7.2 Saponin and soap substitute

Blighia sapida (akee or ackee fruit) an evergreen tree which belongs to the family *sapindaceae*-soapberry family is a species that is also known for its use as a soap substitute or in the preparation of some kinds of soaps. In Nigeria (Yoruba) it is called *Isin* and in Hausa language it is known as *Fisa*. The ashes of the dried husks and the seeds are used in the preparation of a kind of soap. The fruit also contains saponins. The fruit is used to produce soap in some parts of Africa. The rail, the seed and the pod (Fig. 14.7a-g) are used in the production of what is known as Ackee soap. Some towns and villages in Nigeria were using fruits in washing their clothes many years back before having knowledge of the use of modern detergents. The introduction of modern commercial detergents has now openly reduced the use of *Blighia sapida* fruit in rural Nigeria. Note: the traditional use of fruit in washing clothes has cut across most of the houses in rural areas, a recent interview schedule in some villages of Niger State in Nigeria revealed that people of all ages are familiar with the use of the fruit, and in most families, someone has used the *Blighia sapida* fruit for washing clothes at some time (Warra 2012).

14.8 Palm tree crop

The oil palm (*Elaeis guineensis*), a perennial crop plant that grows between 10 degrees north and 10 degrees south latitude, is native to the humid tropics of West and Central Africa. In these areas, oil palm grows naturally in the wild, in isolated stands (Fig. 14.8a-f), or as oil palm groves (Nkongho et al. 2015). Because the oil palm does not develop suckers as some other palm species do, it only grows at one location and cannot be propagated vegetatively to create clones. However, tissue culture, which involves growing tiny fragments of tissue (explants) on particular nutritional solutions, makes it possible to create clones (Corley and Tinker 2003). The genetic improvement in this crop is mainly through breeding and a mix of improved agronomy and management, coupled with breeding selection have quadrupled the oil yield of the crop (Soh et al. 2017).

14.8.1. Chemical properties of palm oil

Equal amounts of saturated and unsaturated fatty acids are present in palm oil. Palm oil's saturated fatty acid content is made up of 44% palmitic acid and 5% stearic acid. Oleic acid, which makes up about 40% of the content of palm, is the mono-unsaturated component. Linoleic acid, a polyunsaturated fatty acid that is an important fatty acid, makes up 10% of palm oil. Palm oil has a deep orange color because it contains substantial amounts of beta-carotene, which is the precursor to vitamin A (Almustafa et al. 1995).

14.8.2 Uses of palm oil

Most people use palm oil for eating, cooking, and other edible purposes. It is also valuable as fuel for internal combustion engines and in the manufacturing of candles, soaps, lubricants, and the plantation business (Opeke 2005). The oil can also be used as a source of vitamin A in place of cold liver (Kochhar 1998).

14.9 Date palm (*Phoenix dactylifera* L.) crop

The date palm (Phoenix dactylifera L.), which has more than 20 variants worldwide (Fig. 14.9a-f), is one of the earliest cultivated trees in desertic areas. The date seed is the primary byproduct of the date fruit industry which is typically discarded, despite the fact that it can be used as a component in animal feed or turned into decaffeinated coffee. The oil from the seed is used in the creation of biodiesel and cosmetic products (Abdul Afiq 2013).

14.9.1 Genetic resources, propagation and conservation

For the growth of date production and its significance for food security in many countries, the conservation of date palm genetic resources is considered a very important phenomenon. The conservation of date palm varieties in their original habitats is regarded as one of the vital aspects of the preservation of its genetic resources (Bekheet and Taha 2013). The preservation of field gene banks was found to be very successful in maintaining and providing the long-term conservation of date palm cultivars. However, it was regarded as comparatively very costly and required a large space. Although the in vitro germplasm conservation method is advantageous and requires little space and simple maintenance, there may be genetic or somaclonal diversity among the regenerated plants. In situ/on-farm conservation, protocols have once again been attempted using effective but expensive methods, which are only available on a small scale (Abul-Soad et al. 2017). Some studies dwelled on the current status of date palm (*Phoenix dactylifera* L.) genetic resources (Ataga et al. 2012). In date palms, sexual propagation caused by heterozygosity hinders the propagation of true-to-type genotypes. The offshoots from axillary buds located at the base of the trunk during the juvenile life of the palm tree were used for vegetative growth. In vitro culture techniques could be helpful due to the slow offshoot generation, which results in restricted quantities and tediousness that can't keep up with the quickly increasing need of kinds; nonetheless, genetic effect inhibits the efficient application. To expedite the genetic improvement of the date palm, somatic embryos are produced in enormous quantities using bioreactors (Jain 2012). In order to plant high-quality fruit-producing and genetically diverse cultivars and prevent any disease outbreaks in the future, it has been shown to be helpful to understand the genetic diversity and cultivar classification for date palms. It was possible to identify date palm cultivars using a genetic resource. This resource coupled with reduced representation genotyping by sequencing data was able to identify minimal genetic variability and the majority of cultivars in the area analyzed. Municipalities had no bearing on the diversity estimations. The development of agricultural resources may be aided by the diversification of the indigenous flora employing molecular genetic markers for date palm cultivar identification (Thareja et al. 2018).

14.9.2 Potential bio-products from date palm

The fruits of date palms compose of carbohydrates, lipids, protein, vitamins, dietary fibers, and mineral matter. Date fruit contains carbohydrates (73–85% by weight), which can be hydrolyzed to sugar. The carbohydrates are changed to bioethanol through anaerobic fermentation by the action of microorganisms. Utilization of date biomass waste and date seed as a bio-fuel resource was reported (Demirbas 2017). It has been observed that date palm (Phoenix dactylifera L.) cell suspension culture can produce polyphenols. Further research was recommended for enhancing cell suspension culture media conditions for the efficient large-scale production of beneficial polyphenols from the date palm and scale-up process (Naik and Al-Khayri 2018). The date palm tree (Phoenix dactylifera L.) is a source of natural fibers that may be harvested from a variety of date palm sections, including the midribs, mesh, leaflets, and spadix stems. The high population of date palms produces a lot of by-products that can be harvested annually. Natural fibers are now in demand for many different purposes, including making rope for use in ships, geotechnical applications like geotextiles, and construction uses like reinforcing asphalt concrete and gypsum plaster in 2019 (Elseify et al. 2019, 2023). The trash from palm trees can be used to make inverted sugar. A method for producing inverted sugar and using date palm fibers as a new carrier for invertase adsorption was reported (Mahmoud 2016). It was demonstrated that Date Palm Pollen (DPP), a natural product well-known in folk medicine in the Arab world, could be used in the agri-food and pharmaceutical field (Sebii et al. 2019). Date and their processing by-products are used as substrates for the production of bioactive compounds, exopolysaccharides, organic acids, bakery yeast, date-flavored probiotic fermented dairy, and items with added value, like antibiotics etc. can be produced (Tang 2014).

14.10 Coconut (*Cocos nucifera* L.) crop

Coconut is of the Arecaceae family and is a unique species of the genus Cocos. A classification by agriculturists refers to coconut as a drupe which is a fruit, a nut and also a seed. Known as the "tree of life", coconut is attributed to have many benefits. The coir obtained from the husks has a natural elastic fiber that is used to make strings, floor mats, ropes and brushes, and the leaves are utilized in making baskets, brooms, thatches roofing and temporary sheds. Coconut lumber is employed in making furniture, and building houses, husks and coconut shells are used in the production of fuel and charcoal. Since 2007, annually over 60 million metric tons of coconuts have been produced globally. Coconut and various coconut products are utilized as cosmetics, pharmaceuticals, beverages etc. Coconuts are also utilized in the production of delicacies like candy, crepes, chocolate, etc. Valuable products of coconut parts are summarized in Fig. 14.10 (Agricdemy 2018).

14.10.1 Value-added products

Coconut palm (*Cocos nucifera*) is a very vital industrial crop in many tropical countries that generates large amounts of residue. However, this residue, the coconut husk can be used to generate second-generation (2G) ethanol (Bolivar-Telleria et al. 2018). Coconut wastes as a bioresource for sustainable energy: quantifying wastes, calorific values and emission was reported (Obeng et al. 2020). The coconut is treasured for the variety of possibilities that may be found in each of the fruit's components. One of the products that include all nine essential amino acids is coconut milk. Coconut milk can be made into products with various fat contents to cater to the needs of consumers who prefer non-dairy milk. It can also be fermented into healthful and delicious yogurts and kefirs. You may buy gluten-free, coco flour, and low-GI flour. The Macapuno and Aromatic varieties of the coconut are highly prized for their distinct flavors and aromas. Using recent advancements in biotechnology, it is possible to increase the value of each of these goods even further (Dayrit and Nguyen 2020). Coconut water's special chemical constituents of vitamins, amino acids, sugars, minerals, and phytohormones justified wide applications. Scientific proof braces the importance of coconut water for medicinal and health uses. Coconut water is used traditionally as a growth supplement in plant tissue culture/micropropagation (Yong et al. 2009).

14.11 Rubber crop (*Hevea brasiliensis* Muell. Arg.)

An essential crop for the production of natural rubber is *Hevea brasiliensis.* Currently, rubber tree agriculture is active on more than 9.5 million hectares in around 40 nations, producing 6.5 million tons of dry rubber annually. With 12 million tons of natural rubber needed by 2020, the world's supply is just barely keeping up with demand (Venkatachalam et al. 2006). Hevea rubber clones are widely grown in tropical areas of the world as the primary sources for the manufacture of Natural Rubber (NR), which is a crucial raw material for many industries (Venkatachalam et al. 2007). It is mostly cultivated for its production of latex, a milky plant liquid, which serves as a basis for various rubber products.

14.11.1 Genetic engineering and developments

The transgenic rubber plants created by the International Rubber Research Groups produce foreign proteins with substantial economic value. A human serum albumin and an antibody are examples of such procedures. Compared to other options, harvesting usable proteins from rubber trees looks to be more profitable (Omo-Ikerodah et al. 2009). In order to decide whether to begin a breeding program with the aim of raising the total solids, it was stated that the genetic prospect of a rubber-segregating population was determined by using repeatability as a measure (García et al. 2018). The highly complex rubber tree DNA is essential for producing Hevea's distinctive traits. Yet, there is still a dearth of information regarding the molecular processes involved in the biosynthesis of rubber, disease resistance, etc., in superior rubber clones. The use of transgenic approaches and marker-assisted selection has greatly increased the breeding effectiveness for latex yield and disease resistance, among other traits (Supriyaa and Priyadarshan 2019).

14.11.2 Potential for products development

GCMS, phytochemical, and FTIR analysis indicated that rubber leaf extract which contains the carbonyl groups, aromatic rings, and double bonds is a good natural plant extract as a corrosion inhibitor (Okewale and Olaitan 2017). Natural rubber skims latex concentrate/montmorillonite as environmentally-friendly nanocomposite value-added material was reported (Ismail and Veerasamy 2011).

Figure 14.1. (a-c) *Azadirachta indica* A. Juss. (Neem) tree and riped fruits.

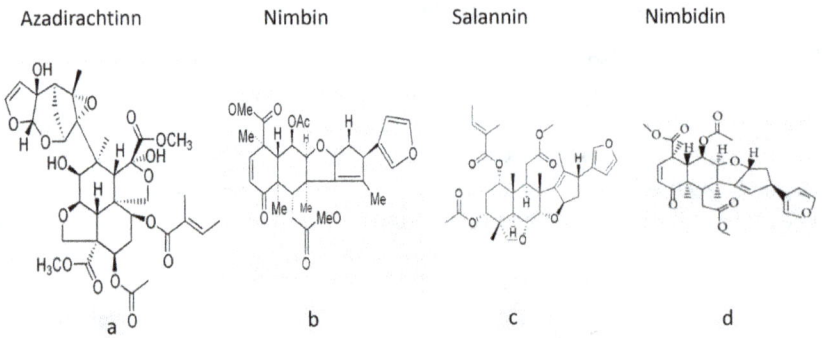

Azadirachtinn   Nimbin   Salannin   Nimbidin

Figure 14.2. The structures of (a) Azadirachtin, (b) Nimbin, (c) Salannin and (d) Nimbidin.

Figure 14.3. (a-b) Flowers and fruits of *Melia azedarach* L.

Figure 14.4. (a) Shea (*Vitellaria paradoxa*) tree. (b) Shea fruit unripe. (c) Shea fruit riped, (d) Shea nuts. (e) Shea kernel. (f) Shea butter.

Figure 14.5. (a) Baobab (*Adansonia digitata*) tree. (b) *Adansonia digitata* seeds. (c) *Adansonia digitata* seed oil.

Figure 14.6. Bitter kola (*Garcinia kola*).

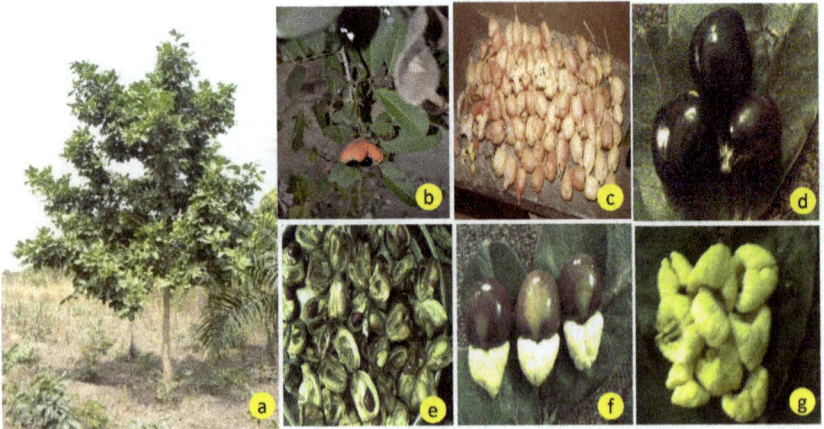

Figure 14.7. The major sections of *Blighia sapida* (the vegetable brain) (a-g) The tree, fruit, ripe fruits, seeds, husks, seeds with arilli and arilli.

Figures 14.8. Palm tree (*Arecaceae*) (a-f) Seedling, tree, plantation, leaves, puds and nuts.

Figures 14.9. Date palm (*Phoenix dactylifera* L.) crop (a) Date trees farm. (b) Town date tree plantation. (c-e) Date varieties. (f) Date seeds.

Figure 14.10. (a) Coconut (*Cocos nucifera* L.) tree. (b) Coconuts. (c) Coconut cross-sectional. (d) Coconut oil.

CHAPTER

15

Desert Crops

15.1 Eucalyptus tree crop

Australia is home to a large number of eucalyptus species that grow there naturally. In the Philippines, Papua New Guinea, Indonesia, and Timor, a few species of eucalyptus are naturally abundant. Some species include E. globules, E. saligna, and E. camaldulensis, among others. Eucalyptus thrives in a variety of ecological settings. Some are hardy species that grow in semi-arid regions, while others may grow in marshy and swampy environments. Different soil types such as soils that are loamy and fertile, sands that are infertile and heavy clays can also favor its growth (Oballa et al. 2010). The biological components of leaf essential oil, leaf oil as a natural remedy and the pharmacological, and toxicological qualities of the leaf oil of various Eucalyptus species were all attempted to be gathered globally. About 900 species of the genus Eucalyptus L'Heritier exist, and the leaves of more than 300 of these species produce volatile essential oil. A high percentage of 1,8-cineole (more than 70%) is present in about 20 of these species, which are used commercially for the production of essential oils for the pharmaceutical and cosmetic sectors. However, despite being widely planted for the production of pulp, plywood, and solid wood, eucalyptus has a surprising range of biological activities. These include antiseptic, antioxidant, antimicrobial, chemotherapeutic, respiratory and gastrointestinal disorder treatment, insecticidal/insect repellent activity, nematicidal, wound healing, and herbicidal, acaricidal, and perfumes, soap production, and a tendency to remove grease (Dhakad et al. 2018). Chemistry and medicinal importance of the seed essential oil of *Eucalyptus torelliana* F. and results showed that the essential oil from the seeds of E. torelliana, a muell grown in Nigeria, has antinociceptive, antioxidant, and anti-inflammatory potential, providing the first scientific validation of the oil's use as a phytotherapeutic agent against reactive oxidative, nociceptive, and inflammatory processes (Ololade and Olawore 2013).

15.1.1 *Eucalyptus grandis* Maiden (Rose gum)

The most prevalent *Eucalyptus* species in Nigeria is *Eucalyptus grandis* (Fig. 15.1a-f), which is distinguished by straight, quick-growing trees that can reach heights of up to 50 m and diameter at breast height (dbh) of 2 m. The species can occasionally be difficult to distinguish morphologically since *E. grandis* and *E. saligna* frequently generate hybrids (Oballa et al. 2010). A study showed that Eucalyptus grandis is a very strong wood species, utilizable for load-bearing work in construction like flooring and panelling, and very useful for fuel wood and charcoal production (Network et al. 2021).

15.1.2 Potential for harmful insects' repellency

Seed and leaf extracts of *Eucalyptus* contain compounds toxic to mosquitoes' larvae, filarial mosquito Culex quinquefasciatus vector of bancroftian filariasis, which can be developed and used as alternatives to the chemical insecticides in reducing filarial vector density and in the control of mosquitoes (Elbanna 2006, Mandal 2011). Tested against Lesser grain borer *Rhyzopertha dominica* (F.) (Chandel et al. 2019). Nano-encapsulated *Eucalyptus* extract was reported to be affective against the pest *Myzus persicae*, as a recommendable alternative to synthetic pesticides, also effective against House Rat, Rattus rattus (Khoshraftar et al. 2019, Sigla et al. 2014).

15.1.3 Antimicrobial and Antifungal action

A study showed that species of *Eucalyptus* can be utilized as natural antibiotics for the treatment of various infectious diseases caused by bacterial strains namely, *Staphylococcus aureus, Streptococcus, Lactobacillus, E. coli and Pseudomonas* and could have potential in understanding the relationship between traditional and modern medicines (Bachheti et al. 2011). Using the disc diffusion agar method, the antifungal activity of Eucalyptus grand essential oil against storage fungi of Arachis hypogea was evaluated. The oil displayed a broad range of fungal toxicity. Hydro distillation was used to extract the leaves essential oils. Due to its broad range of activity and lack of phytotoxicity, it was advised that the oil be effective as a fumigant against storage fungus to prevent the stored groundnut seeds from spoiling (Mumtaz et al. 2012).

15.2 Jojoba crop

Jojoba (*Simmondsia chinensis*) crops usually flourish in arid and semiarid regions (Esmail et al. 2016). The report showed that the plant is capable of protecting the soil against water and wind erosion (Salih et al. 2002). The major usage of the plant is mainly for its unique oil, and utilization of the extracts derived from other parts of the plant. Jojoba liquid wax is currently used as an active substrate for the formulation or as an ingredient, though it showed that the whole plant like the roots, leaves, and hulls can be utilized as a bioactive molecules source (Tietela et al. 2021). The number of sex-specific markers detected in jojoba plants might aid in improving the biofuel production industry (Bafeel and Bahieldi 2020). Jojoba crushed seed and oil cakes are reportedly effective in promoting and increasing most of the growth parameters, leaf nutrient contents and controlling root-knot nematodes (Hafez et al. 2017).

15.2.1 Genetic improvement and biotechnological advances

In terms of the tested characters, such as growth rate, plant shape, leaf size, length of flowering, seed productivity, and even genetic analysis, there was a great deal of variation reported among the genotypes of jojoba. This shows that it is very important for the breeding program and advances jojoba plants (Nahla et al. 2018). The development of effective biotechnological techniques for a good and improved cultivation and production cycle is very vital for farmers cultivating the plants (Al-Obaidi et al. 2017). It has been demonstrated that efficient biotechnology methods can be modified for jojoba propagation and the cloning of genes encoding for economically significant features. According to reports, in vitro seedling culture successfully accomplished jojoba micropropagation (Aly and Basarir 2012).

15.2.2 Industrial raw materials and products

The only plant documented as having to employ liquid wax esters (WEs) as a key seed storage reserve is jojoba (Simmondsia chinensis) (Rajangam et al. 2013). It is a browsing plant as well as an oil source with long-standing promise for use in medicine and food. Analysis of jojoba oil showed that it is similar to sperm whale oil, special liquid wax, that maintains viscosity at very high temperatures ABD and gained interest globally for cosmetics pharmaceutical use; industrial; an ingredient for toiletries, medicines, livestock feed, cosmetics; and as a lubricant. Moreover, jojoba has shown promise for reforestation, reducing desert creep, reducing browse, developing marginal lands, and providing revenue for locals (Khairi et al. 2019).

15.3 Gingerbread plum (*Neocarya macrophylla*) tree

Gingerbread plum (*Neocarya macrophylla*) trees (Fig. 15.2a-e) flourish in arid and semiarid regions, particularly in the Western part of Africa and Central America, especially Panama. It is known in the Hausa Language as 'Gawasa' (National Research Council, 2008). Seed oils from such plants can be used to make soaps (Amza et al. 2011). Quality evaluation of gingerbread plum (*Neocarya macrophylla*) seed oil and preparation of soap from the oil extract was reported (Warra 2012). The possible incorporation of gingerbread plum oils into cosmetics, pharmaceutical, and food products by exploiting the physicochemical properties of refined and unrefined oils of gingerbread plum (*Neocarya macrophylla*) kernels was reported (Warra, 2016). A (+)-catechin-3'-O-rhamnopyranoside, flavan-3-ol glycoside isolated from the stem bark extract of N. macrophylla was evaluated for its antimicrobial activities, analgesic, anti-antivenom, and inflammatory properties as reported. The results showed that catechin-3-O-rhamnopyranoside had significant analgesic, antimicrobial activities, anti-inflammatory, and antivenom providing new theory in the search for better therapeutic agents in the treatment of pain, inflammation, snakebite, and microbial infections. It was isolated from the ethyl acetate soluble fraction of the stem bark of N. macrophylla. The study showed a novel comprehensive biological activity of this compound (Yusuf et al. 2019a). The venom of Naja nigricollis was tested on the leaf of N. macrophylla using bioassay-guided isolation of antivenom components. Chemical tests were used to determine the substance's identity, and by comparing its 1H-NMR data to published data, this isolation of the component from the plant's leaf appears to be the first report. The research suggested that N. macrophylla leaves have significant anti-snake venom efficacy, providing the scientific rationale for its usage in traditional snake bite remedies (Yusuf et al. 2019b).

15.3.1 Products and materials development

Producing carbonaceous materials from plant matter constitutes an important resource. It has been discovered that gingerbread plum shells, an agricultural waste in Niger, can produce active carbons. Suggestions were provided on the resourcefulness of the active carbons derived from Parinari macrophylla shells to be examined via their performances as adsorbents or as catalytic supports. The high economic and ecological value will result from this, as well as a way to recycle agricultural waste (Siragi et al. 2021).

Figure 15.1. (a-b) Eucalyptus (*Eucalyptus grandis*) tree and branches. (c) Eucalyptus leaves (d) Fresh seeds of Eucalyptus. (e) Eucalyptus seeds powder. (f) Dried Eucalyptus seeds.

Figure 15.2. (a) Gingerbread plum (*Neocarya macrophylla*) tree. (b) Gingerbread plum flower. (c) Gingerbread plum nut. (d) Gingerbread plum seeds. (e) Gingerbread plum seed oil.

Plant-derived Biopharmaceuticals

Plant biosystems make it simple and affordable to manufacture treatments manufactured from plants, and they do not require refrigeration or costly medical equipment. These opportunities provide plant-based pharmaceuticals with the potential to treat ailments for which people historically had access to treatment in nations with limited resources (Bamogo et al. 2019). Through biotechnology, plants are engineered by utilizing living organisms with the aid of different transformation protocols like Agrobacterium-mediated transformation, biolistic gene gun etc. to develop biopharmaceutical products (Wani and Sah 2015). With development of this genetic engineering, plants have the potential to produce active proteins of pharmacological importance, like cytokines, antibodies from mammals, hormones, substitutes of vaccines, blood products and a lot of other therapeutic agents (Goldstein and Thomas 2004). In recent decades, plants have been used as a platform for the effective synthesis of important recombinant proteins due to a variety of benefits, including as quick production and scalability, the capacity to produce novel glycoforms, and the requirements for food crop safety (Moon et al. 2020). The efficient use of plants as production systems for recombinant proteins is recommended thereby providing different alternatives for transgene targeting and modification. Prospects of plants as a production system for therapeutic and prophylactic biopharmaceuticals with respect to posttranslational modifications reviewed (Warzecha 2008).

16.1 Biopharming for industrial crops

Turning plants into factories using molecular machinery in order to produce a desired protein with economic interest is nowadays a breakthrough in the manufacturing of valuable pharmaceutical products. Crop plant biotechnology has recently made strides beyond conventional agricultural production of fiber, food, or feed and has risen to address more challenging modern health, social, and industrial issues. The creation of new medicinal items, speciality and fine chemicals, phytoremediation, and renewable energy sources to replace non-renewable fossil fuels are all part of the new era. Plants have offered a true, low-cost replacement production mechanism for high-value goods. Antibodies, feed additives, vaccine antigens, hormones for human and animal health, and industrial proteins are currently the main products created from plants. In fact, biopharmaceuticals have special advantages due to their scalability, affordability, product safety, impact on the environment, and concerns with politics, socioeconomics, and regulation. Nonetheless, adoption is still limited, particularly in the poor world (Chakauyaa et al. 2006).

16.2 Concept of biopharming for transgenic pharmaceutical crops

Biopharming or plant molecular farming (PMF) is the genetic modification of plants to develop new industrial pharmaceuticals. It has the potential to offer revolutionary gains, but it also raises a host of daunting challenges. It is an example of the "third generation" of manipulated plant genetics. The second generation of genetically modified plants sought to enhance their agronomic properties (such as cold and drought tolerance) and nutritional qualities. The first generation of genetically engineered plants was pesticide and insect resistance (Guo 2004, Huot 2003). Biopharming in a broader sense is the harvest of bioactive molecules from mass-cultured organisms and crops (also called molecular farming), for utilization in industrial products ingredients and pharmaceuticals. Bioprospecting is different from biopharming in that it obtains its materials from natural populations. Although the industry has long extracted antibiotics and enzymes from mass-produced microorganisms, biopharming is not a completely novel idea. A number of drugs such as opium (Papaver somniferum) alkaloids, vincristine (Catharanthus roseus, Vinca rosea), camptothecin (Camptotheca acuminata and Nothapodytes Nimmoniana), digoxin (Digitalis species), placitaxel (Taxus species), rauvolfia and reserpine (Rauvolfia serpentina), etc., are extracted from wild or cultivated non-edible plant species. The cultivated edible plant species provide bioactive substances as piperine (Piper nigrum), curcumin (Curcuma domestica, Curcuma longa), papain (Carica papaya), bromelin (Ananas comosus), etc. The conventional ways of using genetically modified (GE) transgenic crop plants and domesticated animals are different from modern biopharming. The ability to manufacture vaccines and antibodies in agricultural plants without the use of embryonated eggs or cell cultures is a significant benefit of contemporary biopharmaceutical research (Rao 2008). The word "biopharming" refers to the experimental application of biotechnology in which crop plants are genetically altered to produce medicinal proteins and compounds that they do not naturally produce. The practice known as "biopharming" or "plant molecular farming" (PMF) entails using genetically modified (GM) plants to produce compounds with pharmacological or commercial value. The expense of biopharmaceuticals now restricts access to them. Biopharmaceuticals made from plants are less expensive to make and store, easier to scale up for large manufacturing, and safer than those made from animals. Discussions included advancements in this area and potential environmental issues (Daniell et al. 2001). Research and business interest in advanced biotechnologies to produce preferred Biopharmaceuticals (BPC) has increased. Yet, among the biggest obstacles are the BPC's cost, safety, and accessibility. Yet, producing and storing plant BPC might be cheap and possibly secure. The appropriate examples are vegetables and legume-producing plants. Leguminous crops, like alfalfa (Medicago sativa), have the major benefit of having a huge biomass production since they can be harvested frequently throughout the year. Phytoestrogens found in legume-producing plants include flavonoids, isoflavonoids, coumestans, and lignans. For the purpose of human and animal treatments, plant-derived low-cost monoclonal antibodies (plantibodies) can be made from legumes like peas, soybeans, and alfalfa. More research, in particular involving nutrigenomics and metabolomics, is necessary to specify such impacts on human health indicators of particular BPC due to the existence of a variety of bioactive chemicals in legumes (Nikkhah 2016, Nikkhah 2012). The creation of significant pharmaceutically and commercially valuable proteins in plants is known as molecular pharming. The goal is to find a low-cost, safe method for mass-producing recombinant medicinal proteins. Recombinant proteins could be produced in practically infinite amounts through molecular farming in plants and used in medicine and the life sciences as diagnostic and therapeutic agents. Plants generate a significant quantity of biomass, and the production of proteins can be boosted by the field growth of stable transgenic plant lines or plant suspension cell culture in fermenters. Furthermore, recombinant protein-rich organs can be produced by transgenic plants for long-term protein storage. This highlights the potential of employing transgenic plants as bioreactors for the molecular farming of recombinant treatments, such as vaccines, diagnostics, such as recombinant antibodies, plasma proteins, cytokines, and growth factors, and these include vaccinations (Fisher and Emans 2000). The creation of valuable molecules from agricultural plants for uses other than food, feed, or fiber (also called molecular farming, pharming, or biopharming). Already, plants are grown to create useful chemicals, such as many medications. Because plants are genetically modified (GE) to generate the substances we want them to, biopharming is distinctive (Mason and Amtzen 1995). Pharmaceuticals, commonly known as plant-made medications, are produced by genetically altering plants (PMP). With the help of biopharming, it was possible to produce some pharmaceuticals more quickly and affordably, giving patients access to more economical medication (Macaulay 2003). Producing pharmaceutical compounds and other valuable proteins on a big scale is made easy and affordable by molecular farming in plants. Plants provide a number of biosafety advantages over conventional production methods, including the lack of endotoxins, human infections, and oncogenic DNA sequences. Some biosafety issues, however, still need to be resolved. They show the final product's quality and safety as well as the broader consequences of molecular farming on human health and the environment. The technological underpinnings of molecular farming in plants and the identification of potential biosafety risks, such as the spread of transgenes in the environment, the accumulation of recombinant proteins in the ecosystem, the contamination of food and feed chains with transgenes and their products, and the quality and safety of the produced goods, were discussed (Commandeur et al. 2003). The technology of biopharming, also known as "molecular farming," may eventually result in the widespread commercialization of drug therapies that are grown using agricultural practices. Biopharmaceutical (biopharma) crops are plants genetically engineered to produce drugs for the pharmaceutical industry. During the past 20 years, one of the promised advantages of plant genetic engineering has been the production of medicinal compounds in crops. This application of biotechnology, also referred to as "pharming," "biopharming," or "molecular farming," has progressed from research to experimentation in greenhouses and fields. Pharmaceutical medications, such as vaccinations for infectious diseases and therapeutic proteins for the treatment of conditions like cancer and heart disease, will be more readily available and less expensively priced thanks to biopharming. Plants are genetically modified to generate certain substances, typically proteins, which are then removed and refined after harvest to create "plant-made pharmaceuticals" (PMPs) (Byrne 2008). Figure 16.1 provides a simplified illustration of molecular farming (Sahu et al. 2014).

16.3 Vaccine production

Now considered sources of innovative pharmaceutical proteins are medicinal entities such as vaccines, antibodies, blood replacements, and others. Plant-derived antibodies, vaccines, and other proteins are far more important than mammalian-derived rDNA pharmaceuticals since they are free of mammalian viral vectors and human infections. Moreover, plants have low capital and operating expenses, superior scalability, eukaryotic posttranslational modifications, a relatively high protein output, low cost of cultivation and high biomass production, and a relatively quick "gene to protein" time (Jelaska et al. 2005). Compared to fermentation-based manufacturing techniques, transgenic plants that express foreign proteins with industrial or medicinal value are more cost-effective. The transient or steady expression of foreign genes has led to the production of specific vaccines in plants. It's been established that a number of different types of bacterial infections can be prevented by using a variety of preventative measures, including the use of antibiotics. When given as food, transgenic potato tubers that expressed a bacterial antigen-induced humoral and mucosal immune reactions. These findings offer "proof of concept" for the production of vaccines using plants (Mason and Arntzen,1995). The production of recombinant proteins in plants has many potential advantages for generating biopharmaceuticals relevant to clinical medicine. One of these is the need for purification, which can be removed when the recombinant protein-containing plant tissue is consumed as food (edible vaccines). Production of edible vaccines in selected crop plants e.g. corn-based edible vaccines was extensively reviewed

(Daniell et al. 2001). In contrast to commercially available vaccines, efforts are being made to develop inexpensive edible vaccines that offer a number of advantages. For this aim, recombinant proteins, such as viral and bacterial antigens and antibodies, can be expressed by transgenic plants. Vaccines against illnesses like hepatitis B, cholera, HIV, etc. have been created using common food plants like banana, tomatoes, rice, carrot, etc. Hence, the up- and down-regulation of desired genes that are employed to modify plants play a significant part in the advancement of genetic crops. Recombinant DNA technology's role in creating transgenic lines and cultivars of various crops with improved nutrient quality, biofuel production, increased production of vaccines and antibodies, increased resistance to pests, diseases, and abiotic stresses, as well as the safety precautions for their commercialization, has been thoroughly discussed (Ahmad et al. 2012). Advances in the production of edible vaccines in the carrot system were suggested as an alternative in the protection against infectious diseases in animal models, and eventually humans. Even though there are problems associated with non-transgenic carrot plant-made edible vaccines, which have been approved in clinical trials, limitations to using the carrot system have a low protein expression. Despite this, edible vaccine development and production are in urgent need. Low production costs and ease of delivery are some challenges that need extensive studies (Issaro et al. 2018). With the advancement of science and technology, there has been much advancement in vaccination technology. Biopharming technology is the most recent technology serving as a platform for vaccine development. Plant vaccines, which are currently administered as capsules or injections or, in the future, as edibles, are preferred for developing nations. Automated biopharming factories are working in this regard. The biolistic transformation technique eliminates the need for vectors, but targeting is compromised. Nuclear transformation is the most suited and well-studied of all. Recombinant DNA stability can be maintained by over-expressing the target protein, and optimization is achieved by picking an efficient promoter. Expression, whether steady or transient, has benefits. Maximum protein extraction is completed by downstream processing, and cost-effective processing with purification is still a goal to be attained (Iqbal et al. 2018). Significantly, an option that combines safety and effectiveness that enables oral distribution by eating edible plant tissue is provided by transgenic plants that produce nonreplicative subunit vaccines (Hefferon 2010).

16.4 Search for medicinal plants against novel Coronavirus disease (COVID-19)

Recently, medicinal plants and their extracts, such as herbs, have drawn a lot of attention because they have demonstrated promising results in the fight against SARS-CoV-2 infection and have proven important in treating COVID-19 patients as alternatives to conventional medical approaches during this devastating pandemic. Summary of several phytochemical compounds, their mechanisms of action, and potential antiviral effects against significant viral diseases (Anand et al. 2021). Herbal or medicinal plants are known for their effectiveness in the treatment of diseases. A study using clinical data from Wuhan showed that Traditional Chinese Medicine (TCM) was reported to have a systemic theoretical background on the pathological evolution of novel coronavirus pneumonia and played a significant role in combating the novel coronavirus pneumonia (NCP) caused by this fast-spreading virus COVID-19. Work of this nature will serve as a resource for drug discovery against 2019 novel coronavirus (SARS-CoV-2)(Luo et al. 2020, Ling 2020). Traditional Chinese Medicine's use of herbs to treat coronavirus infections was investigated during the SARS-CoV and MERS-CoV outbreaks. Initial investigations in China revealed that the alcoholic extract of sweet wormwood (Artemisia annua) was the second most potent herbal remedy used to treat the 2005 SARS-CoV. Medical researchers in Denmark and Germany would work with the Max Planck Institute of Colloids and Interfaces, Potsdam, Germany, to test Artemisia annua plant extract and artemisinin derivatives in laboratory cell tests against the novel coronavirus illness (COVID-19). According to reports, the plant extract from Artemisia annua and pure derivatives derived from the plant, like artemisinin, were tested on cells in research facilities in Germany and Denmark (The Max Planck Society 2020). Many medicinal properties of *Euphorbia hirta* Linn, a herb. There have been reports of the production of a milky white latex. This herb also shows antimalarial, antioxidant, antiasthmatic, antibacterial, anticancer, anti-inflammatory, galactogenic, antidiarrheal, antifertility, antiamoebic, and antifungal activities (Kumar et al. 2010). Some of the symptoms, such as fever, coughing, and respiratory difficulties, may be helped by it. A research team at the University of Ibadan is investigating the potential of Euphorbia hirta and related plants to treat COVID-19 and other respiratory illnesses as well as chronic flu (University of Badan website 2020). A Nigerian viewpoint on the potential value of African medicinal plants as prospective sources of anti-COVID-19 therapeutic leads were examined (Attah et al. 2021). To treat patients with COVID-19, N. sativa may be utilized as adjuvant therapy in conjunction with modified conventional medications. Thymoquinone (TQ), an active component of black seed oil, has antioxidant, anti-inflammatory, antiviral, antibacterial, immunomodulatory, and anticoagulant properties. TQ also shows antiviral potential against a number of viruses, including murine cytomegalovirus, Epstein-Barr virus, hepatitis C virus, human immunodeficiency virus, and other coronaviruses. The activity of cytokine suppressors, lymphocytes, natural killer cells, and macrophages is boosted by TQ. Recently, TQ has expressed notable antiviral activity against a SARSCoV-2 strain isolated from Egyptian patients and, interestingly, molecular docking studies have also shown that TQ could potentially inhibit COVID-19 development through binding to the receptor-binding domain on the spike and envelope proteins of SARS-CoV-2, which may hinder virus entry into the host cell and inhibit its ion channel and pore-forming activity. Additional research has indicated that TQ may block SARS CoV2 proteases, which could reduce viral replication. It has also exhibited strong antagonistic activity toward angiotensin-converting enzyme 2 receptors, which enables it to prevent virus uptake into the host cell. Asthma, dyslipidemia, diabetes, hypertension, renal dysfunction, and malignancy are only a few of the chronic diseases and conditions against which they may be protective. Recently, TQ has been used in clinical trials to treat a variety of illnesses (Badary et al. 2021, Maideen 2020). Scientists' computational analyses recommended the following Moroccan medicinal plants as potential inhibitors of the major SARS-CoV-2 protease: Prior to the clinical essay, the synthesis of a few compounds and the evaluation of their in vitro and in vivo activity against SARS-Cov-2 major protease were reported to be of interest. The researchers have found three molecules (Crocin, Digitoxigenin, and β-Eudesmol) among 67 which are very interesting either on the chemical side or on the biological side and they proposed these three molecules as inhibitors of SARS-CoV-2 main protease. This was based on the energy types of interaction between these molecules and studied protein (Aanouz et al. 2020). With the current development of several vaccine candidates, antivirals capable of limiting virus transmission or blocking infection are being explored. To generate vaccines and antiviral proteins inexpensively and at a mass scale, plant production platforms are being used (Mahmood et al. 2021).

16.5 Challenges of biopharming for biopharmaceuticals

As upcoming commercial state-of-the-art plant-human-animal biotechnologies, plant-derived biopharmaceuticals are thought to be in the development stage. As such they may have a high developmental size and economies, safety worldwide, storage capacity, and reasonably efficient distribution. These proteins provide excellent chances for the cost-effective synthesis of drugs and vaccines. Yet, there are significant drawbacks, including ambiguous manufacturing practice rules for grown plants and dubious regulatory foundations in platform-raised crops. For the new goods to be widely applicable, careful collaboration and compromises of the academic and commercial property environment are required. Authorized operation permits must be made sure of in order for applications and use to be successful in particular nations and locations. Products will thereby enable saving both human and animal lives where environmental issues are present at the same time (Nikkhah 2014a). Due to its low cost and simplicity of administration, oral vaccination delivery is a very good alternative to injection. Oral vaccines also improve the likelihood of developing mucosal immunity against infectious pathogens that enter the body through a mucosal surface. The stomach and gut's ability to degrade protein components before they can trigger an immune response is a big worry with oral vaccinations, however. Many delivery methods have been created to transport intact proteins to the gut while preventing protein breakdown. They include transgenic plant tissue, bio-encapsulation tools like liposomes, and recombinant strains of attenuated microbes (Daniell et al. 2001).

Wide-ranging uses of delectable crop plants, vegetables, and fruits in the natural prevention and treatment of widespread ailments are anticipated in the not-too-distant future. Concerns about biosafety, dose uniformity, regulatory requirements, and the longevity of herbicides and pesticides are still unclear (Nikkhah 2014b). The potential uses of genetically modified crops for biomanufacturing innovations and new prospects for the agricultural industry hold great promise, but they also carry significant concerns. From the standpoint of their coexistence with conventional agriculture, the management of the externalities and of any unexpected economic impacts that develop in this environment is crucial and raises challenging challenges for regulators of pharmaceuticals and industrial compounds. Such plant-based commercial products and medications rely on fascinating scientific and technological frameworks (Moschini 2006).

16.6 Drug discovery and delivery

Drug discovery from plants requires a multidisciplinary approach that combines biological, botanical, ethnobotanical and phytochemical protocols. New lead compounds for the synthesis of medications for a variety of pharmacological targets, including cancer, HIV/AIDS, malaria, Alzheimer's disease, and pain, are provided by plants (Jachak and Saklani 2007). Several blockbuster pharmaceuticals are obtained directly or indirectly from plants, which are a source of essential novel pharmacologically active chemicals. Despite the present focus on synthetic chemistry as a means of discovering and producing medications, plants continue to make a significant contribution to the treatment and prevention of disease. Even before the start of the twenty-first century, just 11% of the 252 medications that the WHO deemed to be basic and essential were from flowering plants (Veeresham 2012). According to a literature review, the foundation for the development of new plant-based drugs should be the identification of the appropriate candidate plants using Ayurvedic knowledge, traditionally recorded use, tribal non-documented use, and thorough literature search. An in-depth understanding of the predominance of specific Ayurvedic characteristics can be obtained from ingredients in ancient documented formulations and analysis of their Ayurvedic attributes when frequently conducted based on which suitable candidate plants may be chosen for bioactivity-based fractionation. The integration of Ayurvedic knowledge with drug discovery necessitates a paradigm shift from sequential to parallel extraction in the extraction process. Standardized extract or an isolated bioactive drug-able molecule as the new drug may result from bioassay-guided fractionation of the identified plant (Katiyar et al. 2015). Drug development from medicinal plants is now being studied using a variety of methods, including botanical, phytochemical, biological, and molecular approaches. Drug development from medicinal plants continues to yield promising novel treatments for a variety of pharmacological conditions, including cancer, HIV/AIDS, Alzheimer's, malaria, and pain (Balunas and Kinghorn 2005).

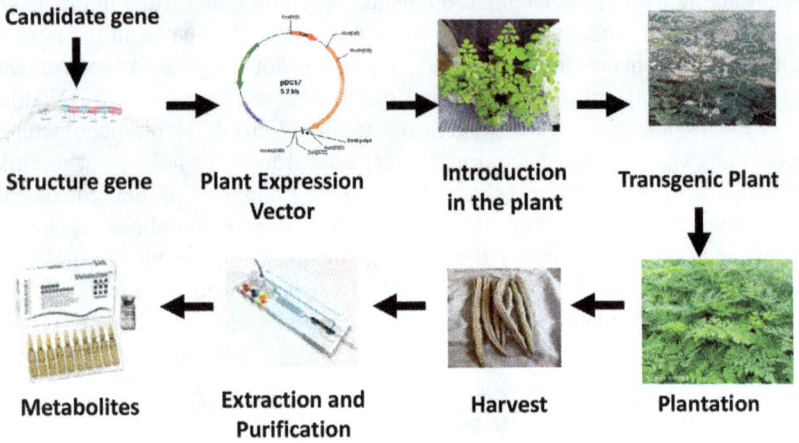

Figure 16.1. Molecular farming – Genetic transformation.
Modified from Sahu et al. (2014).

Bio-based Products, Waste to Wealth and Development of Circular Bioeconomy

The production of biological resources that are naturally renewable and their transformation into bioenergy, food, feed, and biobased goods through fresh innovation and effective industrial biotechnologies comprise the circular bio-economy. It offers solutions to environmental, societal and economic concerns, including mitigation of climate change, food security, and energy and resource efficiency.

17.1 Industrial crops as feedstock for the development of biocomposites

The growing awareness of polymer composites is a result of growing environmental concerns and the urgent need for more environmentally friendly materials, such as biodegradable fillers and reinforcements sourced from renewable sources, particularly those from forests. Some industrially significant composites are typically referred to as "green" composites. Biodegradable polymers made from natural resources are another crucial component of green composites (Singh et al. 2017). Recent studies focused on developing ecologically acceptable and long-lasting "green" composites made of plant fibers and have provided information on the processing, matrix-reinforcement combinations, shape, and characteristics of these materials. These resources are particularly helpful in packaging, building, automotive, medical, and other industries (Satyanarayana 2015).

17.2 Industrial oil crops and production of protein-based plastics

Emerging technologies have made it possible to produce protein-based plastics from industrial oil crops residuals. The modern world depends on plastic materials for many applications, from syringes as medical devices to the packaging of food to keep them fresh. These materials are produced at high temperatures using petrochemicals derived from fossil fuels. Recently the price stability of fossil oil likewise its secure supply has become an issue of discussion. Petrochemical feedstock will either be costly or simply unavailable at some time in the future. If we are to obtain sustainable gains from the products made presently available from fossil oil, we will eventually need to find substitute ways of making the products that currently come from petroleum sources. Another disadvantage of most current petroleum-based plastics is that they degrade slowly placing stress on the environment if not disposed of properly (Newson 2012). The majority of the protein in wheat gluten materials is not hazardous and is entirely biodegradable, regardless of the technical procedure used, according to a review of research on the film-forming capabilities of wheat gluten, maize zein, and soy protein (Domenek et al. 2004, Zhang et al. 2010).

17.3   Development of bio-based chemicals

There are excellent opportunities for the development of bio-based chemicals within the systems of the biobased economy and biorefinery. This frequently happens in conjunction with the creation of bioenergy or biofuels. The worldwide chemical industry might earn $10–15 billion from the development of bio-based products (Jong et al. 2012). Industrial crops such as bioenergy crops provide an advantage of reducing our dependence on petrochemicals that have now become the primary source for making the majority of our consumer and industrial products. Metabolites synthesized in plants are utilizable as biobased platform chemicals for certain substitution of their counterparts derived from petroleum is reviewed in the literature. The goal of these plant metabolic engineering techniques is to increase the number of metabolites in biomass. Recent developments in the control of the isoprenoid and shikimate biosynthetic pathways, both of which are sources of numerous useful chemicals, were highlighted. Enhancing the commercial value of biomass and achieving sustainable lignocellulosic biorefineries may be possible by implementing and improving modified metabolic pathways for the accumulation of coproducts in bioenergy crops (Lin and Eudes, 2020). Plant seeds are reported to have served as feedstock for potential industrial chemicals. Many plant species' seeds contain rare fatty acids and lipids that have proven valuable as a renewable (non-petroleum-based) industrial feedstock or in the cosmetic sector. Seeds typically contain, or have the capacity to create, protective compounds that function as plant growth regulators, fungicides, insecticides, and herbivore repellents, in addition to proteins and energy-storing components like carbohydrates and lipids (Powell 2009).

17.4   Agri/Silvicultural waste or byproducts for industrial applications

Agricultural wastes (AW) are byproducts of cultivating and processing raw agricultural products, in this context they include fruits, vegetables and other crop products. Agricultural wastes can appear in solid, liquid or slurry form depending on the nature of agricultural activities (Foster 2015). A report suggested that high-quality activated carbon from cotton biomass waste materials is potentially useful for various adsorbent applications (Sartova et al. 2018). Agricultural waste fibers have the potential to be used as building materials, decorative items, and flexible raw materials. They are likewise used in composite attributable to its high strength, eco-friendly nature, minimal expense, accessibility and supportability. The possible properties of farming waste strands have started a great deal of exploration to involve these filaments as a material to substitute man-made filaments for safe and harmless ecosystem items. Rural waste can be gotten from oil palm ranches and some other agrarian industries, for example, rice straw, rice husk, pineapple, coconut, sugarcane, and banana. They deliver a lot of biomasses that are delegated as regular strands which as of recently just 10% are used as elective natural substances for some ventures, like auto part, biocomposites, biomedical and others (Dungani et al. 2016). Squanders from agro-enterprises as possible normal wellsprings of cell reinforcements and antimicrobials and the doable mechanical applications in food and aging ventures, particularly the bioethanol business was taken advantage of (Shirahigueand Ceccato-Antonini 2020). Cellulosic polymers to be specifically cellulose, di- and triacetate created from fourteen farming squanders; Branch and fiber left over from oil extraction from oil palm (Elaeis guineensis), raffia, piassava, bamboo mash, bamboo bark from raphia palm (Raphia hookeri), stem and cob of maize plant (Zea mays), natural product fiber from coconut natural product (Cocos nucifera), sawdust from the cotton tree (Cossypium Hirsutum), pear wood (Musa paradisiaca). It has been shown that these materials which were viewed as horticultural squanders and permitted to spoil away in this way as a disturbance to the climate contains cellulose (Israel et al. 2008). Agricultural waste material being profoundly productive, a minimal expense and a sustainable wellspring of biomass can be taken advantage of for weighty metal remediation. These biosorbents can additionally be adjusted for improved effectiveness and various reuses to upgrade their materialism at a modern scale (Garima et al. 2008). Research announced was given on the improvement of practical biobased composite items from agrarian squander/biomass (Muniysamy et al. 2015). Lignocellulosic agribusiness squanders as biomass feedstocks for second-age bioethanol creation were accounted for (Saini et al. 2015).

17.5 Bio-based adhesives from crop plant-derived products

Bio-based adhesives synthesized from cottonseed oil residue (COR) which could give a theoretical framework and scientific road map for the appropriate processing technique and application technology development of COR were reported (Liu et al. 2020). Alfalfa leaf protein was used as a bioresource in the development of eco-accommodating a hay leaf protein-based wood adhesive. This will grow the determination of unrefined components for protein-based wood cement (Zhang et al. 2020). A basic survey has featured future open doors for bio-based glues – benefits past inexhaustibility which covered new functionalities by means of new breakthroughs, sub-atomic models, the upsides of vegetable oils, for example, hydrophobicity, decreased human and ecological poisonousness and the presentation of bio-based contrasted with oil-based types of cement (Heinrich 2019). Protein concentrates showed that better cement execution was most likely according to optimizing protein sub-atomic weight dissemination and hydrophobicity. A way to deal with further developing soy protein-based bond execution and furthermore diminish the general expense of the plant protein-based composites as types of cement were illustrated. Soy protein has been the most broadly explored protein-based glue; because of its cutthroat purposes in both food and feed ventures, soy protein stays a costly item to relieve such a contest, protein composites were researched by blending low-worth sorghum and canola proteins into soy proteins. The study of Li et al. (2020) revealed a straightforward, easy method for increasing the performance of soy protein-based adhesives while simultaneously lowering the total cost of plant protein-based composite adhesives. In starch-based wood glue, many new methodologies have been approached for its compelling use in wood/ wood composite cement giving practically identical execution as manufactured types of cement. Starch-based glues made with an emphasis on starch change strategies for further developing properties of starch-based cement was given as a survey (Gadhave et al. 2017).

17.6 Phytomass value chain and value additions

Due to large quantities of biomass that can be produced annually a great interest arose in products from Miscanthus. Potential uses include use as bedding for animals, mulch for horticulture applications, and insulation to improve energy conservation. Miscanthus has excellent natural absorbent qualities which makes it very attractive for spill management and as a bedding material. Compostable food service ware has been produced from Miscanthus to replace products from plastic that do not biodegrade. Building applications include fiberboard, particleboard, and composites. Miscanthus has high-quality cellulose for material applications and is an excellent source of cellulose where high quality is important. Nanocellulose applications from this crop are of interest and this is an active area of research, and cellulose from Miscanthus for paper production is one of the applications that was reported. An important issue was raised on how best to use Miscanthus biomass not only for energy production but also for conversion to different bioproducts. Miscanthus serve as an important raw material for producing fiber-based materials such as construction or paper industry products due to its high yields with prominent content of lignocellulose, a low requirement for nutrient inputs, and low susceptibility to pests and diseases (Pidlisnyuk et al. 2021).

17.7 Valorization and pyrolysis of crop plant resources for products development

Different alternatives for the valorization of citrus wastes were provided for sustainable management of large amounts of residues from citrus processing industries. Several studies have evaluated the possibility of valorization of citrus peel for biomethane or bioethanol production in biochemical processes. Notwithstanding, much exploration and research is expected to get rid of the poisonous impacts of rejuvenating ointments on the microbial local area. Placing into thought the high added worth of the recuperated compounds, use of strip squander in nourishment (for creation of gelatin, dietary strands, and so on), corrective and pharmaceutical enterprises (extraction of flavonoids, enhancing specialists and citrus extract) despite the fact that still non monetarily economical, yet as a rule is promising (Calabro 2017). The worth of lingering biomass has these days attracted worldwide consideration because of its sustainable structure and overflow combined with worldwide financial and ecological issues related to the substantial utilization of petro-synthetics. By refocusing their interests on biomass-based fillers, ligninolytic proteins, synthetics, and biocompatible compounds that may be obtained from a variety of lignocellulosic waste materials, professionals are converting lignocellulosic bio-innovation into an attractive field. Lignocellulosic materials' potential use in biotechnology, including the development of proteins, creature feed, synthetic substances, bio-fills, mash and paper, and composites was explored (Iqbal et al. 2013). The capability of banana strips (Musa sapientum) for bio-valorization was audited. The usage of banana strips (BP) in biotechnological applications has entered many items which have the potential for industrialization and commercialization. The utilization of crude BP might have been effective as creature feed yet this has not been accomplished at this point with bio-transformation with further developed protein content (Jamal et al. 2012). A study supported methodologies for another idea of horticultural waste administration before biorefinery, in light of thorough materials science examinations. Effective pretreatments for the extraction, partition and fractionation of farming waste related to comprehension of the subtleties of its microstructure and properties which can be fundamental for high-productivity biorefinery was investigated to act as a significant and crucial reason for scientists and businesses in the area of straw biomass biorefinery. To accomplish the greatest productivity conceivable in agrarian waste valorization, it is significant to comprehend that not all pieces of the straw are similarly important and can be treated in the equivalent biorefinery process. The wheat straw stem, which is made out of hubs and internodes, has been displayed to have unmistakable properties and attributes. Detachment of these physical parts before the biorefinery interaction presents a novel region for future exploration ventures, as it can prompt better execution of the expected item. For instance, the hub has higher extractives and debris content, which end up being a lessening factor for biocomposite or bioenergy creation. This has demonstrated that the usage of rural waste as unrefined substances decidedly influences ecological and financial angles by creating extra pay for the ranchers, yet in addition producing practical high-performing bioproducts (Ghaffar 2020). The likely waste-to-energy advancements that can be utilized in the valorization of banana squander were checked on. It was realized that the energy content of banana squanders, whenever bridled through fitting waste-to-energy innovations, wouldn't just address the energy prerequisite for the handling of banana mash, however, would likewise offer an extra advantage of keeping away from petroleum products using sustainable power. Involving Uganda's banana industrialization is a model which is rustic based with restricted specialized information and the monetary ability to arrange present-day sun-powered advances and thermo-changes for drying banana organic product mash. The benefits of different waste-to-energy advances as well as their deficits were investigated. Anaerobic absorption stands apart as the most possible and suitable waste-to-energy innovation for addressing the energy shortage and waste weight in the banana industry. Expected choices for the improvement of anaerobic assimilation of banana squander were likewise explained (Gumisiriza et al. 2020). Carrot strip is a genuine illustration of phytonutrient-rich agro modern side-effects created from the handling of carrots. The customary techniques for the extraction of phytonutrients require a huge volume of natural solvents, complex methodology and costly hardware. Thus, the advancement of green and less difficult extraction strategy is worthwhile to the valorization of agroindustrial squandering as far as financial and manageability. A review assessed the materialness of carotene-gelatin hydrocolloid complexation to the co-extraction of carotenoids and gelatin from a carrot strip squander. Carrot strip squander is a possible feedstock for this extraction strategy since it is wealthy in carotenoids and gelatin, which could frame the colloidal complex prompted by water. In contrast with the traditional dissolvable extraction technique, the quantity of working strides in carotene-gelatin hydrocolloid complexation is altogether lower and the cancer prevention agent action of β-carotenoids was higher. The carotene-gelatin hydrocolloid complexation is viewed as a green extraction strategy that empowers the valorization of farming waste to recuperate carotenoids (Jayesree et al. 2021). A cooperative undertaking with the essential area for the necessary valorization of the waste produced in the crisp handling vegetable lines of a rural helpful is at present being grown, especially centered on cabbage, carrot, celery, and leek. Valorization of vegetable crisp handling buildups as useful powdered fixings with a center around the likely effect of pretreatments and drying techniques on bioactive mixtures and their bioaccessibility were audited (Ramírez-Pulido et al. 2021).

17.8   Biochar development

Biochar is typically thought of as a charcoal-like substance that is produced by pyrolyzing natural material from agricultural and forest service wastes (also known as biomass). To reduce contamination and safely retain carbon, biochar is made using a specific interaction. Natural materials, such as wood chips, leaf litter, or dead plants, are burned in a container with extremely little oxygen during pyrolysis. As the materials are used up, they nearly never emit degrading emissions. The natural material is converted into biochar during the pyrolysis cycle, a stable form of carbon that can only be released into the environment with great effort. Pyrolysis produces energy or intensity, which can be captured and used as a clean energy source. Biochar is excellent at converting carbon to a stable structure (Recovery Global 2021). Biochar is any type of biomass including, yet not restricted to, wood and wood materials, bark, grasses, farming and modern side-effects, which have been made through the pyrolysis method (Brasher 2021). Biochar creation from areca nutsquanders was accounted for. Aside from soil carbon sequestration, warm change of biomass like Areca catechu to biochar can turn into a feasible procedure for biomass the board (Vijayanand et al. 2016). A slow pyrolysis strategy has been applied to the straw and tail of rapeseed plants, and the impacts of temperature and warming rate on the yields and qualities of the strong items (biochars) have been researched. The biochars acquired are carbon-rich, receptive, and generally contamination-free potential strong biofuels (Karaosmanoğlu et al. 2000).

17.9   Lignin platform

Around 20–35% of the mass of a tree or plant is made out of lignin. It is the most plentiful natural fragrant. The biosphere comprises absolute lignin which surpasses 300 billion tons and every year increments by around 20 billion tons. The exceptional properties of lignin including its profoundly fragrant nature and lower oxygen content contrasted with polysaccharides render lignin an exceptionally fascinating bio-polymer to be changed over into a valuable, sustainable compound structure that impedes and related energizes and materials (Bruijnincx et al. 2016). A survey has featured research improvement in lignin biosynthesis, lignin hereditary designing and different organic and substance methods for depolymerization used to change over lignin into biofuels and bioproducts. The bioproducts can diminish our dependence on petrol-based items, which is much needed as oil saves are non-sustainable and restricted. Additionally, hereditary change of the lignin construction will decrease synthetic substances utilized for its breakdown in the bioethanol business and in this way diminish the pretreatment expenses and compound waste (Welker et al. 2015).

17.10   Pectin production

The extraction of gelatin from orange strips (Fig. 17.1a) was reported. The investigation was completed on citrus orange strips, a side-effect of citrus orange handling, as a wellspring of gelatin. This is for increment benefits for citrus orange cultivators and processors. The effort focused on the enhancement of the cycle component needed for the extraction of substantial value-added goods like gelatin from orange strips, which represents a misuse of the handling of squeezed oranges. The result of the work featured that the sweet orange strips are a great wellspring of gelatin and can possibly become a significant unrefined substance for food handling enterprises (Alamineh 2018). Production by extraction and portrayal of gelatin from locally accessible citrus assortment to be specifically lemon natural product strip has been accounted for which showed that lemon strip waste can be utilized as a superior wellspring of gelatin which can be utilized for making the food items in food businesses (Akhtar et al. 2020). Previously numerous specialists are dealing with the improvement of the piece of the cycle innovation required for the extraction of significant worth added items for example orange oil and gelatin from orange strips, which is a misuse of the squeezed orange handling industry. Gelatin and natural ointment are separated by two techniques: steam refining and draining. The gelatin yield from steam refining is more than the filtering process, revealed numerous specialists. Gelatin is significantly utilized in food handling ventures, clinical, dairy, wholesome, well-being, corrective items, and drug and is a conventional gelling specialist for jams and jellies. It was recommended that as innovative work (Research and Development) as of late expanding to find more yield of gelatin, in future there is the assumption for some creative and energizing applications in the field of extraction of gelatin. This seems, by all accounts, to be a chance for youthful specialists that are committed to practical business (Kanse et al. 2017). One of the expected hotspots for gelatin creation is the pineapple strip because of the great substance of gelatin in its dietary fiber organization. Gelatin is utilized as a food thickener, emulsifier, stabilizer and gelling specialist in the food industry. Microwave-helped extraction of gelatin from pineapple strips (Fig. 17.1b) was accounted for (Zakaria et al. 2021). The utilization of Punica granatum strips for the extraction of gelatin was additionally revealed (Sathish et al. 2018).

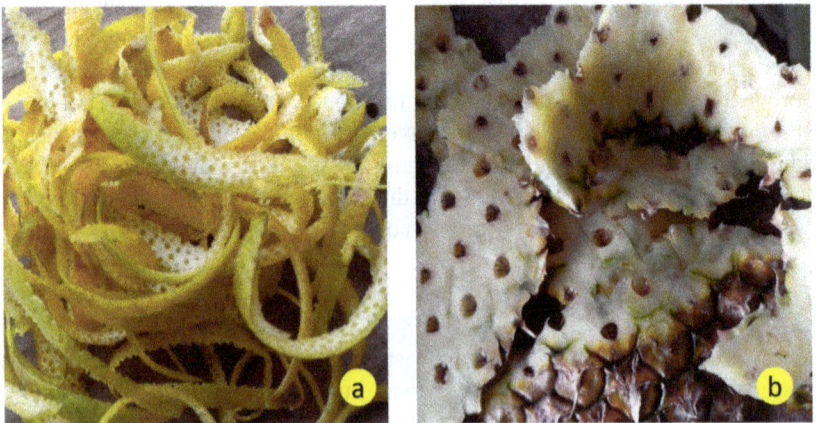

Figure 17.1. (a) Sweet orange peels. (b) Pineapple peels.

Industrial Crops as Complementary for Food Security and Climate Change Mitigation

Many different efforts that attempt to provide solutions or remedies, complement food security measures and reduce or prevent the emission of greenhouse gases were suggested by experts worldwide. Here we present an overview of road maps for their realization.

18.1 Industrial crops as complementary for food security

Cassava, a hardy crop that can withstand challenging growth circumstances and long-term storage, significantly contributes to food security. An effort was made to increase cassava production, with two distinct and intriguing areas of production: large-scale production for industrial starch, and its expansion as a crop for small-scale farmers to ensure food security, especially considering climate change (Amelework et al. 2021). The potential of cereal crops with a Smart Village Based GIS modeling approach to support food security was reported (Utomo and Jumiatun 2022). Fears exist of less food and nutritional security in growing industrial crops and engaging in their farming and processing. However, it was argued that such fears may be exaggerated. Food security is frequently predicted to increase when considering the synergies and higher revenues between cultivating food and industrial crops. Even with industrial farming, considerable progress is expected to achieve. 108 of the 119 million acres worldwide that are planted for industrial crops are made up of oil palm, cocoa, cotton, sugar cane, rubber, and coffee. Out of the 347 million ha that was planted with crops in Africa in 2011–2013, only 21 million ha were planted with industrial crops. A few crops are once more dominant: on nearly 19 of the 21 million acres, oil palm, cotton, cocoa, coffee, and sugar cane were planted. In the places where they were planted, industrial crops have not expanded quickly either. Africa's cultivable land increased by more than 3.8 times between 1961–1963 and 2011–2013, although the area used for industrial crops only increased by 43% during that time. This implies that industrial crops are not insignificant. They might occupy a sizable portion of the local terrain, yet it does assist in putting some recent worries into perspective. There is no widespread transition from food to industrial crops in the farmlands of Africa or the rest of the world (Wiggins et al. 2015). Findings showed a positive connection between cocoa cultivating, family crop pay, and food security questioning the writing that money crops sabotage food security, causing exceptional notice of complementarities between cocoa, food harvest, and cashew creation. Even though ensuring food security pay from cocoa alone was not adequate, rather it improved the monetary capacity of ranchers to enhance cashew and grow food crop creation. This minimization of reliance on the market for food staples improved yearly yield pays, and guaranteed a consistent progression of pay.

Overall, the benefits of cocoa growing for food security were predicted by cashew diversification, the substitution of food crops with cocoa, socioeconomic characteristics such as land ownership, livestock ownership, and formal education, and very slightly by total yearly crop income (Hashmiu et al. 2022).

18.2 The potential of industrial crops in climate mitigation

With regards to environmental change and relief, biotechnology can contribute well to bringing biological system weakness down to its ramifications. To address environmental change variation and moderation for upgraded rural flexibility, efficiency, and food security, as well as to add to the decrease of ozone-depleting substances, an assortment of regular and current biotechnology approaches was investigated. The extremely adverse consequences of environmental change on modern yields could be emphatically tended to by utilizing customary agrarian biotechnology strategies like energy-effective cultivating, the utilization of biofertilizers, tissue culture, and reproducing for versatile assortments, which would uphold endeavors to sequester carbon (Mtui 2011). Substantial expansions in the seriousness with which saltiness, waterlogging and immersion influence crop creation in large numbers of the world's farming districts can be caused by environmental change. Three potential ways to deal with the advancement of harvests for saline, waterlogged and immersed soils are proposed: (i) crop species choice from the inside; (ii) the development of crossbreeds between adjusted wild species and yield plants; and (iii) the training of halophytes (Mullan and Barrett-Lennard 2010). Row crop and forest systems have been the subject of numerous studies aimed at lowering greenhouse gas emissions and carbon (C) sequestration; however, it was noted that little attention has been paid to contributions from niche crop industries like ornamental gardening. Landscapes in rural, suburban, and urban areas are impacted by the ornamental horticulture business. Even though aesthetic gardening has certain unfavorable effects on the environment (such as CO_2 and trace gas outflow), it also has the ability to lower greenhouse gas emissions and boost carbon sequestration. Possible areas in ornamental horticulture container-grown plant production where techniques could be changed to promote carbon absorption and reduce greenhouse gas emissions have been highlighted (Marble et al. 2011). New research from Michigan State University showed that biofuel crops can assist in mitigating climate change when grown on land of otherwise little agricultural value (Michigan State University 2022). Case Studies for coffee and rice as adaptation climate-sensitive crops in agroforestry were reported (Lin et al. 2015). Perennial lignocellulosic crops grown on marginal lands were found to be potential boosters in climate change mitigation (Martinez-Feria et al. 2022). By deploying improved weathering, global croplands and bioenergy crops have the ability to mitigate climate change. Enhanced weathering (EW) in bioenergy crops provides the opportunity to sequester CO_2 while reducing fossil fuel combustion. This is as biofuel use increases (Kantola et al. 2017). As a low greenhouse gas emitter compared to cereals and oilseed rape, sunflower crops may contribute to the mitigation strategy. To lessen the uncertainty in oil yield and projections, sunflower crop models should be taken into consideration as an absorber for these new environmental elements (Debaeke et al. 2017). Genetically modified (GM) crops, especially industrial crops can support climate change mitigation by reducing agricultural greenhouse gas (GHG) emissions. Also, possible decreases in production emissions, GM yield gains also mitigate land-use change and related emissions (Kovak et al. 2022).

Entrepreneurial Approaches for Sustainable Greener Ecosystems

In addition to bio-fortified food crops, diet foods for diabetic patients, and edible vaccines derived from plant sources, biotechnology has made it possible for crops to mature more quickly, produce higher yields, to be more nutritious, have longer shelf lives, and improved flavor and color. The development of industries through biotechnology initiatives will provide employment possibilities and reduce poverty. Biotechnology, entrepreneurship, and reducing poverty all interact (Enwean 2013).

A discussion of some of the accomplishments made in relation to the manufacturing and commercialization of transgenic crops was given, along with an overview of the plant transformation processes. Crop plants can be modified to increase productivity by adding genes that are of interest, such as those that improve soil nutrient uptake, insect resistance, resistance to viral and bacterial diseases, improved nutritional and health qualities, drought tolerance, extreme heat, high salinity, and heavy metal soil contamination. Transgenic crops can produce commercially viable biotechnological goods like biodegradable polymers and metabolites with medicinal qualities (Khan and Liu 2009).

Natural resources are abundant in the world, especially crops that can be genetically modified for home, environmental, and industrial uses. People all around the world are extremely motivated to start business startups in order to boost global economic growth. This might be required to address the escalating issues in the open market. In light of the changing employment pace in communities and schools as well as the rapid changes in technology, it has become necessary to develop an ecosystem through biotechnology that is rich in entrepreneurial ideas and an environment that encourages students and researchers. Transgenic industrial crops can be used to advance entrepreneurial biotechnology, creating an environment that will allow for the successful establishment of biotechnology firms and the global exploitation of research and investment (Tonukari 2004).

It is anticipated that individuals with scientific aptitude and creative abilities who can translate their work into marketable goods and services will produce any potential industrial crop products for these businesses (Bell 2010). In several nations, transgenic crops are currently grown commercially. According to reports, 13 million hectares worldwide have been dedicated to transgenic pest-resistant cultivars as of 2001. The regulatory biosafety organizations mandate that these types of cultivars be evaluated for their environmental impact prior to being certified for commercial use because they offer substantial advantages but also carry possible dangers (Fontes et al. 2002). With grants or funding backing from governments or organizations, many novel biotechnologies in industrial crops encountered fierce competition in academia. Academic institutions cannot afford to cover the high cost of product development, but investors and businesses are willing to in the name of profit. Contrarily, investors and businesses struggle to manage research and create findings, making it difficult or impossible to bring forth conclusions (Life Science Austria 2017). Also, when scientific evidence is sufficient to support a development plan, it is imperative to transfer technology or endeavor from a university to a business or industry. However, it is impossible to ignore the risk that the startup's investors may lose money if the product's final value fails (Shimasaki and Craig 2009). In order to obtain research agreements in the academic-industry, there are many areas of joint research that might foster synergy between biotech businesses and university laboratories. Given that those successful biotechnology companies produce biotechnology products, the development of brilliant persons with scientific and technological expertise who can translate that information into an idea for a good or service that is deserving of commercialization or converting into real goods is essential. It has become crucial to develop talented individuals with scientific and technical know-how who can transform their ideas into products or services worthy of commercialization or turning into economic goods, an individual whose business is in the realm of high technology. This is because biotech companies are successfully producing biotechnology products. A compilation of papers about the quickly developing biotechnology sector was published in a variety of literature (Steven et al. 2014). Inter-institutional scientific cooperation in the field of biotechnology is now seen as the engine propelling the sector forward. For academic and industry scientists doing biotechnology research and functioning as entrepreneurs, concrete synergy is still essential. This is done by demonstrating a commitment to the major commercial worth of their intellectual capital (Oliver 2004). This can be successfully used in the business of growing industrial crops.

19.1 Industrial crops and biochemical entrepreneurship

There are opportunities for biochemists to become entrepreneurs. Biotechnology links with bioprospecting genomic studies to make it easier to solve issues pertaining to the sectors needed for sustained growth. Recombinant DNA technology, however, has slowed down research and development endeavors, particularly in poor nations, and has now created limitations on commitments to entrepreneurial developments due to a lack of facilities and technical expertise (Bogoro 2015). A significant indicator that biotechnology is being accorded a professional place in international development is the advancement of the biotechnology package and industrial crop value chain globally. The ability of young entrepreneurs to come up with ideas, resolve to make them a reality, and have the vision to create money and jobs will determine the direction of the global economy in the future. It's critical to protect business ideas so that rivals cannot easily steal them. It is a good idea to secure legal protection for the product by patenting it. This is especially true in the field of life sciences. Investors will want to make certain that a concept won't become worthless through copying (Kolchinsky 2004).

19.2 Opportunities in plant science and industrial biotechnology enterprise

Startup ecosystem or self-employment opportunities include herbal cosmetic formulations and marketing, medicinal plant identification, preservation processes through refining for medical and pharmaceutical delivery, entrepreneurial horticulture and products development, waste-to-wealth development, energy production from crops biomass, etc. The utilization of plant enzymes to make bio-based products like chemicals, ingredients, detergents, materials, and biofuels through industrial biotechnology is one of the great potentials of industrial crops. It enables microbial production of enzymes (specialized proteins) as biocatalysts that facilitate complex biochemical reactions. Fermenting sugars derived from plants to ethanol produce biofuel, the same process is utilized in making wine and conversion of plant oils to biodiesel. This requires crops such as sugar cane, corn, wheat, oil seed, rape or sugar beet. Such products come either directly from cells or are made using enzymes taken from cells. In a way, cells bio-factories production lines of enzymes in the manufacture of the desired product. A whole factory or just specific workers are desirable in the production of whole cells or isolated enzymes. They can be utilized as tools in making biotech products, cells, and enzymes serve as biotech products themselves, for example, probiotic yogurts and non-soy veggie burgers contain microbial cells; and enzymes are utilized in washing detergents, food processing, cosmetics, etc. Cereal crops are a source of the industrial sugar used for microbial fermentation, though only a small proportion of the crop is used as most sugars are inaccessible to traditional processes, the remaining fraction serves as lignocellulosic biomass and is generally discarded. Development is ongoing to access the sugars locked up in waste-derived feedstocks such as agricultural residues, forestry residues, and post-consumer waste (CPI 2023).

References

Aanouz, I., Belhassan, A., El-Khatabi, K., Lakhlifi, T., El-ldrissi, M. and Bouachrine, M. 2020. Moroccan Medicinal plants as inhibitors against SARS-CoV-2 main protease: Computational investigations, Journal of Biomolecular Structure and Dynamics, 1–20.

Aasim, M., Khawar, K.M., Yalcin, G. and Bakhsh, A. 2014. Current trends in fenugreek biotechnology and approaches towards its improvement. American Journal of Social Issues and Humanities, Special Issue, 128136

Abbasi, R. and Baheti, V. 2018. Preparation of nanocellulose from jute fiber waste. Journal of Textile Engineering & Fashion Technology, 4(1): 101–104.

Abdallah, I., Yehia, R. and Mohamed Abdel-hady Kandil, M.A. 2020. Biofumigation potential of Indian mustard (*Brassica juncea*) to manage Rhizoctonia solani. Egyptian Journal of Biological Pest Control, 30(99): 8.

Abdellatef, E. and Khalafallah, M.M. 2008. Influence of growth regulators on callus induction from hypocotyls of medium staple cotton (*Gossypium hirsutum* L.). Cultivar barac B-67. Journal of Soil and Nature, 2(1): 17–22.

Abdel-Fattah, S.M, Yehia, H.A, Fouzy, A.S.M., Ramadan, M.M. and Nooh, A. 2015. Antifungal efficacy and chemical composition of essential oil from the Egyptian sweet orange peel (*Citrus sinensis*, L.). International Journal of Advanced Research, 3(10): 1257–1269.

Abd Karim, N.A., Ibrahim, M.D., Kntayya, S.B., Rukayadi, Y., Abd Hamid, H. and Ahmad Faizal Abdull Razi, A.F. 2016. Moringa oleifera Lam: Targeting chemoprevention. Asian Pacific Journal of Cancer Prevention, 17(8): 3675–3686.

AbdUlgadir, K.S., Suliman, S.I., Zakria, I.A. and Hassan, N.E.A. 2015. Antimicrobial potential of methanolic extracts of *Hibiscus sabdariffa* and *Ricinus communis*. Advancement in Medicinal Plant Research, 3(1): 18–22.

Abdul-Awal, S.M., Nazmir, S., Nasrin, S., Nurunnabi, T.R. and Jamal Uddin, S. 2016. Evaluation of pharmacological activity of Hibiscus tiliaceus. Springer Plus, 5: 1209.

Abdul Afiq, M.J., Abdul Rahman, R., Che Man, Y.B., AL-Kahtani, H.A. and Mansor, T.S.T. 2013. Date seed and date seed oil. International Food Research Journal, 20(5): 2035–2043.

Abiodun, O.A., Adepeju, A.B., Dauda, A.O. and Adebisi, T.T. 2015. Potentials of under-utilized Blighia sapida seed as a source of starch for industrial utilization. American Journal of Food Science and Nutrition, 2(6): 108–112.

Aboaba, S.A., Adebayo, M.A., Ogunwande, I.A., Tajudeen O. and Olayiwola, T.O. 2015. Volatile constituents of *Jatropha gossypifolia* L. grown in Nigeria. American Journal of Essential Oils and Natural Products, 2(4): 08–11.

Abreu, C.A., Cantoni, M., Coscione, A.R. and Paz-Ferreiro, J. 2012. Organic matter and barium absorption by plant species grown in an area polluted with scrap metal residue. Applied Environ. Soil Science, http://dx.doi.org/10.1155/2012/476821.

Abreu, C.A., Coscione, A.R., Pires, A.M. and Paz-Ferreiro, J. 2012. Phytoremediation of a soil contaminated by heavy metals and boron using castor oil plants and organic matter amendments. Journal of Geochem. Explor., 123: 3–7.

Abro, G.H., Syed, T.S., Tunio, G.M. and Khudro, M.A. 2004. Performance of transgenic Gt cotton against insect pest infestation. Biotechnology, 3(1): 75–81.

Abubakar, U.S., Yusuf, K.M., Abdullahi, M.S., Abdu, G.T., Abdulrazak, A., Muhammad, S., Binta, I.K., Osodi, F.A. and Aliyu, I. 2018. Cultivation, phytochemical and *in vitro* antiplasmodium activity of *Artemisia annua* L. (Asteraceae). Journal of Medicinal Plants Studies, 6(4): 151–155.

Abul-Soad, A.A., Jain, S.M. and Jatoi, M.A. 2017. Biodiversity and conservation of date palm. pp. 313–353. *In:* M.R. Ahuja and S.M. Jain (eds.), Biodiversity and Conservation of Woody Plants, Sustainable Development and Biodiversity 17. Springer International Publishing AG.

Abu Reidah, I.M., Arráez-Román, D., Warad, I., Fernández-Gutiérrez, A. and Segura-Carretero, A. 2017. Global characterization of bioactive compounds of *Viciafaba* L. by-product by means of UHPLCMS2. 2nd International Electronic Conference on Metabolomics. 20–27 November, 2017. Universidad de graneda.

Addala, A., Belattar, N. and Elektorowicz, M. 2018. *Nigella sativa* L. Seeds biomass as a potential sorbent in sorption of lead from aqueous solutions and wastewaters. Oriental Journal of Chemistry, 34(2): 638–647.

Adebayo, I.A., Arsad, H., Mohd Razip and Samian, M.R. 2018. Total phenolics, total flavonoids, antioxidant capacities, and volatile compounds gas chromatography-mass spectrometry profiling of Moringa oleifera ripe seed polar fractions. Pharmacogn Magazine, 14(54): 191–194.

Adebayo, I.A., Balogun, W.G. and Arsad, H. 2017. Moringa oleifera: An apoptosis inducer in cancer cells. Tropical Journal of Pharmaceutical Research, 16(8): 2289–2296.

Adeoye, B.K., Akinbode, B.A., Awe, J.D. and Akpa, C.T. 2017. Evaluation of biosorptive capacity of waste watermelon seed for lead (II) removal from aqueous solution. American Journal of Environmental Engineering, 10(1): 1–8.

Adebisi, F., Afolayan Michael, A., Patrick, I. Olutayo, O., Adedayo, A., Olakunle, F., Ademola, A. and Abayomi, O. 2013. Isolation and characterization studies of Moringa oleifera root starch as a potential pharmaceutical and industrial biomaterial. International Journal of Chemistry and Applications, 5(2): 117–126.

Adhikari, T. and Kumar, A. 2012. Phytoaccumulation and tolerance of *Ricinus communis* L. to nickel. Int. Journal Phytoremediat., 14(5): 481–492.

Adjatin, A., Aboudou, R., Loko, L.Y., Bonou-Gbo, Z., Sanoussi, F., Orobiyi, A., Djedatin, G., Yedomohan, H. and Dansi, A. 2018. Ethnobotanical investigation and diversity of sweet potato (*Ipomea batatas* L.) landraces grown in Northern Benin. Int. J. Adv. Res. Biol. Sci., 5(8): 59–73.

Adole, A.M., Yatim, J.M., Ramli, S.A., Othman, A. and Mizal, N.A. 2019. Kenaf fibre and its bio-based composites: A conspectus. Pertanika Journal of Science and Technolohy, 27(1): 297–329.

Agricdemy. 2018. Business opportunities in the coconut value chain. https://agricdemy. com/post/coconut-business-nigeria 2/11/2020

Agyenim-Boateng, K.G., Lu, J., Shi, Y., Zhang, D. and Yin, X. 2019. SRAP analysis of the genetic diversity of wild castor (*Ricinus communis* L.) in South China. PLoS ONE, 14(7): e0219667.

Agency for Renewable Resources. 2014. Bioplastics. Fachagentur Nachwachsende Rohstoffe e. V. (FNR), Germany. pp. 24–42.

Ahaotu, E.O., Osuji, C.N., Nwabueze, E.U. and Ibeh, C.C. 2013. Benefits of *Jatropha gossipifolia* in Nigeria. International Journal of Agriculture and Biosciences, 2(6): 349-355.

Aher, R.R., Belge, S.A., Kadam, S.R., Kharade, S.S., Misal, A.V. and Yeole, P.T. 2016. Therapeutic importance of fenugreek (*Trigonella foenum-graecum* L.): A review. Journal of Plant Science & Research, 3(1): 1–4.

Ahmad, P., Ashraf, M., Younis, M., Hu, X., Kumar, A., Akram, N.A. and Al-Qurainy, F. 2012. Role of transgenic plants in agriculture and biopharming. Biotechnology Advances, 30(3): 524–540.

Ahmad, K.A., Abdullah, M.E., Hassan, N.A., Usman, N. and Ambak, K. 2017. Investigating the feasibility of using Jatropha Curcas oil (JCO) as bio based rejuvenator in reclaimed asphalt pavement (RAP). MATEC Web of Conferences, 103: 1–9, 09013.

Ahmad, A., Alghamdi, S.S., Mahmood, K. and Afzal, M. 2016. Fenugreek a multipurpose crop: Potentialities and improvements. Saudi Journal of Biological Sciences, 23: 300–310.

Ahsan, M.Z., Majidano, M.S., Channa, A.R., Panhwar, F.H., Soomro, A.W., Khaskheli, F.I. and Kashif Rashid, K. 2014. Regeneration of cotton (Gossypium hirsutum L.) through asexual methods, a review. American-Eurasian Journal of Agricultural and Environmental Science, 14(12): 1478–1486.

Aikpokpodion, P.O. 2012. Defining genetic diversity in the chocolate tree, *Theobroma cacao* L. grown in west and central Africa. pp. 185–212. *In:* Prof. Mahmut Caliskan (ed.), Genetic Diversity in Plants, ISBN: 978-953-510185-7. InTech. Available from: http:// www.intechopen.com/books/genetic-diversity-in-plants/defining-geneticdiversity-in- the-chocolate-tree-theobroma-cacao-l-grown-in-west-and-central-africa

Aisha, Z.M., Ayodeji, O.A., Datsugwai, M.S.S. and Aliyu, Isah Moriki. 2017. Microbiological analysis of Baobab yoghurt produced using *Lactobacillus bulgaricus*. International Journal of Microbiology and Biotechnology, 2(2): 93–10.

Akande, T.O., Odunsi, A.A., Olabode, O.S. and Ojediran, T.K. 2012. Physical and nutrient characterization of raw and processed castor (*Ricinus communis* L.) seeds in Nigeria. World Journal of Agricultural Sciences, 8(1): 89–95.

Akbar, F., Rabbani, M.A., Masood, M.S. and Shinwari, Z.K. 2011. Genetic diversity of sesame (*Sesamum indicum* L.) germplasm from Pakistan using RAPD markers. Pakistan Journal of Botany, 43(4): 2153–2160.

Akhtar, W.J., Abrha, M.G., Omre, P.K. and Gebru, G.G. 2020. Extraction and characterization of pectin from lemon peels. Research Journal of Chemistry and Environmental Science, 8(1): 25–37.

Akinlabi, E.T., Anane-Fenin, K. and Akwada, D.R. 2017. Bamboo the Multipurpose Plant. Springer International Publishing AG. Switzerland. pp. 149–220.

Akinsanoye, O.A. and Omotoso, M.A. 2018. Oleochemicals from water melon (*Citrullus lanatus*) seed oils. The Pharmaceutical and Chemical Journal, 5(4): 97–114.

Akoh, C.C. and Min, D.B. (Eds.). 2008. Food Lipids: Chemistry, Nutrition, and Biotechnology, 3rd ed. CRC Press. https://doi.org/10.1201/9781420046649.

Akoetey, W., Britain, M.M. and Morawicki, R.O. 2017. Potential use of byproducts from cultivation and processing of sweet potatoes. Ciência Rural, Santa Maria, 47(5): 1–8.

Akpomie, Olubunmi Olufunmi, A.O. 2010. The preservative potentials of sweet orange seed oil on leather products in Nigeria. African Journal of Biotechnology, 9(5): 678–681.

Alamineh, E.A. 2018. Extraction of pectin from orange peels and characterizing its physical and chemical properties. American Journal of Applied Chemistry, 6(2): 51–56.

Alcheikh, A. 2015. Advantages and challenges of hemp biodiesel production: A comparison of hemp vs. other crops commonly used for biodiesel production. MSc thesis. Faculty of Science and Sustainable Development. p. 3.

Ali, E.N. and Kemat, S.Z. 2016. Bioethanol Produced from Moringa oleifera Seeds Husk. 29th Symposium of Malaysian Chemical Engineers (SOMChE). pp. 11.

Ali, G.M., Yasumoto, S. and Seki-Katsuta, M. 2007. Assessment of genetic diversity in sesame (*Sesamum indicum* L.) detected by amplified fragment length polymorphism markers. Electronic Journal of Biotechnology, 10(1): 1–10.

Ali, N. 2010. Soybean processing and utilization. pp. 345–427. *In:* Singh, G. (ed.), The Soybean Botany, Production and Uses. CAB International. Wallingford, UK.

Al-Asmari, A.K., Albalawi, S.M., Athar, M.T., Khan, A.Q., Al-Shahrani, H. and Islam, M. 2015. *Moringa oleifera* as an anti-cancer agent against breast and colorectal cancer cell lines. PLoS ONE, 10(8): e0135814

Al-Harbawy, A.W. and AL-Mallah, M.K. 2014. Production and characterization of biodiesel from seed oil of castor (*Ricinus communis* L.) plants. International Journal of Science and Technology, 3(9): 508–513.

Al-Obaidi, J.R., Halabi, M.F., AlKhalifah, N., Asanar, S., Al-Soqeer, A.A. and Attia, M.F. 2017. A review on plant importance, biotechnological aspects, and cultivation challenges of jojoba plant. Biological Research, 50.

Al-Rmalli, W.S., Dhamani, A.A., Abuein, M.M. and Gleza, A.A. 2008. Biosorption of mercury from aqueous solutions by powdered leaves of castor tree (*Ricinus communis* L.). J. Hazard. Mat., 152: 955–959.

Al-Sebaeai, M.A., Alfawaz, M., Chauhan, A.K., AL-Farga, A. and Fatma, S. 2017. Physicochemical characteristics and nutritional value of fenugreek seeds and seed oil. Int. J. Food Sci. Nutr., 2(6): 52–55.

Alves, E.E.N., Souza, C. de F. and Inoue, K.R.A. 2012. Production of biogas and biofertilizer from biodigestion of castor bean cake with and without animal manure. Engenharia na Agricultura, 20(6): 493–500.

Aly, M.A.M. and Basarir, A. 2012. Biotechnology approaches and economic analysis of jojoba natural products. Biotechnological Production of Plant Secondary Metabolites, 1: 159. https://doi.org/10.2174/9781608051441201010159.

Ambrose, D.C.P. and Naik, R. 2016. Development of a process for utilisation of banana waste. International Journal of Research in Applied, Natural and Social Sciences, 4(6): 83–88.

Aminul Islam, A.K.M., Mominul Islam, A.K.M., Nadhirah, N.A., Anuar, N. and Yaakob, Z. 2015. Propagation of *Jatropha curcas* through seeds, vegetative cuttings and tissue culture. pp. 131–158. *In:* Gregory Medina (ed.), *Jatropha curcas*: Biology, Cultivation and Potential Uses. Published by NOVA Science Publisher, USA. ISBN: 978-1-63483-089-8.

Amouri, M., Mohellebi, F., Zaid, T.A. and Aziza, M. 2016. Sustainability assessment of *Ricinus communis* biodiesel using LCA approach. Clean Techn. Environ. Policy, 2016. DOI 10.1007/s10098-016-1262-4.

Anand, A.V., Balamuralikrishnan, B., Kaviya, M., Bharathi, K., Parithathvi, A., Arun, M., Senthilkumar, N., Velayuthaprabhu, S., Saradhadevi, M., Al-Dhabi, N.A. and Arasu, M.V. 2021. Medicinal plants, phytochemicals, and herbs to combat viral pathogens including SARS-CoV-2. Molecules, 26(6): 1775.

Ananthi, T.A.S. and Manikandan, P.N.A. 2013. Potential of rhizobacteria for improving lead phytoextraction in *Ricinus communis*. Remediation, 24(1): 99–106.

Ananthi, T.A.S., Meerabai, R.S. and Krishnasamy, R. 2012. Potential of *Ricinus communis* L. and *Brassica juncea* (L.) Czern. under natural and induced Pb phytoextraction. Universal J. Environ. Res. Tech., 2(5): 429–438.

Anastasi, U., Sortino, O., Cosentino, S.L. and Patanè, C. 2015. Seed yield and oil quality of perennial castor bean in a Mediterranean environment. International Journal of Plant Production, 9(1): 99–116.

Andreazza, R., Bortolon, L., Pieniz, S. and Camargo, F.A.O. 2013. Use of high-yielding bioenergy plant castor bean (*Ricinus communis* L.) as a potential phytoremediator for copper-contaminated soils. Pedosphere, 23(5): 651–661.

Annapurna, D., Rajkumar, M. and Prasad, M.N.V. 2016. Potential of castor bean (*Ricinus communis* L.) for phytoremediation of metalliferous waste assisted by plant growth-promoting bacteria: Possible cogeneration of economic products. pp. 149–178. *In:* Prasad, M.N.V. (ed.), Bioremediation and Bioeconomy, Amsterdam: Elsevier.

Armendáriz, J., Lapuerta, M., Zavala, F., Garcia-Zambrano, E. and del Carmen Ojeda, M. 2015. Evaluation of eleven genotypes of castor oil plant (*Ricinus communis* L.) for the production of biodiesel. Ind. Crops Prod., 77: 484–490.

Arriel, N.H.C., Mauro, A.O.D., Arriel, E.F., Unêda-Trevisoli, S.H., Costa, M.M., Bárbaro, I.M. and Muniz, F.R.S. 2007. Genetic divergence in sesame based on morphological and agronomic traits. Crop Breeding and Applied Biotechnology, 7: 253–261.

Arun, A., Prabha, P., Kumar, K.K. and Balasubramanian, P. 2016. Development of putative transgenic lines of cassava variety H-226. African Journal of Biotechnology, 15(13): 497–504.

Asare, P.A. Kpankpari, R., Adu, M.O., Afutu, E. and Adeyinka, A.S. 2020. Phenotypic characterization of Tiger Nuts (*Cyperus esculentus* L.) from major growing areas in Ghana. Scientific World Journal Volume 2020: 11.

Assogbadjo, A.E., Kyndt, T., Sinsin, B., Gheysen, G. and Van Damme, P. 2006. Patterns of genetic and morphometric diversity in Baobab (*Adansonia digitata*) populations across different climatic zones of Benin (West Africa). Annals of Botany, 97: 819–830.

Ataga, C.D., Mohammed, A.H. and Yusuf, A.O. 2012. Status of date palm (*Phoenix dactylifera* L.) genetic resources in Nigeria. International Journal of Life Sciences and Pharma Research, 2(2): 1–6.

Attah, A.F., Fagbemi, A.A., Olubiyi, O., Dada-Adegbola, H., Oluwadotun, A., Elujoba, A. and Babalola, C.P. 202.1 Therapeutic potentials of antiviral medicine with COVID-19 in focus: A Nigerian perspective. Frontiers in Pharmacology, 12: 596855.

Atiku, F.A., Warra, A.A. and Enimola, M.R. 2014a. FTIR spectroscopic analysis and fuel properties of wild castor (*Ricinus communis* L.) seed oil. Open Science Journal of Analytical Chemistry, 1(1): 6–9.

Atiku, F.A., Warra A.A. and Bashar M.M. 2014b. FTIR analysis and fuel properties of castor (*Ricinus communis* L.) bean oil. World Research Journal of Organic Chemistry, 2(1): 31–34.

Atlaw, T.K. 2018. Preparation and utilization of natural aloe vera to enhance quality of mango fruit. Journal of Food and Nutrition Sciences, 6(3): 76–81.

Austin-Phillips, S. and Ziegelhoffer,T. 2001. The production of transgenie alfalfa value-added proteins. pp. 285–301. *In:* G. Spangenberg (ed.), Molecular Breeding of Forage Crops. Springer Science+Business Media Dordrecht.

Azam, M.M., Mamun-Or-Rashid, A.N.M., Towfique, N.M., Sen, M.K. and Nasrin, S. 2013. Pharmacological potentials of *Melia azedarach* L. – A review. American Journal of BioScience, 1(2): 44–49.

Aziera, Z.N. and Majid, N.M. 2015. Uptake and translocation of zinc and cadmium by *Ricinus communis* planted in sewage sludge contaminated soil. UKM Journal, Publisher Penerbit Univeriti Kebangsaan, Malaysia.

Babalola, F.D. and Agbeja, B.O. 2010. Marketing and distribution of Garcinia kola (Bitter kola) in southwest Nigeria: Opportunity for development of a biological product. Egyptian Journal of Biology, 12: 12–17.

Babita, M., Maheswari, M., Rao, L.M., Shanker, A.K. and Rao, D.G. 2010. Osmotic adjustment, drought tolerance and yield in castor (*Ricinus communis* L.) hybrid. Environ. Experim. Bot., 69(3): 243–249.

Babu, K.N., Samsudeen, K., Minoo, D., Geetha, S.P. and Ravindran, P.N. 2000. Tissue culture and biotechnology of ginger. pp. 181–209. *In:* P.N. Ravindran, K. Nirmal Babu (eds.), Ginger the Genus Zingiber. CRC Press, 2000 N.W. Corporate Blvd., Boca Raton, Florida.

Bachheti, R.K., Joshi, A. and Singh, A. 2011. Oil content variation and antimicrobial activity of eucalyptus leaves oils of three different species of Dehradun, Uttarakhand, India. International Journal of ChemTech Research, 3(2): 625–628.

Badary, O.A., Hamza, M.S. and Tikamdas, R. 2021. Thymoquinone: A promising natural compound with potential benefits for COVID-19 prevention and cure. Drug Design Development and Therapy, 15: 1819–1833.

Bafeel, S. and Bahieldi, A. 2020. Validation and detection of sex-specific markers in jojoba (*Simmondsia chinensis*) plants in Saudi Arabia. Applied Ecology and Environmental Research, 18(4): 5023-5035.

Bagde, P.P., Dhenge, S. and Bhivgade, S. 2017. Extraction of pectin from orange peel and lemon peel. International Journal of Engineering Technology Science and Research, 4(3): 1–7.

Baishya, M. and Kalita M.C. 2015. Phytoremediation of crude oil contaminated soil using two local varieties of castor oil plant (*Ricinus communis*) of Assam. Int. J. Pharma Bio. Sci., 6(4): 1173–1182.

Bajay, M.M., Pinheiro, J.B., Batista, C.E.A., Nobrega, M.B.M. and Zucchi, M.I. 2009. Development and characterization of microsatellite markers for castor (*Ricinus communis* L.), an important oleaginous species for biodiesel production. Conservation of Genetic Resources. DOI 10.1007/s12686-009-9058

Bakhsh, A., Anayo, E., Khabbazi, S.D., Karakoc, O.C., Cengiz Sancak, C. and Sebahattin Ozcan, S. 2016. Development of insect-resistant cotton lines with targeted expression of insecticidal gene. Archive of Biological Science, 68(4): 773–780.

Baladhiya, C.S. and Joshi, D.C. 2016. Effect of castor cake on biogas production by adding it with cattle dung. International Journal of Science, Environment and Technology, 5(2): 547–551.

Balan, V., Kumar, S., Bals, B., Chundawat, S.P.S., Jin, M. and Dale, B.E. 2012. Biochemical and thermochemical conversion of switchgrass to biofuels. *In:* A. Monti (ed.), Switchgrass (pp. 153-185).

Baldanzi, M., Myczkowski, M.L., Salvini, M. and Macchia, M. 2015 Description of 90 inbred lines of castor plant (*Ricinus communis* L.). Euphytica, 202: 13–33.

Bale, A.T., Adebayo, R.T., Ogundele, D.T. and Bodunde, V.T. 2013. Fatty acid composition and physicochemical properties of castor (*Ricinus communis* L.) seed obtained from Malete, Moro local government area, Kwara State. Nigeria. Chemistry and Materials Research, 3(12): 11–13.

Baloch, A.W., Baloch, M., Jatoi, S.H., Baloch, M.J., Baloch, G.M., MugherI, M.A., Depar, M.S., Mallano, I.A., Baloch, A.M., Gandahi, N., Baloch, I.A. and Ali, M. 2015. Genetic diversity analysis in genetically modified cotton (*Gossypium hirsutum* L.) genotypes. Sindh University Research Journal, 47(3): 527–530.

Balunas, M.J. and Kinghorn, A.D. 2005. Drug discovery from medicinal plants. Life Sciences, 78: 431–441.

Balunas, S.M. and Kinghorn, A. 2007. Challenges and opportunities in drug discovery from plants. Current Science, 92(9): 1251-1257.

Bamogo, P.K.A., Brugidou, C., Sérémé, D., Tiendrébéogo, F., Djigma, F.W., Simpore, J. and Lacombe, S. 2019. Virus-based pharmaceutical production in plants: An opportunity to reduce health problems in Africa. Virology Journal, 16: 167.

Barkah, N.N., Wiryawan, K.G., Retnani, Y., Wibawan, W.T. and Elizabeth Wina, E. 2021. Physicochemical properties of products and waste of black seed produced by cold press method. IOP Conf. Series: Earth and Environmental Science, 756: 8.

Basak, S., Samanta, K.K., Saxena, S., Chattopadhyay, S., Narkar, R., Mahangade, R. and Hadge, G.B. 2015. Flame resistant cellulosic substrate using banana pseudostem sap. Polish Journal of Chemical Technology, 17: 123–133.

Bashir, K.A., Waziri, A.F. and Musa, D.D. 2016. *Moringa oleifera*, a potential miracle tree: A review. IOSR Journal of Pharmacy and Biological Sciences, 11(6): 25–30.

Bata, M. and Rahayu, S. 2016. Study of Hibiscus tiliaceus leaf extract carrier as additive in the diets for fattening of local cattle (in vitro). Pakistan Journal of Nutrition, 15: 969–974.

Bateni, H., Karimi, K., Zamani, A. and Benakashani, F. 2014. Castor plant for biodiesel, biogas, and ethanol production with a biorefinery processing perspective. https://doi.org/10.1016/j.apenergy.2014.09.005

Bauddh, K. and Singh, R.P. 2012a. Growth, tolerance efficiency and phytoremediation potential of *Ricinus communis* (L.) and *Brassica juncea* (L.) in salinity and drought affected cadmium contaminated soil. Ecotox. Environ. Safe., 85: 13–22.

Bauddh, K. and Singh, R.P. 2012b. Cadmium tolerance and its phytoremediation by two oil yielding plants *Ricinus communis* (L.) and *Brassica juncea* (L.) from the contaminated soil. Int. J. Phytomed., 14(8): 772–785.

Bauddh, K. 2014. *Ricinus communis* (castor bean): A multipurpose crop for the sustainable environment. Dream-2047, 16(11): 31–32.

Bauddh, K. and Singh, R.P. 2014. Studies on bio-accumulation and partitioning of Cd in *Brassica juncea* and *Ricinus communis* in presence of vermicompost, chemical fertilizers, biofertilizers and customized fertilizers. Ecol. Eng., 74: 93–100.

Bauddh, K. and Singh, R.P. 2015. Effects of organic and inorganic amendments on bio-accumulation and partitioning of Cd in *Brassica juncea* and *Ricinus communis*. Ecol. Eng., 74: 93–100.

Bauddh, K., Singh, K., Singh, B. and Singh R.P. 2015. *Ricinus communis*: A robust plant for bio-energy and phytoremediation of toxic metals from contaminated soil. Ecol. Eng., 84: 640–652.

Bauddh, K., Singh, K. and Singh, R.P. 2016. *Ricinus communis* L., a value added crop for remediation of cadmium contaminated soil. Bull. Environ. Contam. Toxicol., 96(2): 265–269.

Begum, K., Ahmad, H. and Iqbal, A. 2019. Characterization of wheat germplasm with seed quality parameters and glutenin subunits. Pakistan Journal of Botany, 51(3): 1–8.

Behera, M.K. and Mohapatra, N.P. 2015. Biomass accumulation and carbon stocks in 13 different clones of teak (*Tectona grandis* Linn. F.) in Odisha, India. Current World Environment, 10(3): 1011–1016.

Behera, T.K., Staub, J.E., Bhera, S. and Simon, P.W. 2007. Bitter gourd and human health. Medicinal and Aromatic Plant Science and Biotechnology, 1(2): 224–226.

Behmaram, R., Saleh, G., Foroughi, M., Noori, Z., Panandam, J.M. and Jalaluddin Harun, J. 2014. Genetic control of fiber yield and quality in kenaf (*Hibiscus cannabinus* L.). Iranian Journal of Genetics and Plant Breeding, 3(1): 41–31.

Bekheet, S.A. and Taha, H.S. 2013. Complementary strategy for conservation of date palm germplasm. Global Journal of Biodiversity Science and Management, 3(1): 96–107.

Belay, G. 2018. Biotech could spur Africa's industrialization. The Herald 4th March, 2018. Accessed 4/3/2018 https://www.herald.co.zw/biotech-could-spur-africas-industrialisation/

Bell, J.R. 2010. Handbook of bioentrepreneurship (Book Review). New England Journal of Entrepreneurship, 13(1): 1–2.

Bello, E.I., Ayodeji, O.Z., Ogunbayo, S. and Bello, K. 2019. Characterization and glycerine analysis of mustard (*Brassica juncea* L.) seed oil and biodiesel. Journal of Advances in Biology & Biotechnology, 22(2): 1–8.

Berenji, J., Sikora, V., Fournier, G. and Beherec, O. 2013. Genetics and selection of hemp. pp. 48–71. *In:* Pierre Bouloc, Serge Allegret and Laurent Arnaud (eds.); Glen Cousquer (translator), Hemp: Industrial Production and Uses. CAB International. UK.

Bertacchi, S., Bettiga, M. and Danilo Porro, D. 2020. Camelina sativa meal hydrolysate as sustainable biomass for the production of carotenoids by Rhodosporidium toruloides. Biotechnology for Biofuels, 13(1): 47.

Betancor, M.B., Sprague, M., Usher, S., Sayanova, O., Campbell, P.J., Napier, J.A. and Tocher, D.R. 2015. A nutritionally-enhanced oil from transgenic Camelina sativa effectively replaces fish oil as a source of eicosapentaenoic acid for fish. Scientific Reports, 5: 8104.

Betancora, M.B., MacEwana, A., Spraguea, M., Gonga, X., Monterob, D., Hanc, L., Napierc, J.A., Norambuenad, F., Izquierdob, M. and Tochera, D.R. 2021. Oil from transgenic Camelina sativa as a source of EPA and DHA in feed for European sea bass (*Dicentrarchus labrax* L.). Aquaculture, 530: 735-759.

Betha, U.K. and Lavanya, C. 2019. Molecular diversity and population structure in breeding lines of castor (*Ricinus communis* L.). International Journal of Current Microbiology and Applied Sciences, 8(2): 2024–2038.

Bhadkariya, R.K., Jain, V.K., Chak, G.P.S. and Gupta, S.K. 2014. Remediation of cadmium by indian mustard (*Brassica juncea* L.) from cadmium contaminated soil: A phytoextraction study. International Journal of Environment, 3(2): 229–237.

Bhalerao, S.A. 2013. Arbuscular mycorrhizal fungi: A potential biotechnological tool for phytoremediation of heavy metal contaminated soils. Int. J. Sci. Nat., 4(1): 1–15.

Bhaludra, C.S.S., Yadla, H., Cyprian, F.S., Bethapudi, R.R., Basha, S.D. and Anupalli, R.R. 2014. Genetic diversity analysis in the genus *Aloe vera* (L.) using RAPD and ISSR markers. International Journal of Pharmacology, 10: 479–486.

Bhargav, A.R.A. and Joga, R.H. 2018. Effectiveness of the mustard seed powder and cake for the coagulation of synthetic municipal wastewater. Research Journal of Pharmaceutical, Biological and Chemical Sciences, 9(4): 1422–1428.

Bhaskar, T., Chang, J.S., Khanal, S., Lee, D-J.S.K., Mohan, S.V. and Rittmann, B.E. 2016. Waste biorefinery – Advocating circular economy. Bioresource Technology, 215: 1.

Bhat, T.K., Kannan, A., Singh, B. and Sharma, O.P. 2013. Value addition of feed and fodder by alleviating the antinutritional effects of tannins. Agricultural Resources, 2(3): 189–206.

Bhatti, I. and Akhlaq, M. 2013. The scientific importance of *Nigella sativa* (kalonji) and honey in accordance with TIB-E-NABVI. Gomal University Journal of Research, 29(1): 27–30.

Bichi, M.H. 2013. A review of the applications of *Moringa oleifera* seeds extract in water treatment. Civil and Environmental Research, 3(8): 1–10.

Bilia, A.R., Santomauro, F., Sacco, C., Bergonzi, M.C. and Donato, R. 2014. Essential oil of *Artemisia annua* L.: An extraordinary component with numerous antimicrobial properties. Evidence-based Complementary and Alternative Medicine, 2014: 7.

Bimal, R. and Singh, A.K. 2017. Biology and biotechnology of mango (*Mangifera indica* L.) with reference to *in vitro* cell and tissue culture in endangered and endemic cultivars: A review. Journal of Cell and Tissue Research, 17(3): 6285–6292.

Bisht, I.S., Bhat, K.V., Lakhanpaul, S. and Biswas, B.K. 2004. Broadening the genetic base of sesame (*Sesamum indicum* L.) through germplasm enhancement. Plant Genetic Resources, 2(3): 143–151.

Biswas, K., Chattopadhyay, I., Banerjee, R.K. and Bandyopadhyay, U. 2002. Biological activities and medicinal properties of neem (*Azadirachta indica*). Current Sci., 82: 1336-1345.

Biswas, M.K., Ridwan-Ul-Risty, K.M. and Datta, A. 2019. A proposal of a sustainable and integrated plant for jute fiber extraction in an eco-friendly manner. International Journal of Scientific & Engineering Research, 10(1): 801-809.

Blanca, J., Esteras, C., Ziarsolo, P., Pérez, D., Collado, C., RodrÃguez de Pablos, R., Ballester, A., Roig, C., Cañizares, J. and Picó, B. 2012. Transcriptome sequencing for SNP discovery across *Cucumis melo*. BMC Genomics, 13(1): 1–18.

Blessing, N.A., Agbagwa, I.O. and Okoli, B.E. 2011. Comparative physicochemical screening of Jatropha L. species in the Niger Delta. Research Journal of Phytochemistry, 5: 107–114.

Blickman, T. 2017. Morocco and cannabis, reduction, containment or acceptance. Drug policy briefing, 49. March 2017 – Transnational Institute. Available from https://www.tni.org/files/publication-downloads/dpb_49_eng_web.pdf. Accessed 3/10/2019

Bogoro, S.E. 2015. Entrepreneurship for Development. Convocation Lecture delivered at the 2nd Convocation Ceremony of the Kaduna State University, Kaduna on December 11, 2015.

Bolaji, Z.S., Gana, Andrew K.G. and Benson, A.O. 2014. Castor oil plant (*Ricinus communis* L.): Botany, ecology and uses. International Journal of Science and Research, 3(5): 1333–1341.

Bolivar-Telleria, M., Turbay, C., Favarato, L., Carneiro, T., deBiasi, R.S., Fernandes, A.A.R., Santos, A.M.C. and Fernandes, P.M.B. 2018. Second-generation bioethanol from coconut husk. BioMed Research International, 20 pp.

Bonanno, G. 2014. *Ricinus communis* as an element biomonitor of atmospheric pollution in urban areas. Water Air Soil Poll., 225(2): 1852.

Borugadda, V.B. and Goud, V.V. 2014. Epoxidation of castor oil fatty acid methyl esters (COFAME) as a lubricant base stock using heterogeneous ion-exchange resin (IR-120) as a catalyst. Energy Procedia, 54: 75–84.

Bosiacki, M., Kleiber, T. and Kaczmarek, J. 2013. Evaluation of suitability of *Amaranthus caudatus* L. and *Ricinus communis* L. in phytoextraction of cadmium and lead from contaminated substrates. Arch. Environ. Prot., 39(3): 47–59.

Boumchita, S., Lahrichi, A., Benjelloun, Y., Lairini, S., Nenov, V. and Zerrouq, F. 2016. Application of peanut shell as a low-cost adsorbent for the removal of anionic dye from aqueous solutions. Journal of Materials and Environmental Sciences, 6(7): 2353–2364.

Brasher, K. 2021. Using Biochar as Container Substrate for Plant Growth. Available at https://www.mafes.msstate.edu/discovers/index.asp Accessed 21/2/2021.

Brosse, N., Dufour, A., Meng, X., Sun, Q. and Ragauskas, A. 2012. Miscanthus: A fast-growing crop for biofuels and chemicals production. Biofuels, Bioproduction and Biorefinery, 19 pp.

Brown, S.K. and Maloney, K.E. 2015. Apple breeding, genetics and genomics. New York Fruit Quarterly, 23(3): 5–8.

Bruijnincx, P., Weckhuysen, B., Gruter, G., Westenbroek, A. and Edith Engelen-Smeets, E. 2016. Lignin Valorisation the Importance of a Full Value Chain Approach. Utrecht University. 22 pp.

Brunerová, A., Roubík, H. and Brožek, M. 2018. Bamboo fiber and sugarcane skin as a bio-briquette fuel. Energies, 11(218): 1–20.

Buba, C.I., Okhale, S.E. and Muazzam, I. 2016. Garcinia kola: The phytochemistry, pharmacology and therapeutic applications. International Journal of Pharmacognosy, 3(2): 67–81.

Butkutė, B., Lemežienė, N., Cesevičienė, J.M., Liatukas, J. and Dabkevičienė, G. 2013. Carbohydrate and lignin partitioning in switchgrass (*Panicum virgatum* L.) biomass as a bioenergy feedstock. Zemdirbyste-Agriculture, 100(3): 251–260.

Byrne, P. 2008. Bio-pharming. Fact Sheet. No. 0.307. Colorado State University. Available from https://mountainscholar.org/bitstream/handle. Accessed 12/11/2019 Iranian Journal of Genetics and Plant Breeding, 3(1): 32–21.

Cabral, L., Soares, C.R.F.S., Giachini, A.J. and Siqueira, J.O. 2015. Arbuscular mycorrhizal fungi in phytoremediation of contaminated areas by trace elements: Mechanisms and major benefits of their applications. World J. Micro Biol. Biotechnol., 31(11): 1655–1664.

Cabrales, R.A.R., Marrugo, J.L.N. and Plaza, G.A.T. 2011. Evaluation of seed yield and oil contents in four materials of *Ricinus communis* L. Agronomía Colombiana, 29(1): 43–48.

Calabro, P.S., Folino, A., Tamburino, V., Zappia, G., Zema, D.A. and Zimbone, S.M. 2017. Valorisation of citrus processing waste: A review. Proceedings Sardinia 2017/ Sixteenth International Waste Management and Landfill Symposium/ 2-6 October 2017. S. Margherita di Pula, Cagliari, Italy / © 2017 by CISA Publisher, Italy.

Caliceti, C., Capriotti, A.L., Calabria, D., Bonvicini, F., Chiozzi, R.Z., Montone, C.M., Piovesana, S., Zangheri, M., Mirasoli, M., Simoni, P., Laganà, A. and Roda, A. 2019. Peptides from cauliflower by-products, obtained by an efficient, ecosustainable, and semi-industrial method, exert protective effects on endothelial function. Oxidative Medicine and Cellular Longevity, 2019: 13, Article ID 1046504.

Çalişkan, O., Bayazit, S., Öktem, M. and Ergül, A. 2017. Evaluation of the genetic diversity of pomegranate accessions from Turkey using new microsatellite markers. Turkish Journal of Agriculture and Forestry, 41: 142–153.

Campbell, D.N. 2013. Determining the agronomic and physiological characteristics of the castor plant (*Ricinus communis* L.): Developing a sustainable cropping system for Florida. MSc. thesis, University of Florida, pp. 1–90.

Campestrini, L.H., Kuhnen, S., Lemos, P.M.M., Bach, D.B., Dias, P.F. and Maraschin, M. 2006. Cloning protocol of Aloe vera as a study-case for "Tailor-made" biotechnology to small farmers. Journal of Technology Management and Innovation, 1(5): 76–79.

Capuani, S., Fernandes, D.M., Rigon, J.P.G. and Ribeiro, L.C. 2015. Combination between acidity amendments and sewage sludge with phosphorus on soil chemical characteristics and on development of Castor bean. Communications in Soil Sci. Plant Analysis, 46(22): 2901–2912.

Cárdenas-Aguiar, E., Suárez, G., Paz-Ferreiro, J., Askeland, M.P.J., Méndez, A. and Gascó, G. 2020. Remediation of mining soils by combining *Brassica napus* growth and amendment with chars from manure waste. Chemosphere, 261: 127798.

Carpinella, C., Ferrayoli, C., Valladares, G., Defago, M. and Sara Palacios, S. 2002. Potent Limonoid insect antifeedant from *Melia azedarach*. Bioscience, Biotechnology, and Biochemistry, 66(8): 1731–1736.

Carreno, L.V.N., Garcia, I.T.S., Raubach, W.C., Krolow, M., Santos, C.C.G., Probst, L.F.D. and Fajardo, H.V. 2009. Nickel-carbon nanocomposites prepared using castor oil as precursor: A novel catalyst for ethanol steam reforming. J. Power Sources, 188: 527–531.

Cecchi, C.G.S. and Zanchi, C. 2005. Phytoremediation of soil polluted by nickel using agricultural crops. Environ. Manag., 36: 675–681.

Chadare, F.J., Linnemann, A.R., Hounhouigan, J.D., Nout, M.J.R. and Van Boekel, M.A.J.S. 2008. Baobab food products: A review on their composition and nutritional value. Critical Reviews in Food Science and Nutrition, 49(3): 254–274.

Chakauyaa, E., Chikwamba, R. and Rybicki, E.P. 2006. Riding the tide of biopharming in Africa: Considerations for risk assessment. South African Journal of Science, 102: 284–288.

Chakraborthy, G.S., Sharma, G. and Kaushik, K.N. 2008. *Sesamum indicum*: A review. Journal of Herbal Medicine & Toxicology, 2(2): 15–19.

Chan, A.P., Crabtree, J., Qi Zhao, Q., Lorenzi, H., Orvis, J., Puiu, D., Melake-Berhan, A., Jones, K.M., Redman, J., Chen, G., Cahoon, E.B., Gedil, M., Stanke, M., Haas, B.J., Wortman, J.R., Fraser-Liggett, C.M., Ravel, J. and Rabinowicz, P.D. 2010. Draft genome sequence of the oilseed species *Ricinus communis*. Nature Biotechnology, 28(9): 951–958.

Chandel, A.K., Antunes, F.A.F., N-Hilares, R.T., Cota, J., Ellila, S., Silveira, M.H.L., dos Santos, J.C. and da Silva, S.S. 2018. Bioconversion of hemicellulose into ethanol and value-added products: Commercialization, trends, and future opportunities. pp. 97–134. *In:* Chandel, A.K., Silveira, M.H.L. (eds.), Advances in Sugarcane Biorefinery Technologies, Commercialization, Policy Issues and Paradigm Shift for Bioethanol and By-Products. Elsevier Radarweg 29, 1000 AE Amsterdam, Netherlands.

Chandel, R., Nebapure, S.M., Sharma, M., Subramanian, S., Srivastava, C. and Khurana, S.M.P. 2019. Insecticidal and repellent activities of eucalyptus oil against lesser grain borer *Rhyzopertha dominica* (Fabricius). Journal of Microbiology Biotechnology Food Science, 9(3): 525–529.

Chandra, S., Lata, H., Ikhlas, A. Khan, I.A. and El-Sohly, M.A. 2017. *Cannabis sativa* L.: Botany and horticulture. pp. 79-100. *In:* S. Chandra, H. Lata and M.A. E-l Sohly (eds.), *Cannabis sativa* L. – Botany and Biotechnology. Springer International Publishing AG.

Chandrasekaran, U., Xu, W. and Liu, A. 2014. Transcriptome profiling identifies ABA mediated regulatory changes towards storage filling in developing seeds of castor bean (*Ricinus communis* L.). Cell & Bioscience, 4: 33.

Charoensi, S. 2014. Antioxidant and anticancer activities of *Moringa oleifera* leaves. Journal of Medicinal Plant Research, 8(7): 318–325.

Chatzakis, M.K., Tzanakakis, V.A., Mara, D.D. and Angelakis, A.N. 2011. Irrigation of castor bean (*Ricinus communis* L.) and sunflower (*Helianthus annus* L.) plant species with municipal wastewater effluent: Impacts on soil properties and seed yield. Water, 3: 1112–1127.

Chaudhary, S., Chaudhary, P.S., Chikara, S.K., Sharma, M.C. and Iriti, M. 2018. Review on fenugreek (*Trigonella foenum-graecum* L.) and its important secondary metabolite diosgenin. Notulae Botanicae Horti Agrobotanici Cluj-Napoca, 46(1): 22–31.

Chavarriaga-Aguirre, P., Brand, A., Medina, A., Prías, M., Escobar, R., Martinez, J., Díaz, P., Camilo López, C., Roc, W.M. and Tohme, J. 2016. The potential of using biotechnology to improve cassava: A review. In Vitro Cell Developmental Biology—Plant, 52: 461–478.

Chen, G.Q., Johnson, K. and Morale, E. 2018. Recurrent selection for improved oil content in castor bean. pp. 67–74. *In:* C. Kole and P. Rabinowicz (eds.), The Castor Bean Genome, Compendium of Plant Genome. Springer Nature Switzerland AG.

Chen, Y. and Tai, W. 2018. Castor oil-based polyurethane resin for low-density composites with bamboo charcoal. Polymers, 10: 1100.

Chen, Y., Liu, X., Wang, M. and Yan, X. 2014. Cadmium tolerance, accumulation and relationship with Cd subcellular distribution in *Ricinus communis* L., Acta Scientiae Circumstantiae, 34(9): 2440–2446.

Cheng, Z., Lu, B., Baldwin, B.S., Sameshima, K. and Chen, J. 2002. Comparative studies of genetic diversity in kenaf (*Hibiscus cannabinus* L.) varieties based on analysis of agronomic and RAPD data. Hereditas, 136: 231–239.

Chessa, I. 2010. Cactus pear genetic resources conservation, evaluation and uses. pp. 43–53. *In:* A. Nefzaoui, P. Inglese and T. Belay (eds.). Improved Utilization of Cactus Pear for Food, Feed, Soil and Water Conservation and Other Products in Africa. Proceedings of International Workshop, Mekelle (Ethiopia), 19-21 October, 2009.

Chhajro, M.A., Rizwan, M.S., Guoyong, H., Jun, Z. and Kubar, K.A. 2015. Enhanced accumulation of Cd in castor (*Ricinus communis* L.) by soil-applied chelators. Int. J. Phytoremed., 18: 664–670.

Chhikara, S., Abdullah, H.M., Akbari, P., Schnell, D. and Dhankher, O.P. 2018. Engineering *Camelina sativa* (L.) Crantz for enhanced oil and seed yields by combining diacylglycerol acyltransferase1 and glycerol-3-phosphate dehydrogenase expression. Plant Biotechnology Journal, 16: 1034–1045.

Chindo, I.Y., Osuide, J.O. and Yongabi, K.A. 2011. *Azadirachta indica* (neem) seed oil as adjuvant for antimicrobial activity. Int. Res. Journal Appl. Basic Sci., 2(8): 299–302.

Chipojola, F.M., Mwase, W.F., Kwapata, M.B., Bokosi, J.M., Njoloma, J.P. and Maliro, M.F. 2009. Morphological characterization of cashew (*Anacardium occidentale* L.) in four populations in Malawi. African Journal of Biotechnology, 8(20): 5173–5181.

Choudhri, P., Rani, M., Sangwan, R.S., Kumar, R., Kumar, A. and Chhokar, V. 2018. De novo sequencing, assembly and characterisation of Aloe vera transcriptome and analysis of expression profiles of genes related to saponin and anthraquinone metabolism. BMC Genomics, 19: 427.

Chow, C.K. (Ed.). 2007. Fatty Acids in Foods and Their Health Implications. CRC Press.

Christianto, V. and Smarandache, F. 2019. On the efficacy of *Moringa oleifera* as anticancer treatment: A literature survey. BAOJ Cancer Research & Therapy, 5(2): 1–5.

Chukwuneke, J.L., Ewulonu, M.C., Chukwujike, I.C. and Okolie, P.C. 2019. Physico-chemical analysis of pyrolyzed bio-oil from *Swietenia macrophylla* (mahogany) wood. Heliyon, 5: e01790.

Clark, D.G. 2004. Applications of plant biotechnology to ornamental crops. *In:* Christou, P., Klee, H. (eds.), Handbook of Plant Biotechnology. John Wiley & Sons, Ltd.

Colombo, R. and Papetti, A. 2019. Avocado (*Persea americana* Mill.) by-products and their impact: From bioactive compounds to biomass energy and sorbent material for removing contaminants: A review. International Journal of Food Science and Technology. 54(4): 943-951.

Commandeur, U., Twyman, R.M. and Fischer, R. 2003. The biosafety of molecular farming in plants. AgBiotechNet 2003, April, ABN, 5: 1–9.

Conradie, T.T. 2011. Genetic engineering of sugarcane for increased sucrose and consumer acceptance. Master of Science thesis in Plant Biotechnology at the University of Stellenbosch.

Corley, R.H.V. and Tinker, P.V. 2003. The Oil Palm. 4th Edition. Blackwell Science Ltd, a Blackwell Publishing Company, Oxford. pp. 201.

Corrales-Garcia, J. and Sáenz, C. 2013. Use of cladodes in food products. pp. 45-55. *In:* C. Sáenz, H. Berger, A. Rodriguez-Felix, L. Galleti, J. Corrales-Garcia, E. Sepulveda and C. Rosell (eds.), Agro-industrial Utilization of Cactus Pear. Food and Agriculture Organization of the United Nations (FAO).

Coscione, A.R. and Berton, R.S. 2009. Barium extraction potential by mustard, sunflower and castor bean. Scientia Agricola, 66: 59–63.

Costa, E.T., de, S., Guilherme, L.R.G., de Melo, E.E.C., Ribeiro, B.T., Inacio, E.S.B., Severiano, E.C., Faquin, V. and Hale, B.A. 2012. Assessing the tolerance of castor bean to Cd and Pb for phytoremediation purposes. Biol. Trace Elem. Res., 145: 93–100.

Coulibaly, K., Aka, R.A., Camara, B., Kassin, E., Kouakou, K., Kébé, B.I., Koffi, N.K., Tahi, M.G., Walet, N.P., Guiraud, S.B., Assi, M.E. , Kone, B., N'Guessan, K.F.D and Koné, D. 2018. Molecular identification of Phytophthora palmivora in the cocoa tree orchard of Côte d'Ivoire and assessment of the quantitative component of pathogenicity. International Journal of Sciences, 7(8): 7–15.

CPI. 2023.10 Everyday uses of Biotechnology. Available at https://www.uk-cpi.com/blog/10-everyday-uses-of-biotechnology. Accessed 27/3/2023

Cseke, L.J., Kirakosyan, A., Kaufman, P.B.., Warber, S.L., Duke, J.A. and Brielmann, H.L. 2006. Natural Products from Plants. 2ndEdition. Taylor & Francis Group, LLC. CRC Press, Boca Raton, pp. 427.

Csurhes, S. 2008. Pest plant risk assessment: Neem tree (*Azadiracta indica*). Queensland Government Department of Primary Industries and Fisheries. Brisbane, Qld 4001 https://www.uk-cpi.com/blog/10-everyday-uses-of-biotechnology. Accessed 27/3/2023

Daniel, P. 2016. Developing the insulation sheet of Luffa Cylindrica for Mitticool Fridge. International Journal of Scientific Research and Management, 4(8): 4514–4524.

Daniell, D., Streatfield, S.J. and Wycoff, K. 2001. Medical molecular farming: Production of antibodies, biopharmaceuticals and edible vaccines in plants. Trends in Plant Science, 6(5): 219–226.

Danquah, J.A., Appiah, M., Osman, A. and Pappinen, A. 2019. Geographic distribution of global economic important mahogany complex: A review. Annual Research & Review in Biology, 34(3): 1–22.

Das, A., Moquammel Haque, S.K., Ghosh, B., Nandagopal, K. and Jha, T.B. 2015. Morphological and genetic characterization of micropropagated field grown plants of *Aloe vera* L. Plant Tissue Culture & Biotechnology, 25(2): 231–246.

Das, D.R., Sachan, A.K., Mohd. Shuaib, M. and Mohd. Imtiyaz, M. 2014. Chemical charecterization of volatile oil components of *Citrus reticulata* by GC-MS analysis. World Journal of Pharmacy and Pharmaceutical Sciences, 3(7): 197–204.

Dávila, J.A., Rosenberg, M., Castro, E. and Cardona, C.A. 2017. A model biorefinery for avocado (*Persea americana* mill.) processing. Bioresource Technology, 243: 17–29.

Davis, L.C., Pidlisnyuk, V., Mamirova, A., Shapoval, P. and Stefanovska, T. 2021. Establishing miscanthus, production of biomass, and application to contaminated sites. pp. 77–114. *In:* Larry E. Erickson and Valentina Pidlisnyuk (eds.), Phytotechnology with Biomass Production: Sustainable Management of Contaminated Sites. First edition. CRC Press, Boca Raton.

Dayrit, F.M. and Nguyen, Q. 2020. Improving the value of the coconut with biotechnology. pp. 29–50. *In:* S. Adkins et al. (eds.), Coconut Biotechnology: Towards the Sustainability of the Tree of Life. Springer Nature Switzerland AG.

De Caluwé, E., Kateřina Halamová, K. and Van Damme, P. 2010. *Digitata* L. – A review of traditional uses, phytochemistry and pharmacology. Afrika Focus, (3)1: 11–51.

De Castro, O., Gargiulo, R., Guacchio, E.D., Caputo, P. and De Luca, P. 2015. A molecular survey concerning the origin of *Cyperus esculentus* (Cyperaceae, Poales): Two sides of the same coin (weed vs. crop). Annals of Botany, 115: 733–745.

de Evan, T., Vintimilla, A., Molina-Alcaide, E., Ranilla, M.J. and Carro, M.D. 2020. Potential of recycling cauliflower and romanesco wastes in ruminant feeding: In vitro studies. Animals, 10(8): 1247.

de Lima, G.S., Gheyi, H.R., Nobre, R.G., dos Anjos Soares, L.A., Xavier, D.A. and dos Santos Junior, J.A. 2015. Water relations and gas exchange in castor bean irrigated with saline water of distinct cationic nature. African Journal of Agricultural Research, 10(13): 1581–1594.

de Medeiros, L.L., da Silva, F.L.H., de Queiroz, A.L.M., de Oliveira, Y.S.L., de Souza Junior, E.F., Madruga, M.S. and da Conceição, M.M. 2020. Structural-chemical characterization and potential of sisal bagasse for the production of polyols of industrial interest. Brazilian Journal of Chemical Engineering, 37: 451–461.

Debaeke, P., Casadebaig, P., Flenet, F. and Langlade, N. 2017. Sunflower crop and climate change: Vulnerability, adaptation, and mitigation potential from case-studies in Europe. OCL Oilseeds and Fats Crops and Lipids, 24(1): 15.

Demirbas, A. 2017. Utilization of date biomass waste and date seed as bio-fuels source. Energy Sources Part A: Recovery Utilization and Environmental Effects, 39(8): 1–7.

Dendena, B. and Stefano Corsi, S. 2014. Cashew, from seed to market: A review. Agronomy for Sustainable Development, 34(4): 753–772. Springer Verlag/EDP Sciences/INRA.

Dequigiovanni, G., Ramos, S.L.F., Lopes, M.T.G., Clement, C.R., Rodrigues, D.P., Eliane Fabri, G., Zucchi, M.I. and Veasey, E.A. 2018. New microsatellite loci for annatto (*Bixa orellana*), a source of natural dyes from Brazilian Amazonia. Crop Breeding and Applied Biotechnology, 18: 116–122.

Deshmukh, S.N., Shrivastava, B., Sharma, P., Jain, H.K. and Ganesh, N. 2016. Pharmacognostical and phytochemical investigation of leaves of *Bixa orellana*

Linn. International Journal of Pharmaceutical Sciences Review and Research, 22(1): 247–252.

Devadiga, A., Shetty, K.V. and Saidutta, M.B. 2015. Timber industry waste-teak (*Tectona grandis* Linn.) leaf extract mediated synthesis of antibacterial silver nanoparticles. International Nanotechnology Letters, 5: 205–214.

Devappa, R.K., Harinder, P.S. and Makkar, K.B. 2010. Jatropha toxicity—A review. Journal of Toxicology and Environmental Health, Part B, 13: 476–507.

Devi, S.V., Kole, P.R., Gowthami, R. and Sehrawat, N. 2019. Genetically modified plants: Developments and industrial aspects. pp. 145–172. *In:* Mukesh Yadav, Vikas Kumar, and Nirmala Sehrawat (eds.). Walter de Gruyter GmbH, Berlin/Munich/Boston.

Dhakad, A.K., Pandey, V.V., Beg, S., Rawat, J.M. and Singha, A. 2018. Biological, medicinal and toxicological significance of Eucalyptus leaf essential oil: A review. Journal of Science Food and Agriculture, 98(3): 833–848.

Dhanesh, B.T. and Kochhar, A. 2013. Peanut Processing and It's Potential Food Applications. International Journal of Science and Research, 4(6): 2701–2706.

Dhivya, K., Sathish, S., Balakrishnan, N., Udayasuriyan, V. and Duraialagaraja Sudhakar, D. 2016. Genetic engineering of cotton with a novel cry2AX1 gene to impart insect resistance against *Helicoverpa armigera.* Crop Breeding and Applied Biotechnology, 16: 205-212.

Diambra, L.A. 2011. Genome sequence and analysis of the tuber crop potato. Nature, 475: 189–196.

Di Guardo, M., Bink, C.A.M., Guerra, W., Letschka, T., Lozano, L., Busatto, N., Poles, L., Tadiello, A., Bianco, L., Visser, R.G.F., Weg, E. and Costa, F. 2017. Deciphering the genetic control of fruit texture in apple by multiple family-based analysis and genome-wide association. Journal of Experimental Botany, 68(7): 1451–1466.

Diemeleou, C.A., Zoue, L.T. and Niamke, S.L. 2014. Physicochemical and nutritive characterization of high value nonconventional oil from seeds of *Amaranthus hybridus* Linn. European Scientific Journal, 10(24): 100–115.

Diniz, A.L., Ferreira, S.S., Felipeten-Caten, Gabriel, R.A., dos Santos, M.J.M., Barbosa, V.S.V., Carneiro, M.S. and Souza, G.M. 2019. Genomic resources for energy cane breeding in the post genomics era. Computational and Structural Biotechnology Journal, 17: 1404–1414.

Director General Department of Alternative Energy and Efficiency Royal Government of Thailand. nd. The role of Jatropha Curcas in support of the Thai Government's National Policy for Bio Diesel. https://en.calameo.com/books/0013941983f3fc927dcf9 14/10/2020.

Divakara, B.N., Upadhyaya, H.D., Wani, S.P. and Gowda, C.L. 2010. Biology and genetic improvement of Jatropha curcas L.: A review. Applied Energy, 87(3): 732-742.

Dixit, A. 2019. Banana stem: An underutilised part of the plant. International Journal of Scientific Research, 8(10): 24–25.

Djekota, C., Diouf, D., Sane, S., Mbaye, M.S. and Noba, K. 2014. Morphological characterization of shea tree (*Vitellaria paradoxa* subsp. paradoxa) populations in the region of Mandoul in Chad. International Journal of Biodiversity and Conservation, 6(2): 184–193.

Dölle, K. and Kurzmann, D.E. 2019. Cannabis, the plant of the unlimited possibilities. Advances in Research, 20(3): 1–22.

Domenek, S., Feuilloley, P., Gratraud, J., Morel, M.H. and Guilbert, S. 2004. Biodegradability of wheat gluten based bioplastics. Chemosphere, 54(4): 551–559.

Dongre, A., Parkhii, V. and Gahukar, S. 2004. Characterization of cotton (*Gossypium hirsutum* L.) germplasm by ISSR, RAPD markers and agronomic values. Indian Journal of Biotechnology, 3: 388–393.

Dotaniya, M.L., Datta, S.C., Biswas, D.R., Dotaniya, C.K., Meena, B.L., Rajendiran, S., Regar, K.L. and Manju Lata, M. 2016. Use of sugarcane industrial by-products for improving sugarcane productivity and soil health. International Journal of Recycling Organic Waste in Agriculture, 5: 185–194.

Drabu, S., Khatri, S. and Babu, S. 2012. Neem: Healer of all ailments. Res. J. Pharm. Biol. Sci., 3(1): 120–123.

Duan, N., Bai, Y., Sun, H., Wang, N., Ma,Y., Li, M., Wang, X., Jiao, C., Legall, N., Mao, L., Wan, S., Wang, K., He, T., Feng, S., Zhang, Z., Mao, Z., Shen, X., Chen, X., Jiang, Y., Wu, S., Yin, C., Ge, S., Yang, L., Jiang, S., Xu, H., Liu, J., Wang, D., Qu, C., Wang, Y., Weifang Zuo, W., Xiang, L., Liu, C., Zhang, D., Gao, Y., Xu, Y., Xu, K., Chao, T., Fazio, G., Shu, H., Zhong, G., Cheng, L., Fei, Z. and Chen, X. 2017. Genome re-sequencing reveals the history of apple and supports a two-stage model for fruit enlargement. Nature Communications, 8(249): 1–11.

Duc, P.A., Dharanipriya, P.,Velmurugan, B.K. and Shanmugavadivu, M. 2019. Groundnut shell – A beneficial bio-waste. Biocatalysis and Agricultural Biotechnology, 20: 101206. https://doi.org/10.1007/978-3-642-22144-6_175.

Dungani, R., Karina, M., Sulaeman, S.A., Hermawan, D. and Hadiyane, A. 2016. Agricultural waste fibers towards sustainability and advanced utilization: A review. Asian Journal of Plant Science, 15: 42–55.

Duvall, C.S. 2016. Drug laws, bioprospecting and the agricultural heritage of *Cannabis* in Africa. Space and Polity, 20(1): 10–25.

Dwi, A.A., Faidliyah, N.M., Harimbi, S. and Sriliani, S. 2018. Moringa leaf potential (*Moringa oleifera*) for the manufacture of instant drink powder with variations in tween 80 volume and drying temperature as an antioxidant drink. International Journal of ChemTech Research, 11(05): 295–300.

Dwivedi, N.K., Indiradevi, A., Asha, K.I., Nair, R.A. and Suma, A. 2014. A protocol for micropropagation of *Aloe vera* L. (Indian Aloe) – A miracle plant. Research in Biotechnology, 5(1): 01–05.

Edo, G.I. 2022. Antibacterial, phytochemical and GC-MS analysis of Thevetia peruviana extracts: An approach in drug formulation. Natural Resources for Human Health, 1–9.

Efisue, A.A. 2015. Genetic Fingerprinting of sweet potato [*Ipomoea batatas* (L.) Lam] as revealed by isozyme electrophoresis analysis. Nigerian Journal of Biotechnology, 30: 36–47.

Egharevba, H.O., Oladosun, P. and Izebe, K.S. 2016. Chemical composition and anti-tubercular activity of the essential oil of orange (*Citrus sinensis* L.) peel from north central Nigeria. International Journal of Pharmacognosy and Phytochemical Research, 8(1): 91–94.

El Finti, A., Belayadi, M., El Boullani, R., Msanda, F. and El Mousadik, A. 2010. Genetic structure of cactus pear (*Opuntia ficus-indica*) in Moroccan collection. *In:* VII International Congress on Cactus Pear and Cochineal 995 (pp. 57-61).

Elbanna, S.M. 2006. Larvicidal effects of eucalyptus extract on the larvae of Culex pipiens mosquito. International Journal of Agriculture & Biology, 8(6): 896–897.

Elevitch, C.R. and Thomson, L.A.J. 2006. *Hibiscus tiliaceus* (Beach hibiscus), ver. 1.2. pp. 1-4. *In:* Elevitch, C.R. (ed.), Species for Pacific Island Agroforestry. Permanent Agricultre Reources (PAR). Holualoa, Hawa'I, U.S.A. http://www.traditionaltree.org

Elias, M., Bayala, J. and Dianda, M. 2006. Impediments and innovations in knowledge sharing: The case of the African Shea sector. KM4D Journal, 2(1): 52–67.

Elseify, L.A., Midani, M., Shihata, L.A. and El-Mously, H. 2019. Review on cellulosic fibers extracted from date palms (*Phoenix dactylifera* L.) and their applications. Cellulose. https://doi.org/10.1007/s10570-019-02259-6(0123456789-6().,-volV() 0123458697().,-volV)

Elseify, L.A., Midani, M., El-Badawy, A.A., Awad, S. and Jawaid, M. 2023. Comparative study of long date palm (*Phoenix dactylifera* L.) midrib and spadix fibers with other commercial leaf fibers. Cellulose, 30(3): 1927-1942.

Enaberuel, L.O., Obisesan, I.O., Okolo, E.C., Akinwale, R.O., Aisueni, N.O. and Ataga, C.D. 2014. Genetic diversity of shea butter tree (*Vitellaria paradoxa* C.F. Gaernt) in the Guinea Savanna of Nigeria based on morphological markers. American-Eurasian Journal of Agricultural & Environmental Science, 14(7): 615–623.

Ene, C.O., Ogbonna, P.E., Agbo, C.U., Uche, P. and Chukwudi, U.P. 2016. Evaluation of sixteen cucumber (*Cucumis sativus* L.) genotypes in derived savannah environment using path coefficient analysis. Notule Scientia Biologicae, 8(1): 85–92.

Energy Alternatives India. 2016. Emerging Business Opportunities in Castor Oil Derivatives. Comprehensive Castor Oil Report. The Definitive Guide for Entrepreneurs and Investors. Available from http://www.consult.eai.in/wp-content/uploads/2016/05/Emerging-Business-Opportunities-in-Castor-Oil-Derivatives.pdf. Accessed 22/1/2019

Engadget, J.F. 2018. Grapes genetically resistant to mildew rot could cut pesticide use and French wine prices at the same time. GLP 2017-2018 Annual Report, pp. 13.

Enontiemonria, E.V., Ayodeji, A., Lucky, A.P.A. and Ogheneofego, O. 2012. The effects of trans-esterification of castor seed oil using ethanol, methanol and their blends on the properties and yields of biodiesel. International Journal of Engineering and Technology, 2(10): 1734–1742.

Enwean, I.B. 2013. Biotechnology, Entrepreneurship and Poverty Alleviation. Proceedings of the 2nd Biennial International Conference on Biotechnology and National Development. pp. 46–54.

Ermumcu, M.S.K. and Şanlıer, N. 2017. Black cumin (*Nigella sativa*) and its active component of Thymoquinone: Effects on health. Journal of Food and Health Science, 3(4): 170–183.

Esmail, A.E.G., Amira, K.G.A. and Adss, I.A.A. 2016. Increase the economic value of the jojoba (*Simmondsia chinensis*) yield using evaluation of distinctive clones grown under the Egyptian environmental conditions. International Journal of Agricultural Technology, 12(1): 145–165.

Etemadi, F., Hashemi, M., Barker, A.V., Zandvakili, O.R. and Liu, X. 2019. Agronomy, nutritional value, and medicinal application of Faba bean (*Vicia faba* L.). Horticultural Plant Journal, 5(4): 170–182.

Eynck, C. and Falk, K.C. 2013. Camelina (*Camelina sativa*). pp. 389–391. In: B.P. Singh (ed.), Biofuel Crops: Production, Physiology and Genetics. CAB International.

Ezejiofor, T.I.N., Eke, N.V., Okechukwu, R.I., Nwoguikpe, R.N. and Duru, C.M. 2011. Waste to wealth: Industrial raw materials potential of peels of Nigerian sweet orange (*Citrus sinensis*). African Journal of Biotechnology, 10(33): 6257–6264.

Fakir, A.D. and Waghmare, J.S. 2017. Watermelon waste: A potential source of omega-6 fatty acid and proteins. International Journal of ChemTech Research, 10(6): 384–392.

Falasca, S.L., Ulberich, A.C. and Ulberich, E. 2012. Developing an agro-climatic zoning model to determine potential production areas for castor beans (*Ricinus communis* L.). Ind. Crop Prod., 40: 185–191.

Fan, X., Wang, X. and Chen, F. 2011. Biodiesel production from crude cottonseed oil: An optimization process using response surface methodology. The Open Fuels & Energy Science Journal, 4: 1–8.

Fang, E.F., Hassanien, A.A., Wong, J.H., Bah, C.S.F., Soliman, S.S. and Ng, T.B. 2011. Isolation of a new trypsin inhibitor from the faba bean (*Vicia faba* cv. Giza 843) with potential medicinal applications. Protein & Peptide Letters, 18: 64–72.

Farooq, F., Rai, M., Tiwari, A., Khan, A. and Farooq, S. 2012. Medicinal properties of *Moringa oleifera*: An overview of promising healer. Journal of Medicinal Plants Research, 6(27): 4368–4374.

Faure, J. and Tepfer, M. 2016. Camelina, a Swiss knife for plant lipid biotechnology. Oil Seeds and Fat Crops and Lipids, 9 pp.

Ferreira, J.F.S., Laughlin, J.C., Delabays, N. and de Magalhães, P.M. 2005. Cultivation and genetics of *Artemisia annua* L. for increased production of the antimalarial artemisinin. Plant Genetic Resources, 3(2): 206–229.

Fiorelli, J., Curtolo, D.D., Barrero, N.G., Savastano Jr., H., Eliria Maria de Jesus Agnolon Pallone, E.M. and Johnson, R. 2012. Particulate composite based on coconut fiber and castor oil polyurethane adhesive: An eco-efficient product. Industrial Crops and Products, 40: 69–75.

Fisher, R. and Emans, N. 2000. Molecular farming of pharmaceutical proteins. Transgenic Research, 9(4-5): 279–299.

Fitriana, W.D., Ersam, T., Shimizu, K. and Fatmawati, S. 2016. Antioxidant activity of *Moringa oleifera* extracts. Indones. Journal of Chemistry, 16(3): 297–301.

Flores-López, M.L., Romaní, A., Cerqueira, M.A., Rodríguez-García, R., Jasso de Rodríguez, D. and Vicente, A.A. 2016. Compositional features and bioactive properties of whole fraction from Aloe vera processing. Industrial Crops and Products, 91: 179–185.

Fodil, A. and Arbouche, H.S. 2007. Evaluation of the crop waste products of melon yellow canary as animal feed: Influence of culture zone. Livestock Research for Rural Development, 19(10).

Fontes, E.M.G., Pires, C.S.S., Sujii, E.R. and Panizzi, A.R. 2002. The environmental effects of genetically modified crops resistant to insects. Neotropical Entomology, 31(4): 497–513.

Food and Agriculture Organization of the United Nations (FAO). 2013. Agro-industrial Utilization of Cactus Pear. pp. 1-16. Rome.

Food and Agriculture Organization of the United Nations. 2013. Cassava Save and Grow: Cassava. A guide to sustainable production intensification. pp. 1–4. FAO, 2013.

Foong, F.H.N., Mohammad, A., Solachuddin Jauhari and Arief Ichwan, S.J.I. 2015. Biological properties of cucumber (*Cucumis sativus* L.) extracts. Malaysian Journal of Analytical Sciences, 19(6): 1218–1222.

Foster, C.N. 2015. Agricultural Wastes: Characteristics, Types, and Management. Nova Science Publishers, Inc. New York.

Foster, J.T., Allan G.J., Chan, A.P., Rabinowicz, P.D., Ravel, J., Jackson, P.J. and Keim, P. 2010. Single nucleotide polymorphisms for assessing genetic diversity in castor bean (*Ricinus communis*). BMC Plant Biol., 10(1): 1–11.

Fouassier, J., Morlet-Savary, F., Lalevée, J., Xavier Allonas, X. and Christian Ley, C. 2010. Dyes as photoinitiators or photosensitizers of polymerization reactions. Materials, 3: 5130–5142.

Fowler, P.A., McLauchlin, A.R. and Hall, L.M. 2003. The Potential Industrial Uses of Forage Grasses Including Miscanthus. BioComposites Centre, University of Wales, Bangor, Gwynedd, UK. pp. 27-29.

Friday, O.A., Ajayi, F.F., Folorunsho, O.A. and Ogori, A.F. 2017. Quality characteristics of shortbread biscuit fortified with fermented jack bean flour. International Journal of Agricultural Science and Food Technology, 3(3): 036–041.

Fuller, S. and Stephens, J.M. 2015. Diosgenin, 4-hydroxyisoleucine, and fiber from fenugreek: Mechanisms of actions and potential effects on metabolic syndrome. Advance Nutrytion, 6: 189–197.

Furlan, F.F., Filho, R.T., Pinto, F.H.P.B., Costa, C.B.B., Cruz, A.J.G., Giordano, R.L.G. and Giordano, R.C.G. 2013. Bioelectricity versus bioethanol from sugarcane bagasse: Is it worth being flexible? Biotechnology for Biofuels, 6: 142.

Gabriel, A., Victor, N. and du Preez James, P.C. 2014. Cactus pear biomass, a potential lignocellulose raw material for single cell protein production (SCP): A review. International Journal of Current Microbiology and Applied Sciences, 3(7): 171–197.

Gadhave, R.V., Mahanwar, P.A. and Gadekar, P.T. 2017. Starch-based adhesives for wood/ wood composite bonding: Review. Open Journal of Polymer Chemistry, 7: 19–32.

Galhiane, M.S., Rissato, S.R., Santos, L.D.S., Chierice, G.O., Almeida, M.V.D., Fumis, T., Chechim, I. and Sampaio, A.C. 2012. Evaluation of the performance of a castor-oil based formulation in limiting pesticide residues in strawberry crops. Quím Nova, 35(2): 341–347.

Gami, B., Syed, B.A. and Patel, B. 2015. Assessment of genetic diversity in bamboo accessions of India using molecular markers. International Journal of Applied Science and Biotechnology, 3(2): 330–336.

Ganopoulos, I., Tourvas, N., Xanthopoulou, A., Aravanopoulos, F.A., Avramidou, E., Zambounis, A., Tsaftaris, A., Madesis, P., Thomas Sotiropoulos, T. and Koutinas, N. 2018. Phenotypic and molecular characterization of apple (Malus ×domestica Borkh) genetic resources in Greece. Scientia. Agricola., 75(6): 509–518.

Gantait, S., Debnath, S. and Ali, N. 2014. Genomic profile of the plants with pharmaceutical value. Biotech, 4: 563–578.

Garcia-Mas, J., Benjak, A., Sanseverinoa, W., Bourgeois, M., Mira, G., González, V.M., Hénaff, E., Câmara, F., Cozzuto, L., Lowy, E. and Alioto, T. 2012. The genome of melon (*Cucumis melo* L.). Proceedings of the National Academy of Sciences, 109(29): 11872–11877.

García, D.A.M., Torres, J.M.C., Arboleda, A.A.N. and Gaona, O.J.C. 2018. Repeatability for yield total solids in a segregating population of rubber (*Hevea brasiliensis*) in Colombia. Rev. Fac. Nac. Agron. Medellín, 71(1): 8407–8414.

Garima, D.S., Mahajan, M. and Kaur, P. 2008. Agricultural waste material as potential adsorbent for sequestering heavy metal ions from aqueous solutions – A review. Bioresource Technology, 99(14): 6017–6027.

Gashaw, Z. 2020. Status of black cumin (*Nigella sativa* L.) research and production in Ethiopia: A review. International Journal of Forestry and Horticulture, 6(3): 20–29.

Gbadegesin, M.A., Olaiya, C.O. and Beeching, J.R. 2013. African cassava: Biotechnology and molecular breeding to the rescue. British Biotechnology Journal, 3(3): 305–317.

Gehlot, A., Arya, I.D., Kataria, V., Gupta, R.K. and Arya, S. 2014. Clonal multiplication of multipurpose desert tree *Azadirachta indica* – Neem. Journal of Arid Land Studies, 24(1): 37–40.

Georgiev, V., Ananga, A. and Tsolova, V. 2014. Recent advances and uses of grape flavonoids as nutraceuticals. Nutrients, 6: 391–415.

Ghaffar, S.H. 2020. Wheat straw biorefinery for agricultural waste valorisation. Green Materials, 8(2): 60–67.

Ghareeb, M.A., Shoeb, H.A., Madkour, H.M.F., Refahy, L.A., Mohamed, M.A. and Saad, A.M. 2014. Antioxidant and cytotoxic activities of flavonoidal compounds from *Gmelina arborea* Roxb. Global Journal of Pharmacology, 8(1): 87–97.

Gharib-Eshghi, A., Mozafari, J. and Azizov, I. 2016. Genetic diversity among sesame genotypes under drought stress condition by drought implementation. Agricultural Communications, 4(2): 1–6.

Ghatage, T.S. and Das, P. 2016. High resolution melting curve analysis: An efficient method for genetic purity analysis of cotton (*Gossypium hirsutum*) hybrid and their parental lines. IOSR Journal of Biotechnology and Biochemistry, 2(7): 49–53.

Ghazi, Z., Ramdani, M., Tahri, M., Rmili, R., Elmsellem, H., El Mahi, B. and Fauconnier, M.L. 2015. Chemical composition and antioxidant activity of seeds oils and fruit juice of Opuntia ficus indica and Opuntia dillenii from Morocco. Journal of Materials and Environmental Science, 6(8): 2338–2345.

Gianessi, L.P. and Carpenter, J.E. 2000. Agricultural Biotechnology: Benefits of Transgenic Soybeans. National Center for Food and Agricultural Policy. Washington.

Giri, C.C., Shyamkumar, B. and Anjaneyulu, C. 2004. Progress in tissue culture, genetic transformation and applications of biotechnology to trees: An overview. Trees, 18: 115–135.

Girish, K. and Bhat, S. 2008. Neem – A green treasure. Electronic J. Biol., 4(3): 102–111.

Gnanasambandam, A., Paull, J., Torres, A., Kaur, S., Leonforte, T., Li, H., Zong, X., Yang, T. and Materne, M. 2012. Impact of molecular technologies on faba bean (*Vicia faba* L.) breeding strategies. Agronomy, 2: 132–166.

Goel, G., Makkar, H.P.S., Francis, G. and Becker, K. 2007. Phorbol esters: Structure, biological activity, and toxicity in animals. International Journal of Toxicology, 26: 279–288.

Gokbulut, C. 2010. Sesame oil: Potential interaction with P450 isozymes. Journal of Pharmaceutical Toxicology, 5: 469–472.

Goldberg, R.A. 1952. The Soybean Industry. The University of Minnesota Press, Minneapolis. pp. 37–65.

Goldman, J. 2012. Tasty traits: Introduce genetics with a sensory assessment of apples. The American Biology Teacher, 74(1): 42–44.

Goldman, S.L. and Cole, C. 2014. Compendium of Bioenergy Plants CORN. CRC Press, Boca Raton.

Goldstein, D.A. and Thomas, J.A. 2004. Biopharmaceuticals derived from genetically modified plants. Q. J. Med., 97: 705–716.

Gollakota, K.G. and Jayalakshmi, B. 1983. Biogas (natural gas) production by anaerobic digestion of oil cake by a mixed culture isolated from cow dung. Biochemical and Biophysical Research Communications, 110(1): 32–35.

Gómez-Ortíz, N.M., Vázquez-Maldonado, I.A., Pérez-Espadas, A.R. and Oskam, G. 2010. Dye-sensitized solar cells with natural dyes extracted from achiote seeds. Solar Energy Materials and Solar Cells, 94(1): 40–44.

Gonçalves, R.N., Barbosa, S.D.G.B. and Raquel Elisada Silva-López, R. 2016. Biotechnology Research International. 11 pp.

Gong, Y., Xu, S., Mao, W., Ze-yun Li, Z., Hu, Q., Zhang, G. and Ding, J. 2011. Genetic diversity analysis of faba bean (*Vicia faba* L.) based on EST-SSR markers. Agricultural Sciences in China, 10(6): 838–844.

Gonzalez-Cortés, Reyes-Valdés, M.H., Robledo-Torres, V., Villarreal-Quintanilla, J.A. and Ramírez-Godina, F. 2018. Pre-germination treatments in four prickly pear cactus (*Opuntia* sp.) species from Northeastern Mexico. Australian Journal of Crop Science, 12(10): 1676–1684.

Goodarzi, F., Darvishzadeh, R., Hassani, A. and Hassanzaeh, A. 2011. Study on genetic variation in Iranian castor bean (*Ricinus communis* L.) accessions using multivariate statistical techniques. Journal of Medicinal Plants Research, 5(21): 5254–5261.

Górnaś, P. and Rudzińska, M. 2016. Seeds recovered from industry by-products of nine fruit species with a high potential utility as a source of unconventional oil for biodiesel and cosmetic and pharmaceutical sectors. Industrial Crops and Products, 83: 329–338.

Goswami, S. and Das, S. 2015. A study on cadmium phytoremediation potential of Indian mustard, *Brassica juncea*. International Journal of Phytoremediation, 17(1–6): 583–588.

Goudarzi, M., Fazeli, M., Azad, M., Seyedjavadi, S.S. and Mousavi, R. 2015. Aloe vera gel: Effective therapeutic agent against multidrug-resistant *Pseudomonas aeruginosa* isolates recovered from burn wound infections. Chemotherapy Research and Practice, 2015: 5. Article ID 639806, http://dx.doi.org/10.1155/2015/639806

Gour, V.S. and Datta, M. 2015. Soil carbon sequestration through desert date based forestry in arid and salt affected regions. National Academy Science Letters, 38(2): 127–128.

Goyal, N., Pardha-Saradhi, P. and Sharma, G.P. 2014. Can adaptive modulation of traits to urban environments facilitate *Ricinus communis* L. invasiveness? Environ. Monit. Assess., 186: 7941–7948.

Grassi, G. and McPartland, J.M. 2017. Chemical and morphological phenotypes in breeding of *Cannabis sativa* L. pp. 137–160. *In:* S. Chandra, H. Lata and M.A. ElSohly (eds.), *Cannabis sativa* L. – Botany and Biotechnology. Springer International Publishing AG.

Graudal, L. and Moestrup, S. 2017. Genetic resources conservation and management. *In:* pp. 9-13. W. Kollert and M. Kleine (eds.), The Global Teak Study: Analysis, Evaluation and Future Potential of Teak Resources (Vol. 36). Wien: International Union of Forest Research Organizations (IUFRO). IUFRO World Series.

Grigoriou, A.H. and Ntalos, G.A. 2001. The potential use of *Ricinus communis* L. (Castor) stalks as a lignocellulosic resource for particle boards. Industrial Crops and Products, 13: 209–218.

Grosser, M.D.J. 2015. Using genetically modified biotechnology to improve citrus. Citrus Industry, 10–13.

Gruenwald, J. and Galizia, M. 2005. Market brief in the European Union for selected natural ingredients derived from native species: *Adansonia digitata* L. Baobab. United Nations Conference on Trade and Development. pp. 1–35.

Gudeta, T.B. 2016. Chemical composition, bio-diesel potential and uses of *Jatropha curcas* L. (Euphorbiaceae). American Journal of Agriculture and Forestry, 4(2): 35–48.

Gumisiriza, R., Hawumba, J.F., Okure, M. and Hensel, O. 2017. Biomass waste-to-energy valorisation technologies: A review case for banana processing in Uganda. Biotechnology for Biofuels, 10: 1-29.

Gummadi, S.N., Manoj, N. and Kumar, D.S. 2007. Structural and biochemical properties of pectinases. pp. 99–115. *In:* Polaina, A., MacCab, A.P (eds), Industrial Enzymes Structure, Function and Applications. Springer, Dordrecht, The Netherlands.

Guner, N. and Wehner, T.C. 2004. The genes of watermelon. HortScience, 39(6): 1175–1182.

Gunstone, F.D. 2004. The Chemistry of Oils and Fats: Sources, Composition, Properties and Uses. 1st Edition. Blackwell Publishing Ltd, 9600 Garsington Road, Oxford OX4 2DQ, UK. p. 8.

Gupta, D. and Ranjan, R. 2016. Role of bamboo in sustainable development. ASJ International Journal of Advances in Scientific Research, 2(1): 25–32.

Guy Marcel, B.K., André, K.B., Viviane, Z. and Séraphin, K. 2011. Cashew in Breeding: Research synthesis. International Journal of Agronomy and Agricultural Research, 1(1): 1–8.

Habte, T.Y., Suleiman, I., Adam, H.E. and Krawinkel. 2019. The potential of baobab (*Adansonia digitata* L.) leaves from north and west Kordofan in Sudan as mineral complement of common diets. J. Nutrition and Food Processing, 2(2): 1–11.

Hadi, F., Ali, N. and Fuller, M.P. 2016. Molybdenum (Mo) increases endogenous phenolics, proline and photosynthetic pigments and the phytoremediation potential of the industrially important plant *Ricinus communis* L. for removal of cadmium from contaminated soil. Environ. Sci. Pollut. Res. Int., 23(20): 20408–20430.

Hadi, F., Ul-Arifeen, M.Z., Aziz, T., Nawab, S. and Nabi, G. 2015. Phytoremediation of cadmium by *Ricinus communis* L. in hydrophonic condition. American-Eurasian J. Agric. & Environ. Sci., 15(6): 1155–1162.

Hafez, O.M., Taha, R.A. and El-Nagdi, W.M.A. 2017. Assessing efficacy of jojoba crushed seed and oil cake on growth vigor, nutritional status of superior grapevine cuttings and controlling the root-knot nematode. Agricultural Engineering International: CIGR Journal, Special issue: 111–117.

Hammad, E.A.F. 2008. Chinaberry, *Melia azedarach* L.: A biopesticidal tree. *In:* Capinera, J.L. (ed.), Encyclopedia of Entomology. Springer, Dordrecht. https://doi.org/10.1007/978-1-4020-6359-6_628.

Hamman, J.H. 2008. Composition and applications of *Aloe vera* leaf gel. Molecules, 13: 1599–1616.

Hannan, M.A., Rahman, M.A., Sohag, A.A.M., Uddin, M.J., Dash, R., Sikder, M.H., Rahman, M.S., Timalsina, B., Munni, Y.A., Sarker, P.P. and Alam, M. 2021. Black cumin (*Nigella sativa* L.): A comprehensive review on phytochemistry, health benefits, molecular pharmacology and safety. Nutrients, 13: 1784.

Haque, M. and Ghosh, B. 2013. High frequency microcloning of Aloe vera and their true-to type conformity by molecular cytogenetic assessment of two years old field growing regenerated plants. Botanical Studies, 54: 46.

Haryanto, D.A., Landuma, S. and Purwanto, A. 2014. Fabrication of Dye-sensitized Solar Cell (DSSC) Using Annato Seeds (*Bixa orellana* Linn). Proceedings of the 5th Nanoscience and Nanotechnology Symposium (NNS2013). AIP Conference Proceedings 1586, 104.

Hasan, A.J.M., Hadi, A.M.H. and Ekaab, N.S. 2017. The effect of using castor oil on the pollutants emission in a continuous combustion chamber. Al-Nahrain Journal for Engineering Sciences, 20(4): 937–944.

Hasan, M., Rahmayani, R.F.I. and Munandar. 2018. Bioplastic from Chitosan and Yellow Pumpkin Starch with Castor Oil as Plasticizer. International Conference on Advanced Materials for Better Future. IOP Conf. Series: Materials Science and Engineering 333.

Hasan, N.A., Nawahwi, M.Z. and Ab Malek, H. 2013. Antimicrobial activity of *Nigella sativa* seed extract (Aktiviti Antimikrob Ekstrak *Nigella sativa*). Sains Malaysiana, 42(2): 143–147.

Hashmiu, I., Agbenyega, O. and Dawoe, E. 2022. Cash crops and food security: Evidence from smallholder cocoa and cashew farmers in Ghana. Agriculture & Food Security, 11(12): 21.

Hassanein, A., Salem, J., Faheed, F. and Abdullah El-Nagish, A. 2019. Some important aspects in *Moringa oleifera* Lam. Micropropagation. Acta Agriculturae Slovenica, 113(1): 13–27.

Haung, H., Yu, N., Wang, L., Gupta, D.K., He, Z., Wang, K., Zhu, Z., Yan, X., Li, T., Yang, X.-E. 2011. The phytoremediation potential of bioenergy crop *Ricinus communis* for DDTs and cadmium co-contaminated soil. Bioresour. Technol., 102: 11034–11038.

Hefferon, K.L. 2010. Biopharmaceuticals in Plants Toward the Next Century of Medicine. CRC Press, Taylor & Francis Group. Boca Raton. p. 27.

Hegazy, A.E. and Ammar, M.S. 2019. Utilization of cauliflower (*Brassica oleracea* L. ssp. botrytis) stem flour in improving Balady bread quality. Al-Azhar Journal of Agricultural Research, 44(1): 112–118.

Heinrich, L.A. 2019. Future opportunities for bio-based adhesives – Advantages beyond renewability. Green Chemistry, 21: 1866.

Hennessey-Ramos, L., Murillo-Arango, W. and Guayabo, G.T. 2018. Evaluation of a colorant and oil extracted from avocado waste as functional components of a liquid soap formulation. Revista Facultad Nacional de Agronomia, 72(2): 8855–8862.

Hernández-Nicolás, N.Y., Córdova-Téllez, L., Romero-Manzanares, A., Jiménez-Ramírez, J. and Luna-Cavazos, M. 2018. Traditional uses and seed chemical composition of Jatropha spp. (Euphorbiaceae) in Tehuacán-Cuicatlán, México. Review in Biologia Tropical, 66(1): 266–279.

Heukelbach, J., Oliveira, F.A.S. and Speare, R. 2006. A new shampoo based on neem (*Azadirachta indica*) is highly effective against head lice in vitro. Parasitol Res., 99: 353–356.

Hiltan, M. 2000. Smoking in British Popular Culture Perfect Pleasure. Manchester University Press. pp. 229–241.

Hoang, N.V., Furtado, A., Botha, F.C., Simmons, B.A. and Henry, R.J. 2015. Potential for genetic improvement of sugarcane as a source of biomass for biofuels. Front. Bioeng. Biotechnol., 3: 182.

Hong, S.H., Choi, S.A., Yoon, H. and Cho, K.S. 2011. Screening of *Cucumis sativus* as a new arsenic-accumulating plant and its arsenic accumulation in hydroponic culture. Environ. Geochem. Health, 33: 143–149.

Hu, Z., Wu, Q., Dalal, J., Vasani, N., Lopez, H.O., Sederoff, H.W. and Qu, R. 2017. Accumulation of medium-chain, saturated fatty acyl moieties in seed oils of transgenic *Camelina sativa*. PLoS One, 12(2): e0172296.

Huang, G., Guo, G., Yao, S., Zhang, N. and Hu, H. 2016. Organic acids, amino acids compositions in the root exudates and Cu-accumulation in castor (*Ricinus communis* L.) under Cu stress. Int. J. Phytoremed., 18(1): 33–40.

Huang, H., Yu, N., Wang, L., Gupta, D.K., He, Z., Wang, K., Zhu, Z., Yan, X., Li, T. and Yang, X.-E. 2011. The phytoremediation potential of bioenergy crop *Ricinus communis* for DDTs and cadmium co-contaminated soil. Bioresour. Technol., 102(23): 11034–11038.

Hugo, V. and Zuazo, D. 2008. Mango research and biotechnology. The European Journal of Plant Science and Biotechnology, 2(1): 25–37.

Hui, W., Yang, Y., Wu, G., Wang, Y., Zayed, M.Z. and Chen, X. 2018. Differential gene expression analyses related to fruit yield of *Jatropha curcas* L. using RNA-seq. Biotechnology & Biotechnological Equipment, 32(5): 1126–1133.

Huot, M. 2003. Plant Molecular Farming: Issues and Challenges for Canadian Regulators. Option consommateurs. 2120, rue Sherbrooke est, bur. 604 Montreal, Quebec. pp. 2–4.

Hussain, D.A.S. and Hussain, M.M. 2016. *Nigella sativa* (black seed) is an effective herbal remedy for every disease except death – A prophetic statement which modern scientists confirm unanimously: A review. Advancement in Medicinal Plant Research, 4(2): 27–57.

Hussein, M.M., Abdel-Azeem, A.S., Soraya, T. and El-Damhougy, S.T. 2016. The health benefits of black seed (*Nigella sativa*). Research Journal of Pharmaceutical, Biological and Chemical Sciences, 7(1): 1109–1113.

Huynh, N.T. 2016. Biological treatments of cauliflower (*Brassica oleracea* L. var. Botrytis) outer leaves: Improved extraction and conversion of phenolic compounds. PhD dissertation, Faculty of Bioscience Engineering, Ghent University, Belgium.

Ian, S.E. Bally. 2011. Advances in research and development of mango industry. Rev. Bras. Frutic., Jaboticabal - SP, Volume Especial, E. 057–063.

Ibeagha, O.A. and Onwualu, A.P. 2015. Strategies for improving the value chain of castor as an industrial raw material in Nigeria. Agric. Eng. Int: CIGR Journal, 17(3): 217–230.

Ibrahim, A., Usman, A., Yahaya, A.I., Umar, M.L., Halilu, A., Abubakar, H., Kwanashie, A.J., Mahadi, M.A. and Ibrahim B. 2014. Genetic diversity for nutritional traits in the leaves of baobab, *Adansonia digitata.* African Journal of Biotechnology, 13(2): 301–306. http://ejfa.info/

Ibrahim, H.D. and Ogunwusi A.A. 2017. Prospects for kenaf textiles production in Nigeria. Journal of Natural Sciences Research, 7: 22–28.

Ibrahim, H., Magaji, S., Jibia, S.A., Muhammad, I., Abdulsalam, H. and Abubakar Ismail, A. 2021. Fallen mahogany (*Khaya senegalensis*) leaves, a potential feedstock for biodiesel production. Nigerian Research Journal of Chemical Sciences, 9(1): 37–43.

Ibrahim, H.D., Hammanga, Z., Olugbemi, B.O., Wali, S.L. and Isah, A.B. 2015. Policy brief in the prouction of cocoa in Nigeria. Raw Materials Research and Development Council, Abuja. pp. 5–8.

Ige, P.O. 2018. Above ground biomass and carbon stock estimation of *Gmelina arborea* (Roxb.) stands in Omo forest reserve, Nigeria. Journal of Research in Forestry, Wildlife & Environment, 10(4): 71–80.

Ighalo, J.O., Adeniyi, A.G., Oke, E.O., Adewoye, L.T. and Motolani, F.O. 2020. Evaluation of *Luffa cylindrica* fibers in a biomass packed bed for the treatment of paint industry effluent before environmental release. European Journal of Sustainable Development Research, 4(4): em0132.

Imankulov, N. 2012. Preparation and research on properties of castor oil as a diesel fuel additive. Applied Technologies & Innovations, 6(1): 30–37.

Inácio de Campos, C., Lahr, F.A.R., Christoforo, A.L. and Fátma do Nascimento, M. 2014. Castor oil based polyurethane resin used in the production of medium density fiberboard. International Journal of Composite Materials, 4(4): 185–189.

In Hyun, N., Seung Bum, R., Min Jeong, P., Chul Min, C., Jae Gon, K., Sueng Won, J., Ho Cheo, S. and Min Ho, Y. 2016. Immobilization of heavy metal contaminated mine wastes using *Canavalia ensiformis* extract. Catena, 136: 53–58.

Innova, A. 2009. Castor-oil plants genetically altered to produce new bio-lubricants. Science Daily. Available online. 26/11/2018 https://www.sciencedaily.com/releases/2009/06/090625074514.htm.

International Flora Technologies. 2008. Moringa Oil Product Information Bulletin. p. 1.

Iqbal, H.M.N., Kyazze, G. and Keshavarz, T. 2013. Advances in the valorization of lignocellulosic materials by biotechnology: An overview. BioResources, 8(2): 3157–3176.

Iqbal, Y. 2017. Biomass quality of miscanthus genotypes for different bioconversion routes. Ph.D. Thesis. University of Hohenheim Institute of Crop Science. pp. 1–16.

Iqbal, Y. and Lewandowski, I. 2014. Inter-annual variation in biomass combustion quality traits over five years in fifteen Miscanthus genotypes in south Germany. Fuel Processing Technology, 121: 47–55.

Iqbal, Y. and Lewandowski, I. 2016. Biomass composition and ash melting behaviour of selected miscanthus genotypes in Southern Germany. Fuel, 180: 606–612.

Isaac, O.T. 2012. Characterisation of fuel briquettes from *Gmelina arborea* (Roxb.) sawdust and maize cob particles using *Cissus populnea* gum as binder. Ph.D. Thesis Agricultural and Environmental Engineering, University of Ibadan, Nigeria. 184 pp.

Iscia, A. and Demirerb, G.N. 2007. Biogas production potential from cotton wastes. Renewable Energy, 32: 750–757.

Ismail, N.I.N. and Veerasamy, D. 2011. Value-added natural rubber skim latex concentrate/montmorillonite as environmentally-friendly nanocomposite materials. Journal of Rubber Research, 14(4): 216–229.

Ismail, S., Ahmed, A.S., Anr, R. and Hamdan, S. 2016. Biodiesel production from castor oil by using calcium oxide derived from mud clam shell. Journal of Renewable Energy, 18.

Israel, A.U., Obot, I.B., Umoren, S.A., Mkpenie, V. and Asuquo, J.E. 2008. Production of cellulosic polymers from agricultural wastes. E-Journal of Chemistry, 5(1): 81–85.

Izadi, A., Rashedi, H., Hafarzadegan, R. and Hajiaghaee, R. 2015. Evaluating the effect of enzymatic process on the edible Aloe vera gel viscosity using commercial cellulase. Journal of Applied Biotechnology Reports, 2(3): 299–303.

Jadhav, M.S., Balakrishnan, N., Sudhakar, D. and Udayasuriyan, V. 2015. Genetic transformation of cotton for bollworm *Helicoverpa armigera* (Hubner) resistance using chimeric cry2AX1 gene. Indian Journal of Science and Technology, 8(17): 1–7.

Jagatee, S., Behera, S., Dash, P.K., Sahoo, S. and Mohanty, R.C. 2015. Bioprospecting starchy feedstocks for bioethanol production: A future perspective. Journal of Microbiology Research and Reviews, 3(3): 24–42.

Jain, S.M. 2012. Date palm biotechnology: Current status and prospective – An overview. Emirates Journal of Food and Agriculture, 24(5): 386–399.

Jamal, P., Saheed, O.K. and Alam, Z. 2012. Bio-valorization potential of banana peels (*Musa sapientum*): An overview. Asian Journal of Biotechnology, 4: 14.

Javed, A., Ahmad, A., Tahir, A., Shabbir, U., Nouman, M. and Adeela Hameed, A. 2019. Potato peel waste—Its nutraceutical, industrial and biotechnological applacations. AIMS Agriculture and Food, 4(3): 807–823.

Javed, H. and Jamil, N. 2015. Utilization of mustard oil for the production of polyhydroxyalkanoates by *Pseudomonas aeruginosa*. Journal of Microbiology Biotechnology and Food Science, 4(5): 412–414.

Jayesree, N., Hang, P. Kay, Priyangaa, A., Krishnamurthy, N. Prasad, Ramanan, R. Nagasundara, Turki, M. Aldawoud, Charis, M. Galanakis, Ooi and Wei, C. 2021. Valorisation of carrot peel waste by water-induced hydrocolloidal complexation for extraction of carote and pectin. Chemosphere, 272.

Jelaska, S., Mihaljević, S. and Bauer, N. 2005. Production of biopharmaceuticals, antibodies and edible vaccines in transgenic plants. Current Studies of Biotechnology, IV: 121–127.

Jelodar, N.B., Bhatt, A., Mohamed, K. and Keng, C.L. 2014. New cultivation approaches of *Artemisia annua* L. for a sustainable production of the antimalarial drug artemisin. Journal of Medicinal Plant Research, 8(10): 441–447.

Jensen, E., Robson, P., Norris, J., Cookson, A., Farrar, K., Donnison, I. and Clifton-Brown, J. 2013. Flowering induction in the bioenergy grass *Miscanthus sacchariflorus* is a quantitative short-day response, whilst delayed flowering under long days increases biomass accumulation. Journal of Experimental Botany, 64(2): 541–552.

Jiang, P., Chien, M., Sheu, M., Huang, Y., Chen, M., Su, C. and Liu, D. 2014. Dried fruit of the Luffa sponge as a source of chitin for applications as skin substitutes. BioMed Research International, 2014: 9.

Jiannong, L., Yuzhen, S. and Xuegui, Y. 2018. Review on genetic engineering in castor bean. Advancements Bioequivalence and Bioavailability, 1(4): 1–2.

Jingura, R.M., Musademba, D. and Rutendo Matengaifa, R. 2010. An evaluation of utility of *Jatropha curcas* L. as a source of multiple energy carriers. International Journal of Engineering, Science and Technology, 2(7): 115–122.

Johnson, R. 2013. Hemp as an Agricultural Commodity. Congressional Research Service Report for Congress. p. 1.

Johnson, S., Morgan, E.D. and Peiris, C.N. 1996. Development of the major triterpenoids and oil in the fruit and seeds of neem (*Azadirachta indica*). Annal of Botany, 78: 383–388.

Johnson, W. 2007. Final report on the safety assessment of *Ricinus communis* (castor) seed oil, hydrogenated castor oil, glyceryl ricinoleate, glyceryl ricinoleic acid, potassium ricinoleate, sodium ricinoleate, zinc ricinoleate, ethyl ricinoleate, glycol ricinoleate, isopropyl ricinoleate, methyl ricinoleate, and octyldodecyl ricinoleate. International Journal of Toxicology, 26(3): 31–77.

Jong, E., Higson, A., Walsh, P. and Wellisch, M. 2012. Product developments in the bio-based chemicals arena. Biofuels, Bioproducts and Biorefineries, 6: 606–624.

Jordan, J.H., Easson, M.W., Dien, B., Thompson, S., Brian D. and Condon, B.D. 2019. Extraction and characterization of nanocellulose crystals from cotton gin motes and cotton gin waste. Cellulose, 26(10): 5959-5979.

Joshi, M., Sodhi, K.S., Pandey, R., Singh, J. and Goyal, S. 2014. Transgenic plants as sole source for biopharmaceuticals. International Journal of Recent Trends in Science and Technology, 13(1): 97–106.

Joshi, S.V., Patel, N.T., Pandey, B.I. and Pandey, A.N. 2012. Effect of supplemental Ca^{2+} on NaCl-stressed castor plants (*Ricinus communis* L.). Acta Bot. Croat., 71(1): 13–29.

Josias, H. and Hamman, J.H. 2008. Composition and applications of Aloe vera leaf gel. Molecules, 13: 1599–1616.

Jung, I.L., Lee, J.H. and Kang, S.C. 2015. A potential oral anticancer drug candidate, *Moringa oleifera* leaf extract, induces the apoptosis of human hepatocellular carcinoma cells. Oncology Letters, 10: 1597–1604.

Jytothirmaye, P. and Lingumpelly, R. 2015. Efficiency of *Ricinus communis* extract in modifying cyclophosphamide induced clastogenicity in mice bone marrow cells. World Journal of Pharmacy and Pharmaceutical Sciences, 4(3): 811–822.

Kabbia, M.K., Akromah, R., Asibuo, J.Y., Conteh, A. and Kamara, E.G. 2017. Inheritance of seed quality traits in groundnut (*Arachis hypogaea* L.). International Journal of Development and Sustainability, 6(8): 526–544.

Kaboré, D., Sawadogo-Lingani, H., Diawara, B., Compaoré, C.S., Dicko, M.H. and Jakobsen, M. 2011. A review of baobab (*Adansonia digitata*) products: Effect of processing techniques, medicinal properties and uses. African Journal of Food Science, 5(16): 833–844.

Kadambi, K. and Dabral, S.N. 1955. The silviculture of *Ricinus communis* Linn. Indian Forester, 81(1): 53–58.

Kahramanoğlu, I. and Usanmaz, S. 2016. Pomegranate Production and Marketing. CRC Press, Taylor & Francis Group 6000 Broken Sound Parkway NW, Suite 300 Boca Raton, New York. pp. 1–7.

Kale, S. and Megha, G. 2011. Formulation and in vitro evaluation for sun protection factor of *Moringa oleifera* Lam (family – moringaceae) oil sunscreen cream. International Journal of Pharmacy and Pharmaceutical Sciences, 3(4): 371–375.

Kalibbala, H.M., Wahlberg, O. and Hawumba, T.J. 2009. The impact of *Moringa oleifera* as a coagulant aid on the removal of trihalomethane (THM) precursors and iron from drinking water. Water Science & Technology: Water Supply, 9(6): 707–714.

Kallamadi, P.R., Nadigatla, V.P.R.G.R. and Mulpuri, S. 2015. Molecular diversity in castor (*Ricinus communis* L.). Industrial Crops and Products, 66: 271–281.

Kamenetsky, R., Faigenboim, A., Mayer, E.S., Michael, T.B., Gershberg, C., Kimhi, S., Esquira, I., Shalom, S.R., Eshel, D., Rabinowitch, H.D. and Sherman, A. 2015. Integrated transcriptome catalogue and organ-specific profiling of gene expression in fertile garlic (*Allium sativum* L.). BMC Genomics, 16: 12.

Kammerbauer, J. and Dick, T. 2000. Monitoring of urban traffic emissions using some physiological indicators in *Ricinus communis* L. plants. Arch. Environ. Contam. Toxicol., 39: 161–166.

Kandel, R., Yang, X., Song, J. and Wang, J. 2018. Potentials, challenges, and genetic and genomic resources for sugarcane biomass improvement. Frontiers in Plant Science, 9: 151.

Kang, W., Bao, J., Zheng, J., Hu, H. and Du, J. 2015. Distribution and chemical forms of copper in the root cells of castor seedlings and their tolerance to copper phytotoxicity in hydroponic culture. Environ. Sci. Pollut., R22(10): 7726–7734.

Kanse, N.G., Chirag, S., Swapnil, S. and Suryawanshi Visha, S. 2017. Extraction of pectin from orange peel's and its applications: Review. International Journal of Innovative Research in Science, Engineering and Technology, 6(9): 19452-19457.

Kanti, M., Anjani, K., Usha Kiran, B. and Vivekananda, K. 2015. Agro-morphological and molecular diversity in castor (*Ricinus communis* L.) germplasm collected from Andaman and Nicobar Islands, India. Czech Journal of Genetics and Plant Breeding, 51: 96–109.

Kantola, I.S., Masters, M.D., Beerling, D.J., Long, S.P. and DeLucia, E.H. 2017. Potential of global croplands and bioenergy crops for climate change mitigation through deployment for enhanced weathering. Biology Letters, 13(4): 18.

Karaosmanoğlu, F., Işığıgür-Ergüdenler, A. and Aydın Sever, A. 2000. Biochar from the straw-stalk of rapeseed plant. Energy Fuels, 14(2): 336–339.

Karen Guo, K. 2004. Biopharming: Unique challenges & policy proposals. Availabe from https://dash.harvard.edu/bitstream/handle/1/8965608/Guo.html?sequence=2. Accessed 12/11/2019.

Karimi, S., Abbaspour, H., Sinaki, J.M. and Makarian, H. 2012. Evaluation of drought stress and foliar chitosan on biochemical characteristics of castor bean (*Ricinus communis* L.). Research Journal of Biological Sciences, 7(3): 117–122.

Katiyar, C., Gupta, A., Kanjilal, S. and Katiyar, S. 2015. Drug discovery from plant sources: An integrated approach, 23(1): 10–18.

Katwal, R.P.S. 2003. Teak in India: Status, Prospects and Perspectives. International Conference Quality Timber Products of Teak from Sustainable Forest Management, 2-5 December 2003. Peechi, Kerala, India.

Kavera, H.L. Nadaf. 2017. Genetic improvement for yield through induced mutagenesis in groundnut (*Arachis hypogaea* L.). Legume Research, 40(1): 32–35.

Kemppainen, K., Siika-aho, M., Pattathil, S., Giovando, S. and Kruus, K. 2014. Spruce bark as an industrial source of condensed tannins and non-cellulosic sugars. Industrial Crops and Products, 52: 158–168.

Keshwani, D.R. and Cheng, J.J. 2009. Switchgrass for bioethanol and other value-added applications: A review. Bioresource Technology, 100: 1515–1523.

Khader, M. and Eckl, P.M. 2014. Thymoquinone: An emerging natural drug with a wide range of medical applications. Iran Journal of Basic Medical Sciences, 17: 950–957.

Khairi, M.M.A. 2019. Genetics and breeding of jojoba (*Simmondsia chinensis* (Link) Schneider). *In:* Al-Khayri, J., Jain, S., Johnson, D. (eds.), Advances in Plant Breeding Strategies: Industrial and Food Crops. Springer, Cham. https://doi.org/10.1007/978-3-030-23265-8_8

Khaled, S. and Abu Darda. 2016. Promoting sustainable consumption & production of jute diversified products. Final Evaluation Report – SWITCH ASIA I Jute Value Chain Project Implemented by Care Bangladesh. Dhaka - 1212, Bangladesh. p. 44.

Khan, E.U. and Liu, J.H. 2009. Plant biotechnological approaches for the production and commercialization of transgenic crops. Biotechnology & Biotechnological Equipment, 23(3): 1281–1288.

Khan, D., Zulfiqar, A.S. and Zak, M.J. 2014. Axial dimensions of seeds, within-ragma allocation of phytomass and seed packaging cost in a wild castor bean, *Ricinus communis* L. (Euphorbiaceae). Int. J. Biol. Res., 2(1): 13–22.

Khatun, M.M., Tanny, T., Yesmin, S., Salimullahb, M. and Alam, I. 2018. Evaluation of genetic fidelity of in vitro-propagated Aloe vera plants using DNA-based markers. Science Asia, 44: 87–91.

Khedkar, M.A., Nimbalkar, P.R., Chavan, P.V., Chendake, Y.J. and Bankar, S.B. 2017. Cauliflower waste utilization for sustainable biobutanol production: Revelation of drying kinetics and bioprocess development. Bioprocess and Biosystems Engineering, 40(10): 1493–1506.

Khojely, D.M., Seifeldin Elrayah Ibrahim, S.E., Sapey, E. and Tianfu Han, T. 2018. History, current status, and prospects of soybean production and research in sub-Saharan Africa. The Crop Journal, 226–235.

Khor, K.Z., Lim, V., Moses, E.J. and Abdul Samad, N. 2018. The in vitro and in vivo anticancer properties of *Moringa oleifera*. Evidence-based Complementary and Alternative Medicine, 2018: 14.

Khoshraftar, Z., Safekordi, A.K., Shamel, A. and Zaefizadeh, M. 2019. Synthesis of natural nanopesticides with the origin of Eucalyptus globulus extract for pest control. Green Chemistry Letters and Reviews, 12(3): 286–298.

Kibria, A.A., Kamrunnessa, Rahman, M.M. and Kar, A. 2019. Extraction and evaluation of phytochemicals from banana peels (*Musa sapientum*) and banana plants (*Musa paradisiaca*). Malaysian Journal of Halal Research Journal, 2(1): 22–26.

Kim, H.M., In Choi, I.S., Lee, S., Hwang, I.M., Chun, H.H., Wi, S.G., Kim, J., Shin, T.Y., Kim, J.C., Kim, J.S., Kim, J. and Park, H.W. 2019. Advanced strategy to produce insecticidal destruxins from lignocellulosic biomass Miscanthus. Biotechnology for Biofuels, 12: 188.

King, A.J., He, W., Cuevas, J.A., Freudenberger, Ramiaramanana, D. and Graham, I.A. 2009. Potential of *Jatropha curcas* as a source of renewable oil and animal feed: Review paper. Journal of Experimental Botany, 60(10): 2897–2905.

Kishwar, F., Mahmood, I., Mahmood, T. and Qamar-ul-Haq. 2013. Thymol, an active constituent of *Nigella sativa*, could reduce toxicity of some trace metals (Fe (III), Cr(VI), Cu(II), V(IV) and Co(II)). European Academic Research, 1(6): 1110–1125.

Kochhar, S.L. 1998. Economic Botany in the Tropics, 2nd edition. Macmillian India Ltd, Delhi, India.

Kodjo, T.A., Gbénonchi, M., Sadate, A., Komi, A., Yaovi, G., Dieudonné, M. and Komla, S. 2011. Bio-insecticidal effects of plant extracts and oil emulsions of *Ricinus communis* L. (Malpighiales: Euphorbiaceae) on the diamondback, *Plutella xylostella* L. (Lepidoptera: Plutellidae) under laboratory and semi-field conditions. Journal of Applied Biosciences, 43: 2899–2914.

Koelmel, J., Prasad, M.N.V., Velvizhi, G., Butti, S.K. and Mohan, S.V. 2016. Metalliferous waste in India and knowledge explosion in metal recovery techniques and process for the prevention of pollution. pp. 341–392. *In:* Prasad, M.N.V. and Kaimin Shih (eds.), Environmental Materials and Waste Resource Recovery and Pollution Prevention. Academic Press. Elsevier. ISBN 9780128038376.

Koelmel, J., Prasad, M.N.V. and Pershell, K. 2015. Bibliometric analysis of phytotechnologies for remediation: Global Scenario of Research and Applications. Int. J. Phytoremediation, 17: 145–153.

Kolchinsky, P. 2004. The Entrepreneur's Guide to a Biotech Startup. Available from https://www.ctsi.ucla.edu/researcher-resources/files/view/docs/EGBS4_Kolchinsky.pdf Accessed 14/9/2017

Kołodziej, J., Pudełko, K. and Mańkowski, J. 2023. Energy and biomass yield of industrial hemp (*Cannabis sativa* L.) as influenced by seeding rate and harvest time in Polish agro-climatic conditions. Journal of Natural Fibers, 20(1): 2159609.

Konzen, E.R., Perón, R., Ito, M.A., Brondani, G.E. and Tsai, S.M. 2017. Molecular identification of bamboo genera and species based on RAPD-RFLP markers. Silva Fennica, 51(4): 16. Article id 1691.

Kovak, E., Blaustein-Rejto, D. and Qaim, M. 2022. Genetically modified crops support climate change mitigation. Trends in Plant Science, 27(7): 627–629.

Krenzelok, E.P. 2009. The "castor bean" plant much maligned, but beautiful! Clin. Toxicol., 47.

Kreps, F., Burčová, Z., Jablonský, M., Ház·, A., Frecer, V., Kyselka, J., Schmidt, S., Šurina, I. and Vladimír Filip, V. 2017. Bioresource of antioxidant and potential medicinal compounds from waste biomass of spruce. ACS Sustainable Chemical Engineering, 5(9): 8161–8170.

Krishna, H. and Singh, S.K. 2007. Biotechnological advances in mango (*Mangifera indica* L.) and their future implication in crop improvement – A review. Biotechnology Advances, 25(3): 223–243.

Krishna, M.S. and Jayakumaran, N.A. 2010. Antibacterial, cytotoxic and antioxidant potential of different extracts from leaf, bark and wood of *Tectona grandis*. International Journal of Pharmaceutical Sciences and Drug Research, 2(2): 155–158.

Krishna, M.S. and Jayakumaran, N.A. 2011. Anthraquinones from leaves of *Tectona grandis*: A detailed study on its antibacterial activity and other biological properties. International Journal of Phytomedicine, 3: 50–58.

Kuhn, D.N., Livingstone III, D.S., Richardsc, J.H., Manosalva, P., Van den Berg, N. and Chambers, A.H. 2019. Application of genomic tools to avocado (*Persea americana*) breeding: SNP discovery for genotyping and germplasm characterization. Scientia Horticulturae, 246: 1–11.

Kumar, A., Tyagi, V., Rathi, B. and Priyanka, Manisha. 2017. Chronological review on phytochemical, antioxidant, antimicrobial and clinical studies on biodiesel yielding good luck tree (*Thevetia peruviana*). International Journal of Pure and Applied Bioscience, 5(6): 1499–1514.

Kumar, C., Shukla, S.S. and Pandey, R.K. 2017. A review on *Thevetia peruviana*. Research Journal Pharmacology & Pharmacodynamics, 9(2): 93–96.

Kumar, M., Singh, S. and Singh, S. 2011. *In vitro* morphogenesis of a medicinal plant – *Aloe Vera* L. Asian Journal of Plant Science and Research, 1(1): 31–40.

Kumar, R., Naik, P.K., Kumar, A., Aggarawal, H., Kumar, A. and Chhokar, V. 2016. A combined approach using RAPD, ISSR and bioactive compound for the assessment of genetic diversiy of *Aloe vera* (L.) Burm. f. Indian Journal of Biotechnology, 15: 538–545.

Kumar, S., Malhotra, R. and Kumar, D. 2010. *Euphorbia hirta*: Its chemistry, traditional and medicinal uses, and pharmacological activities. Pharmacognosy Review, 4(7): 58–61.

Kumar, S., Singh, R., Kumar, V., Rani, A. and Jain, R. 2017. *Cannabis sativa*: A plant suitable for phytoremediation and bioenergy production. pp. 269–285. *In:* Kuldeep Bauddh, K., Singh, B., John Korstad, J. (eds.). Phytoremediation Potential of Bioenergy Plants. Springer Nature Singapore Pte Ltd. 2017.

Lachapelle, E. and Montpetit, E. 2015. Putting phytoremediation into action. Scientist, 29: 18–19.

Ladda, P.L. and Kamthane, R.B. 2014. *Ricinus communis* (Castor): An overview. Int. J. of Res. in Pharmacology & Pharmaco-therapeutics, 3(2): 136–144.

Lagunes-Fortiz, E., Robledo-Paz, A.M., Gutiérrez-Espinosa, M.A., Mascorro-Gallardo, J.O. and Espitia-Rangel, E. 2013. Genetic transformation of garlic (*Allium sativum* L.) with tobacco chitinase and glucanase genes for tolerance to the fungus *Sclerotium cepivorum*. African Journal of Biotechnology, 12(22): 3482–3492.

Lakhani, H.N., Patel, S.V., Bodar, N.P. and Golakiya, B.A. 2015. RAPD analysis of genetic diversity of castor bean (*Ricinus communis* L.). International Journal of Current Microbiology and Applied Sciences, 4(1): 696–703.

Laliberté, B., Cryer, N.C., Daymond, A.J., End, M.J., Engels, J., Eskes, B., Gilmour, M., Lachenaud, P., Phillips-Mora, W., Turnbull, C.J., Umaharan, P., Zhang, D. and Weise, S. 2012. A Global Strategy for the Conservation and Use of Cacao Genetic Resources, as the Foundation for a Sustainable Cocoa Economy. 17th International Cocoa Research Conference, Yaoundé, Cameroon, 15-20 October 2012, pp. 1–15.

Lamien, B.N., Nygard, J.I., Ouacdraogo, R., Odacn, J.S. and Guinko, P.C. 2006. Mistletoe impact on Shea tree (*Vitellaria paradoxa* C.F. Gaertn.) flowering and fruiting behaviour in savanna area from Burkina Faso. Environmental and Experimental Botany, 55(1–2): 142–148.

Lanaud, C., Risterucci, A.M., Pieretti, I., N'Goran, J.A.K. and Fargeas, D. 2004. Characterisation and genetic mapping of resistance and defence gene analogs in cocoa (*Theobroma cacao* L.). Molecular Breeding, 13(3): 211–227.

Lanjhiyana, S., Garabadu, D., Ahirwar, D., Bigoniya, P., Rana, A.C., Patra, K.C., Lanjhiyana, S.K. and Karuppaih, M. 2011. Antihyperglycemic potential of Aloe vera gel in experimental animal model. Annals of Biological Research, 2(1): 17–31.

Larsen, A.S., Vaillant, A., Verhaegen, D. and Kjær, E.D. 2009. Eighteen SSR primers for tetraploid Adansonia digitata and its relatives. Conservation Genetics Resources, Springer, 1: 325–328.

Laurentin, H., Ratzinger, A. and Karlovsky, P. 2008. Relationship between metabolic and genomic diversity in sesame (*Sesamum indicum* L.). BMC Genomics, 9: 1-11.

Laurentin, H.E.T. 2007. Genetic diversity in sesame (*Sesamum indicum* L.): Molecular markers, metabolic profiles and effect of plant extracts on soil-borne pathogenic fungi. Ph.D. thesis. Faculty of Agricultural Sciences, Georg-August-University Göttingen, Germany.

Lavanya, C., Murthy, I.Y.L.N., Nagaraj, G. and Mukta, N. 2012. Prospects of castor (*Ricinus communis* L.) genotypes for biodiesel production in India. Biomass Bioenergy, 39: 204–209.

Le, V.Q., Belles-Isles, J., Dusabenyagasani, M. and Tremblay, F.M. 2001. An improved procedure for production of white spruce (*Picea glauca*) transgenic plants using *Agrobacterium tumefaciens*. Journal of Experimental Botany, 52(364): 2089–2095.

Lea, M. 2014. Bioremediation of turbid surface water using seed extract from the *Moringa oleifera* Lam. (Drumstick) tree. Current Protocols in Microbiology, 1G.2.1–1G.2.

Lehtomäki, A. 2006. Biogas Production from Energy Crops and Crop Residues. Academic dissertation. Faculty of Mathematics and Science of the University of Jyvaskyla.

Lenkala, P.K., Rani, R.R., Sivaraj, N.K., Reddy, K.R. and Pradav, M.J. 2015. Genetic variability and character association studies in Jack bean (*Canavalia ensiformis* (L.) DC.) for yield and yield contributing traits. Electronic Journal of Plant Breeding, 6(2): 625–629.

Lewandowski, I., Clifton-Brown, J., Trindade, L.M., van der Linden, G.C., Schwarz, K-U., Müller-Sämann, K., Anisimov, A., Chen, C.-L., Dolstra, O., Donnison, I.S., Farrar, K., Fonteyne, S., Harding, G., Hastings, A., Huxley, L.M., Iqbal, Y., Khokhlov, N., Kiesel, A., Lootens, P., Meyer, H., Mos, M., Muylle, H., Nunn, C., Özgüven, M., Roldán-Ruiz, I., Schüle, H., Tarakanov, I., van der Weijde, T., Wagner, M., Xi, Q. and Kalinina, O. 2016. Progress on optimizing miscanthus biomass production for the European bioeconomy: Results of the EU FP7 Project OPTIMISC. Frontiers in Plant Science, 7: 16-20.

Li, G., Wan, S.W., Zhou, J., Yang, Z.Y. and Qin, P. 2010. Leaf chlorophyll fluorescence, hyperspectral reflectance, pigments content, malondialdehyde and proline accumulation responses of castor bean (*Ricinus communis* L.) seedlings to salt stress levels. Ind. Crop Prod., 31(1): 13–19.

Li, G., Zhang, H., Wu, X., Shi, C., Huang, X. and Qin, P. 2011. Canopy reflectance in two castor bean varieties (*Ricinus communis* L.) for growth assessment and yield prediction on coastal saline land of Yancheng District, China. Ind. Crops Prod., 33: 395–402.

Li, J., Lin, H., Bean, S.R., Sun, X.S. and Wang, D. 2020. Evaluation of adhesive performance of a mixture of soy, sorghum and canola proteins. Industrial Crops and Products, 157: 112898.

Liauw, M.Y., Natan, F.A., Widiyanti, P., Ikasari, D., Indraswati, N. and Soetaredjo, F.E. 2008. Extraction of neem oil (*Azadiracta indica* A. Juss) using n-Hexane and ethanol: Studies of oil quality kinetic and thermodynamic. ARPN J. Eng. Appl. Sci., 3(3): 49–54.

Life Science Austria. 2017. The international Biotech & Medtech Business Plan Handbook. Austria Wirtschaftsservice Gesellschaft mbH.

Lima Neto, M.C., Martins, M.O., Ferreira, Silva S.L. and Silveira, J.A.G. 2015. *Jatropha curcas* and *Ricinus communis* display contrasting photosynthetic mechanisms in response to environmental conditions. Sci. Agric., 72(3): 260–269.

Lima, J.R., Araújo, I.M.S., Pinto, C.O., Goiana, M.L., Rodrigues, M.C.P. and Lima, L.V. 2021. Obtaining cashew kernel protein concentrate from nut processing by-product and its use to formulate vegetal burger. Brazilian Journal of Food Technology, 24: 9, e2020232.

Lima, R.L.S., Severino, L.S., Sampaio, L.R., Sofiatti, V., Gomes, J.A. and Beltrão, N.E.M. 2011. Blends of castor meal and castor husks for optimized use as organic fertilizer. Industrial Crops and Products, 33: 364–368.

Lin, B.B., Burgess, A.J. and Murchie, E.H. 2015. Tree–Crop Interactions: Agroforestry in a Changing Climate Adaptation for Climate-sensitive Crops Using Agroforestry: Case Studies for Coffee and Rice. 2nd Edition. CAB International, UK. pp. 278–308.

Lin, C. and Eudes, A. 2020. Strategies for the production of biochemicals in bioenergy crops. Biotechnology for Biofuels, 13(71): 25.

Ling, C. 2020. Traditional Chinese medicine is a resource for drug discovery against 2019 novel coronavirus (SARS-CoV-2). Journal of Integrative Medicine. In press.

Lins, A.B., Lawinscky, P.R., Barbosa, A.M.M., Gaiotto, F.A. and Corrêa, R.X. 2016. Molecular genetic diversity in a core of cocoa (*Theobroma cacao* L.) clones with potential for selection of disease resistance, plant height and fruit production. African Journal of Biotechnology, 15(44): 2517–2523.

Liu, H., Wu, G.G., Wang, J.B., Wu, X., Bai, L., Jiang, W., Lv, B.B., Pan, A.H., Jia, J.W., Li, P. and Zhao, K. 2016. Characterization and comparison of transgenic *Artemisia annua* GYR and wild-type NON-GYR plants in an environmental release trial. Genetics and Molecular Research, 15(3): 1–17.

Liu, S., Zhu, Q., Gua, Q., He, L. and Li, W. 2015. Bioaviation fuel production from hydroprocessing castor oil promoted by the nickel-based bifunctional catalysts. Bioresource Technology, 183: 93–100.

Liu, X., Su, H., Pang, Y., Yang, D., Jiang, Y., Mao, A., Yuan, Y. and Xu, W. 2019. Synthesis and properties of bio-based adhesives derived from plant oil residues. American Journal of Modern Energy, 5(6): 94–99.

Liu, Z., Guo, S., Xu, J., Zhang, Y., Dong, L., Xiao, S., Bai, R., Liao, B., Su, H.E., Cheng, R. and Chen, S. 2018. Genome size estimation of Chinese cultured *Artemisia annua* L. Journal of Plant Biology and Crop Research, 1: 1002.

Llorach, R., Espín, J.C., Tomás-Barberán, F.A. and Ferreres, F. 2003. Valorization of cauliflower (*Brassica oleracea* L. var. *botrytis*) by-products as a source of antioxidant phenolics. Journal of Agriculture and Food Chemistry, 51(8): 2181–2187.

Locatelli, G.O., Finkler, L. and Finkler, C.L.L. 2019. Orange and passion fruit wastes characterization, substrate hydrolysis and cell growth of *Cupriavidus necator*, as proposal to converting of residues in high value added product. Anais da Academia Brasileira de Ciências, 91(1): e20180058.

Lohaus, R.H. 2019. *Camelina sativa*: A promising oilseed for producing biofuels on marginal lands: Field production and characterization of a low-pectin seed mutant. Ph.D thesis. University of Nevada, Reno.

Loss-Morais, G., Turchetto-Zolet, A.C., Etges, M., Cagliari, A., Körbes, A.P., Maraschin, F.D.S., Margis-Pinheiro, M. and Margis, R. 2013. Analysis of castor bean ribosome-inactivating proteins and their gene expression during seed development. Genetics and Molecular Biology, 36(1): 74–86.

Lu, C. and Kang, J. 2008. Generation of transgenic plants of a potential oilseed crop *Camelina sativa* by *Agrobacterium*-mediated transformation. Plant Cell Reports. 27: 73–78.

Lu, X.Y. and He, C.Q. 2005. Tolerance, uptake and accumulation of cadmium by *Ricinus communis* L. J. Agro-Environ. Sci., 24: 674–677.

Ludvíková, M. and Griga, M. 2019. Chapter 16 – Transgenic fiber crops for phytoremediation of metals and metalloids. pp. 341–358. *In:* Majeti Narasimha Vara Prasad (ed.), Transgenic Plant Technology for Remediation of Toxic Metals and Metalloids. Academic Press, 2019.

Luo, L., Zhang, D., Luo, H., Liu, B., Zhao, K., Zhao, Y., Bian, Y. and Wang, Y. 2020. Treatment efficacy analysis of traditional Chinese medicine for novel coronavirus pneumonia (COVID-19): An empirical study from Wuhan, Hubei Province, China. Chinese Medicine, 15: 34.

Luo, Z., Brock, J., Dyer, J.M., Kutchan, T., Schachtman, D., Augustin, M., Ge, Y., Fahlgren, N. and Abdel-Haleem, H. 2019. Genetic diversity and population structure of a *Camelina sativa* Spring Panel. Frontiers of Plant Science, 10: 184.

Lv, J., Qi, J., Shi, Q., Shen, D., Zhang, S., Shao, G., Li, H., Sun, Z., Weng, Y., Shang, Y. and Gu, X. 2012. Genetic diversity and population structure of cucumber (*Cucumis sativus* L.). PLoS ONE, 7(10): 1–9.

Lyu, J., Ramekar, R., Kim, D., Kim, J.M., Lee, M., Hung, N.N., Kim, J., Ahn, J., Si-Yong Kang, S., Choi, I., Park, K. and Soon-Jae Kwon, S. 2020. Characterization of gene isoforms related to cellulose and lignin biosynthesis in kenaf (*Hibiscus cannabinus* L.) mutant. Plants, 9: 631.

Ma, X., Tang, J., Li, C., Liu, Q., Chen, J., Li, H., Guo, L. and Xie, J. 2014. Identification and quantification of ricin in biomedical samples by magnetic immunocapture enrichment and liquid chromatography electrospray ionization tandem mass spectrometry. Anal Bioanal Chem., 406: 5147–5155.

Ma, Y., Rajkumar, M., Vicente, J.A.F. and Freitas, H. 2010. Inoculation of Ni-resistant plant growth promoting bacterium *Psychrobacter* sp. strain *SRS8* for the improvement of nickel phytoextraction by energy crops. Int. J. Phytoremed., 13: 126–139.

Ma, Y., Prasad, M.N.V., Rajkumar, M. and Freitas, H. 2011. Plant growth promoting rhizobacteria and endophytes accelerate phytoremediation of metalliferous soils. Biotechnol. Adv., 29: 248–258.

Ma, Y., Rajkumar, M., Luo, Y. and Freitas, H. 2011b. Inoculation of endophytic bacteria on host and non-host plants – Effects on plant growth and Ni uptake. J. Hazard. Mater., 196: 230–237.

Ma, Y., Rajkumar, M., Rocha, I., Oliveira, R.S. and Freitas, H. 2015. Serpentine bacteria influence metal translocation and bioconcentration of *Brassica juncea* and *Ricinus communis* grown in multi-metal polluted soils. Frontiers in Plant Science, 5. Article 757, doi:10.3389/ fpls. 2014.00757.

Ma, Y., Rajkumar, M., Zhang, C. and Freitas, H. 2016. Beneficial role of bacterial endophytes in heavy metal phytoremediation. J. Environmental Management, 174: 14–25.

Ma, Y., Yin, Z. and Ye, J. 2017. Lipid biosynthesis and regulation in Jatropha, an emerging model for woody energy plants. pp. 113–127. *In:* S. Tsuchimoto (ed.), The Jatropha Genome. Springer.

Macaulay, J. 2003. Biopharming: Growing medicine crops. Food Technology, 57(9).

Macedo, C.L.O., Cerqueira, D., Souza, A.J., Bispo, A.S.R., Coelho, R.R.R. and Nascimento, R.P. 2013. Production of cellulose-degrading enzyme on sisal and other agro-industrial residues using a new brazilian actinobacteria strain *Streptomyces* sp. SLBA-08 E.P. Brazilian Journal of Chemical Engineering, 30(4): 729–735.

Machida-Hirano, R. 2015. Diversity of potato genetic resources. Breeding Science, 65: 26–40.

Mackler, A.M., Heber, D. and Cooper, E.L. 2013. Pomegranate: Its health and biomedical potential. Evidence-based Complementary and Alternative Medicine, 2013: 2.

Madeira Jr., J.V., Macedo, J.A. and Macedo, G.A. 2011. Detoxification of castor bean residues and the simultaneous production of tannase and phytase by solid-state fermentation using Paecilomyces variotii. Bioresource Technology, 102: 7343–7348. Springer International Publishing AG. Cham, Switzerland.

Maduka, N. and Ire, F.S. 2018. Tigernut plant and useful application of tigernut tubers (*Cyperus esculentus*) – A review. Current Journal of Applied Science and Technology, 29(3): 1-23.

Maghuly, F., Vollmann, J. and Laimer, M. 2015. Biotechnology of Euphorbiaceae (*Jatropha curcas, Manihot esculenta, Ricinus communis*). Applied Plant Genomics and Biotechnology, 6: 87–114.

Magriotis, Z.M., Carvalho, M.Z., de Sales, P.F., Alves, F.C., Resende, R.F. and Saczk, A.A. 2014. Castor bean (*Ricinus communis* L.) press cake from Biodiesel production: An efficient low cost adsorbent for removal of textile dyes. Journal of Environmetal and Chemical Engineering, 2(3): 1731–1740.

Mahmood, K.T., Mugal, T. and Ul Haq, I. 2010. *Moringa oleifera*: A natural gift – A review. Journal of Pharmaceutical Sciences and Research, 2(11): 775–781.

Mahmoud, D.A.R. 2016. Utilization of palm wastes for production of invert sugar. Procedia Environmental Sciences, 34: 104–118.

Mahmud, R., Inoue, N., Kasjima, S. and Shahenn, R. 2008. Assessment of potential indigenous plant species for the phytoremediation of arsenic-contaminated areas of Bangladesh. Int. J. Phytoremed., 10: 119–132.

Mahor, G. and Ali, S.A. 2016. Recent update on the medicinal properties and use of Aloe vera in the treatment of various ailments. Bioscience and Biotechnological Research Communication, 9(2): 273–288.

Maideen, N.M.P. 2020. Prophetic medicine – *Nigella sativa* (black cumin seeds) – Potential herb for COVID-19. Journal of Pharmacopuncture, 23(2): 62–67.

Majeed, M., Satyan, K.S. and Prakash, L. 2007. Neem oil limonoids: Product overview. Sabinsa Corporation. pp. 1–8.

Majeed, A., Tayyab Husnain and Riazuddin, S. 2000. Transformation of virus – Resistant genotype of *Gossypium hirsutum* L. with pesticidal gene. Plant Biotechnology, 7(2): 105–110.

Majumder, S., Saha, P., Datta, K., Swapan, K. and Datta, S.K. 2020. Fiber crop, jute improvement by using genomics and genetic engineering. Advancement in Crop Improvement Techniques, 363-383.

Majumder, S., Datta, K., Sarkar, C., Saha, S.C. and Datta, S.K. 2018. The development of *Macrophomina phaseolina* (fungus) resistant and *Glufosinate* (herbicide) tolerant transgenic jute. Frontiers in Plant Science, 9: 92.

Mak-Mensah, E.E. and Firempong, C.K. 2011. Chemical characteristics of toilet soap prepared from neem (*Azadirachta indica* A. Juss) seed oil. Asian J. Plant Sci. Res., 1(4): 1–7.

Makeswari, M. and Santhi, T. 2013. Tannin gel derived from leaves of *Ricinus communis* as an adsorbent for the removal of Cu (II) and Ni (II) ions from aqueous solution. International Journal of Modern Engineering Research, 3(5): 3255–3266.

Makeswari, M. and Santhi, T. 2013. Removal of malachite green dye from aqueous solutions onto microwave assisted zinc chloride chemical activated epicarp of *Ricinus communis.* Journal of Water Resource and Protection, 5(2): 17. Article ID: 28297, DOI:10.4236/ jwarp.2013.52023

Makeswari, M. and Santhi, T. 2014. Adsorption of Cr (VI) from aqueous solutions by using activated carbons prepared from *Ricinus communis* leaves: Binary and ternary systems. Arabian Journal of Chemistry, 57–69.

Makkar, H., Kumar, V., Oyeleye, O., Akinyele, A.O., Agulo-Escalanta, M.A. and Becker, K. 2011. *Jatropha platyphylla*, a new non-toxic Jatropha species: Physical properties and chemical constituents including toxic and antinutritional factors of seeds. Food Chemistry, 125(1): 63–71.

Maksoud, M.A., El-Shamma, M.S., Saleh, M.A., Zaied, N.S. and Hafez, O.M. 2015. Industrial orange wastes as organic amendments in citrus orchards. American-Eurasian J. Agricultural and Environmental Science, 15(6): 1163–1166.

Malarkodi, M., Krishnaswamy, R. and Chitdeswari, T. 2008. Phytoextraction of nickel contaminated soil using castor Phytoextractor. Journal of Plant Nutrition, 31(2): 219–229.

Malaviya, A., Malhotra, K., Agarwal, A. and Katherine Saikia, K. 2019. Artemisinin: A potent antimalarial drug. pp. 347–364. *In:* Saurabh Saran, Vikash Babu and Asha Chaubey (eds.), High Value Fermentation Products, Volume 1. Scrivener Publishing LLC.

Maldonado-Celis, M.E., Yahia, E.M., Bedoya, R., Landázuri, P., Loango, N., Aguillón, J., Restrepo, B. and Ospina, J.C.G. 2019. Chemical composition of mango (*Mangifera indica* L.) fruit: Nutritional and phytochemical compounds. Frontiers in Plant Science, 10: 1073.

Mallek-Ayadi., S., Bahloul, N. and Kechaou, N. 2018. Chemical composition and bioactive compounds of *Cucumis melo* L. seeds: Potential source for new trends of plant oils. Process Safety and Environmental Protection, 1(13): 68–77.

Mandal, S. 2011. Repellent activity of *Eucalyptus* and *Azadirachta indica* seed oil against the filarial mosquito *Culex quinque fasciatus* Say (Diptera: Culicidae) in India. Asian Pacific Journal of Tropical Biomedicine, 109-112.

Maranz, S., Wiesman, Z. and Garti, N. 2003. Phenolic constituents of Shea (*Vitellaria paradoxa*) kernels. Journal of Agriculture and Food Chemistry, 51: 6268–6273.

Marble, S.C., Prior, S.A., Runion, G.B., Torbert, H.A., Gilliam, C.H. and Fain, G.B. 2011. The importance of determining carbon sequestration and greenhouse gas mitigation potential in ornamental horticulture. HortScience, 46(2): 240–244.

María, J.G.M. and Francisco, P.G. 2012. Obtaining and characterization of biodiesel from castor oil (*Ricinus communis*) and sunflower (*Helianthus annuus*) grown in Tabasco, Mexico. International Journal of Applied Science and Technology, 2(9): 58–74.

Marimuthu, M. and Gurumoorthi, P. 2013. Physicochemical and functional properties of starches from Indian Jack bean (*Canavalia ensiformis*), an underutilized wild food legume. Journal of Chemical and Pharmaceutical Research, 5(1): 221–225.

Marin, C.P., Kaschuk, J.J., Frollini, E. and Nitschke, M. 2015. Potential use of the liquor from sisal pulp hydrolysis as substrate for surfactin production. Industrial Crops and Product, 66: 239–245.

Maroto, F.G. and Alonso, D.L. 2018. Oil biosynthesis and biotechnology in the castor bean. pp. 197–213. *In:* C. Kole and P. Rabinowicz (eds.), The Castor Bean Genome, Compendium of Plant Genomes. Springer Nature Switzerland AG.

Marpudi, S.L., Abirani, L.S.S., Puskala, R. and Srividya, N. 2011. Enhancement of storage life and quality maintenance of papaya fruits using Aloe vera based antimicrobial coating. Indian Journal of Biotechnology, 10: 83–89.

Marques, N.N., Garcia, C.S., Madruga, L.Y.C., Villetti, M.A., Filho, M.M., Edson, N. and Ito, R.B. 2019. Turning industrial waste into a valuable bioproduct: Starch from mango kernel derivative to oil industry. J.R.M., 7(2): 139-152.

Martínez, M.I.S. 2015. Analysis of miscanthus biomass for biofuel production. MSc. Minor Thesis Report Department of Plant Breeding of Wageningen University. pp. 123.

Martinez-Feria, R.A., Basso, B. and Kim, S. 2022. Boosting climate change mitigation potential of perennial lignocellulosic crops grown on marginal lands. Environmental Research Letters, 17: 1–13.

Martins, A.E., Pereira, M.S., Jorgetto, A.O., Ma, U.M., Silva, R.I.V., Saeki, M.J. and Castor, G.R. 2013. The reactive surface of Castor leaf (*Ricinus communis* L.) powder as a green adsorbent for the removal of heavy metals from natural river water. Applied Surface Sci., 276: 24–30.

Mason, H.S. and Arntzen, C.J. 1995. Transgenic plants as vaccine production systems. Trends in Biotechnology, 13: 388–392.

Mayee, C.D., Singh, P., Dongre, A.B., Rao, M.R.K. and Sheo Raj, S. 2018. Transgenic Bt Cotton. CICR Technical Bulletin NO: 22. Central Institute for Cotton Research Nagpur. pp. 1-7. Available from www.cicr.org.in/pdf/transgenic_bt_cotton.pdf. Accessed 2/11/2018

Mayowa Akeem Azeez, M.A., Andrew, J.E. and Sitholec, B.B. 2016. A preliminary investigation of Nigerian-grown *Gmelina arborea* and *Bambusa vulgaris* for pulp and paper production. Maderas-CiencTecnol., 18(1): 1–18.

Mazarei, M., Baxter, H.L., Srivastava, A., Li, G., Xi, H., Dumitrache, A., Rodriguez, M. Jr, Natzke, J.M., Zhang, J.-Y., Turner, G.B., Sykes, R.W., Davis, M.F., Udvardi, M.K., Wang, Z.-Y., Davison, B.H., Blancaflor, E.B., Tang, Y., Stewart, C.N. Jr. 2020. Silencing Folylpolyglutamate Synthetase1 (FPGS1) in switchgrass (*Panicum virgatum* L.) improves lignocellulosic biofuel production. Frontiers in Plant Science, 11: 843.

McCalmont, J., Hastings, A., Mcnamara, N., Richter, G.M., Robson, P., Donnison, I. and Clifton-Brown, J. 2017. Environmental costs and benefits of growing miscanthus for bioenergy in the UK. GCB Bioenergy, 9(3): 489–507.

McKeon, T.A., Hayes, D.G., Hildebrand, D.F., Randall, J. and Weselake, R.J. (eds). 2016. Industrial Crops. Academic Press. 474 pp.

Medeiros, A.M.M.S., Machado, F. and Rubim, J.C. 2015. Synthesis and characterization of a magnetic bio-nanocomposite based on magnetic nanoparticles modified by acrylated fatty acids derived from castor oil. European Polymer J., 71: 152–163.

Meena, K., Anjani, K. and Venkat, R.K. 2014. Molecular diversity in castor germplasm collection originated from North-Eastern Hill Province of India. IJRSI I(VI), www.rsisinternational. org/IJRSI.html

Mehmood, F., Khan, A.U.H. and Khan, Z.U.D. 2011. Appraisal of ecological significance of *Ricinus communis* Linn. in the wasteland of Lahore, Pakistan, Biologia (Pakistan), 57(1&2): 97–103.

Mehran, M.J., Zendehbad, S.H. and Malla, S. 2014. Free radical scavenging and antioxidant potential activity of cassava plants. Asian Journal of Pharmaceutical and Clinical Research, 7(1): 66–70.

Mehta, J., Shukla, A., Bukhariya, V. and Charde, R. 2011. The magic remedy of *Moringa oliferia*: An overview. International Journal of Biomedical and Advance Research, 2: 215–227.

Meirman, G., Yerzhanova, S., Yessimbekova, M. and Masonichich-Shotunova. 2013.

Formation of the genetic resources of forage crops Kazakhstan: Luzerna and wheatgrass. Journal of Entomology and Zoology Studies, 1(4): 141–144.

Melo, E.E.C., Costa, E.T.S., Guilherme, L.R.G., Faquin, V. and Nascimento, C.W.O. 2009. Accumulation of arsenic and nutrients by castor bean plants grown on an As-enriched nutrient solution. J. Hazard. Mater., 168: 479–483.

Melo, E.E.C., Guilherme, L.R.G., Nascimento, C.W.A. and Penha, H.G.V. 2012. Availability and accumulation of Arsenic in oilseeds grown in contaminated soils. Water, Air, & Soil Pollut., 223(1): 233–240.

Memon, S., Mari, S.N., Mari, A. and Gaddi, N.H. 2010. Induction of callus through anther and ovule culture in upland cotton (*Gossypium hirsutum* L.). World Applied Sciences Journal, 8: 76–79.

Mendes, M.G., Santos, C.D. Jr., Dias, A.C.C. and Bonetti, A.M. 2009. Castor bean (*Ricinus communis* L.) as a potential environmental bioindicator. Genetics Mol. Res., 14(4): 12880–12887.

Mensah, M.B., Awudza, J.A. and O'Brien, P. 2018. Castor oil: A suitable green source of capping agent for nanoparticle syntheses and facile surface functionalization. Royal Society Open Science, 5(8): 180824.

Meshram, P.D., Gore, A.J. and Usmani, G.A. 2015. Analysis and optimization of solvent free microwave assisted extraction of bio-oil from orange peels using response surface methodology. International Journal on Recent and Innovation Trends in Computing and Communication, 3(2): 15–20.

Michigan State University. 2022. New MSU research shows how biofuel crops can help mitigate climate change when grown on land of otherwise little agricultural value. AgBioResearch. Accessed 9/7/2022 at https://www.canr.msu.edu/news/new-msu-research-shows-how-biofuel-crops-can-help-mitigate-climate-change-when-grown-on-land-of-otherwise-little-agricultural-value

Mignouna, H.D., Abang, M.M. and Asiedu, R. 2008. Genomics of yams, a common source of food and medicine in the tropics. pp. 549–570. *In:* P.H. Moore and R. Ming (eds.), Genomics of Tropical Crop Plants. Springer.

Milani, M. and Nóbrega, M.B.M. 2013. Castor breeding. *In:* S.B. Andersen (ed.), Plant Breeding from Laboratories to Fields. http://dx.doi.org/10.5772/56216.

Miri, S.M. and Roughani, A. 2018. Biotechnology in Floriculture. 2nd International & 3rd National Congress on Flower and Ornamental Plants. Mahallat, Iran.

Mishra, G., Jawla, S. and Srivastava, V. 2013. *Melia azedarach*: A review. International Journal of Medicinal Chemistry & Analysis, 3(2): 53–56.

Misra, M. and Misra, A.N. 2010. Jatropha: The biodiesel plant biology, tissue culture and genetic transformation – A review. International Journal of Pure Applied Science and Technology, 1(1): 11–24.

Mitra, S. and Tarafdar, J. 2011. Nutritional facts and value added products of yams (*Dioscorea* spp.). pp. 340-343. *In:* Ghosh, S.N. (ed.), Book chapter; Conference paper: Proceedings of the International Symposium on Minor Fruits and Medicinal Plants for Health and Ecological Security (ISMF & MP), West Bengal, India. 19-22 December, 2011-2012.

Mohammadi, A., Mohammadi, N., Alijani, A.M. and Peykarestan, B. 2014. Evaluation of allelopathic potential of two plant species, caster (*Ricinus communis*) and neem (*Azadirachta indica*. A. Juss) against seed germination and seedling growth of lentil (*Lens culinaris* Medik). International Journal of Agriculture and Crop Sciences, 4(8): 54–57.

Mohammad, B.T., Al-Shannag, M., Alnaief, M., Singh, L., Singsaas, E. and Alkasrawi, M. 2018. Production of multiple biofuels from whole camelina material: A renewable energy crop. BioRes., 13(3): 4870-4883.

Mohammed, H.T., Lakhmiri., R., Azmani, A. and Hassan, I.I. 2014. Bio-oil from pyrolysis of castor shell. International Journal of Basic & Applied Sciences, 14(6): 1–5.

Mohammed, N. and Abdullah, M. 2015. Comparative studies and optimization of biodiesel production from oils of selected seeds of Nigerian origin. World Academy of Science, Engineering and Technology, International Science Index, Energy and Power Engineering, 3(5): 538.

Mohan, S.V., Modestra, J.A., Amulya, K., Butti, S.K. and Velvizhi, G. 2016. A circular bioeconomy with bio-based products from CO_2 sequestration. Trends in Biotechnology, 34: 506–519.

Mohan, S.V., Nikhil, G.N., Chiranjeevi, P., Reddy, N.C., Rohit, M.V., Kumar, A.N. and Sarkar, O. 2016. Waste biorefinery models towards sustainable circular bioeconomy: Critical review and future perspectives. Bioresource Technology, 215: 2–12.

Moncada, J., Cardona, C.A. and Rincon, L.E. 2015. Design and analysis of a second and third generation biorefinery: The case of castor bean and microalgae. Bioresour. Technol., 198: 836–843.

Mongkholkhajornsilp, D., Douglas, S., Douglas, P.L., Elkamel, A., Teppaitoon, W. and Pongamphai, S. 2005. Supercritical CO_2 extraction of nimbin from neem seeds – A modelling study. J. Food Eng., 71: 331–340.

Moniruzzaman, M., Yaakob, Z. and Khatun, R. 2016. Biotechnology for Jatropha improvement: A worthy exploration. Renewable and Sustainable Energy Reviews, 54: 1262–1277.

Montpetit, E. and Lachapelle, E. 2017. New environmental technology uptake and bias toward the *status quo*: The case of phytoremediaiton. Environmental Tech & Innovation, 7: 102–109.

Moon, K., Park, J., Park, Y., Song, I., Lee, H., Cho, H.S., Jeon, J. and Hyun-Soon Kim, H. 2020. Development of systems for the production of plant-derived biopharmaceuticals. Plants, 9(30): 21.

Moosavy, M.H., Hassanzadeh, P., Mohammadzadeh, E., Mahmoudi, R., Khatibi, S.A. and Mardani, K. 2017. Antioxidant and antimicrobial activities of essential oil of lemon (Citrus limon) peel in vitro and in a food model. Journal of Food Quality and Hazards Control, 4: 42–48.

Moradi kor, N. and Moradi, K. 2013. Physiological and pharmaceutical effects of fenugreek (*Trigonella foenum-graecum* L.) as a multipurpose and valuable medicinal plant. Global Journal of Medicinal Plant Research, 1(2): 199–206.

Morgil, H., Gercek, Y.C. and Tulum, I. 2020. Single nucleotide polymorphisms (SNPs) in plant genetics and breeding. *In:* Mahmut Çalışkan, Osman Erol and Gül Cevahir Öz (eds.), The Recent Topics in Genetic Polymorphisms. IntechOpen. DOI: 10.5772/intechopen.91886. Available from: https://www.intechopen.com/chapters/71577 Accessed 3/9/2021

Morris, J.B. 2002. Food, industrial, nutraceutical, and pharmaceutical uses of sesame genetic resources. pp. 153–156. *In:* J. Janick and A. Whipkey (eds.), Trends in New Crops and New Uses. ASHS Press, Alexandria, VA.

Moschini, G. 2006. Pharmaceutical and industrial traits in genetically modified crops: Co-existence with conventional agriculture. CARD Working Papers. 451. http://lib.dr.iastate.edu/card_workingpapers/451

Msaakpa, T.S. and Obasi, M.O. 2014. Correlated studies between growth and yield characters of castor bean (*Ricinus communis* L.). International Journal of Scientific and Research Publications, 4(7).

Mtui, G.Y.S. 2011. Involvement of biotechnology in climate change adaptation and mitigation: Improving agricultural yield and food security. International Journal for Biotechnology and Molecular Biology Research, 2(13): 222–231.

Mullan, D.J. and Barrett-Lennard, E.G. 2010. Breeding crops for tolerance to salinity, waterlogging and inundation. pp. 92–114. *In:* Matthew P. Reynolds (ed.), Climate Change and Crop Production. Wallingford UK: CABI.

Mulualem, T., Mekbib, F., Hussein, S. and Gebre, E. 2018. Analysis of biochemical composition of yams (*Dioscorea* spp.) landraces from Southwest Ethiopia. Agrotechnology, 7: 177.

Mulualem, T., Mekbib, F., Shimelis, H., Gebre, E. and Amelework, B. 2018. Genetic diversity of yam (*Dioscorea* spp.) landrace collections from Ethiopia using simple sequence repeat markers. Australian Journal of Crop Science, 12(8): 1222–1230.

Mulugeta, G. and Fekadu, A. 2014. Industrial and agricultural potentials of *Moringa*. Journal of Natural Sciences Research, 4(14): 57-63.

Muniappan, S., Bragadeshwaran, A., Kasianantham, N., Rajasekar, V., Chinnadurai, K., Balusamy, S. and Ibrahim, M.I.M. (2020). Development of biofuel from *Nigella sativa* biomass and its suitability for energy application. Biomass Conversion and Biorefinery, 12: 705–721.

Mumtaz, B., Fatima, S. and Kadam V.B. 2012. Antifungal activity of the essential oils of *eucalyptus* against groundnut storage fungi. DAV International Journal of Science, 1(1): 3738.

Muniysamy, S., Mtibe, A., Motaung, T., Ofoseu, O. and Patnaik. A. 2015. Development of sustainable biobased composite products from agricultural wastes/biomass in South Africa. Green Economy Research Report.

Muthoni, J., Shimelis, H. and Melis, R. 2019. Long-term conservation of potato genetic resources: Methods and status of conservation. Australian Journal of Crop Science, 13(5): 717–725.

Muyibi, S.A., Ameen, E.S.M., Megat, M.J., Noor, M. and Ahmadun, F.R. 2003. Sanitary landfill leachate treatment with *Moringa* seed extract and micro-filtration membrane. pp. 339-347. *In:* Ujang, Z. and Henze, M. (eds.), Environmental Biotechnology: Advancement in Water and Waste Water Applications in the Tropics. IWA Publishing, Malaysia.

Myers, V.R. 2020. 40 Species of pines from around the world. The Spruce. https://www. thespruce.com/pine-trees-from-around-the-world-3269718. 3/10/2020

Nagarani, G. and Siddhuraju, A.A.P. 2014. Food prospects and nutraceutical attributes of *Momordica* species: A potential tropical bioresources – A review. Food Science and Human Wellness, 3: 117–126.

Nagardeolekar, A., Wang, K., Jing, C., Dongre, P., Wood, C., Amidon, T. and Bujanovic, B. 2017. Prospects of Complete Utilization of Miscanthus in a Biorefinery Based on Hot Water Extraction. International Forest Biorefining Conference (IFBC). Thunder Bay, Ontario, Canada. May 9–11, 2017.

Nahar, K. and Borna, R.S. 2013. In vitro plant regeneration from shoot tip explants of *Jatropha curcas* L: A biodiesel plant. ARPN Journal of Science and Technology, 3(1): 38–42.

Nahla, A., Adawy, S., Eliraq, M., Abo El-Khasab, A., Abd El-Aziz, Y., El-Bassell, E., Esmaill, A.S., Attaia, M., Darwish, S.H., El-Ashry, H. and Gawish, M.S. 2018.

Selection and agronomical evaluation for some elite genotypes of jojoba. Journal of Applied Life Sciences International, 18(1): 1–12.

Naik, P.M. and Al-Khayri, J.M. 2018. Cell suspension culture as a means to produce polyphenols from date palm (*Phoenix dactylifera* L.). Ciência e Agrotecnologia, 42(5): 464–473.

Nair, P.N., Ajith, S. and Indira, E.P. 2015. Nilambur – Genotypically a unique teak population in India. International Journal of Agriculture, Environment and Biotechnology, 8(4): 885–890.

Nair, S. and Varalakshmi, K.N. 2011. Anticancer, cytotoxic potential of *Moringa oleifera* extracts on HeLa cell line. Journal of Natural Pharmaceuticals, 2(3): 138–142.

Namuli, A., Bazira, J., Casim, T.U. and Engeu, P.O. 2018. A review of various efforts to increase artemisinin and other antimalarial compounds in *Artemisia annua* L. plant. Cogent Biology, 4: 1513312.

Nangbes, J.G., Nvau, J.B., Buba, W.M. and Zukdimma, A.N. 2013. Extraction and characterization of castor (*Ricinus communis*) seed oil. The International Journal of Engineering and Science, 2(9): 105–109.

Nasim-Uz-Zaman, Rahman Liman, L., Abdul Kader, Sarker, A.B., Rafiul Islam and Parveen, I. 2018. An eco-friendly approach of cotton fabric dyeing with natural dye extracted from Bixa orellana seeds employing different metallic mordants. Chemical and Materials Engineering, 6(1): 1–8.

Nath, U.K., Kim, H., Khatun, K., Park, J., Kwon-Kyoo Kang, K. and Nou, I. 2016. Modification of fatty acid profiles of rapeseed (*Brassica napus* L.) oil for using as food, industrial feed-stock and biodiesel. Plant Breeding and Biotechnology, 4(2): 123–134.

National Center for Biotechnology Information. 2019. Available from https://pubchem. ncbi.nlm.nih.gov/compound/Diosgenin. Accessed 22/12/2019

National Research Council. 1992. Neem: A Tree for Solving Global Problems. National Academy Press, Washington, D.C. pp. 34–74.

National Research Council, 2008: Lost Crops of Africa, Vol. III: Fruits. National Academic Press, Washington D.C. pp. 290–298.

Natividad, L.R. and Rafael, R.R. 2014. Carotenoid analyses and antibacterial assay of annato (*Bixa orellana* L.), carrot (*Daucus carota* L.), corn (*Zea mays* L.) and tomato (*Solanum lycopersicum* L.) extracts. Research Journal of Recent Sciences, 3(3): 40–45.

Nautiyal, P.C. 2002. Groundnut: Post-harvest operations. Food and Agriculture of the United Nations, pp. 9–10.

Nayak, L., Nag, D., Das, S., Ray, D.P. and Ammayappan, L. 2011. Utilisation of sisal fibre (*Agave sisalana* L.) – A review. Agricultural Review, 32(2): 150–152.

Nayak, S., Rout, G.R. and Das, P. 2003. Evaluation of the genetic variability in bamboo using RAPD markers. Plant Soil Environment, 49(1): 24–28.

Nazir, A., Malik, R.N., Ajib, M., Khan, N. and Siddiqui, M.F. 2011. Hyperaccumulators of heavy metals of industrial areas of Islamabad and Rawalpindi. Pak. J. Bot., 43(4): 1925–1933.

Ndung'u, J.W., Anino, E., Njuguna, D.K., Mwangangi, R., Jepkorir, M., Mbugua, R.W., Chepng'etich, J., Ngule, C.M. and Mwitari, P. 2018. Phytochemical screening and synergistic antiproliferative activity against selected cancer cell lines of *Moringa oleifera* and *Indigofera arrecta* leaf extracts. European Journal of Medicinal Plants, 23(2): 1–11.

Neis, F.A., de Costa, F., de Araújo Jr., A.T., Fett, J.P. and Fett-Neto, A.G. 2019. Multiple industrial uses of non-wood pine products. Industrial Crops & Products, 130: 248–258.

Nekonam, M.S., Razmjoo, J., Kraimmojeni, H., Sharifnabi, B., Amini, H. and Bahrami, F. 2014. Assessment of some medicinal plants for their allelopathic potential against redroot pigweed (*Amaranthus retroflexus*). Journal of Plant Protection Research, 54(1): 90–95.

Nesi, N., Delourme, R., Brégeon, M., Falentin, C. and Renard, M. 2008. Genetic and molecular approaches to improve nutritional value of *Brassica napus* L. seed. C.R. Biologies, 331: 763–771.

Neto, M.C.L., Lobo, A.K.M., Martins, M.O., Fontenele, A.V. and Silveira, J.A.G. 2014. Dissipation of excess photosynthetic energy contributes to salinity tolerance: A comparative study of salt-tolerant *Ricinus communis* and salt-sensitive *Jatropha curcas*. J. Plant Physiol., 171(1): 23–30.

Neto, M.C.L., Martins, M.D.O., Ferreira-Silva, S.L. and Silveira, J.A.G. 2015. *Jatropha curcas* and *Ricinus communis* display contrasting photosynthetic mechanisms in response to environmental conditions. Sci. Agric., 72(3): 260-269.

Network, P., Omole, A.O., Adetogun, A.C. and Banjoko, O. 2021. Utilization potentials of *Eucalyptus grandis* (Hill ex Maiden) a municipal tree on the University of Ibadan Campus. Afribary. Retrieved from https://afribary.com/works/utilization-potentials-of-eucalyptus-grandis-hill-ex-maiden-a-municipal-tree-on-the-university-of-ibadan-campus

Newkirk, R. 2010. Soybean Feed Industry Guide. Canadian International Grains Institute. pp. 3–6.

Newson, W.R. 2012. Protein Based Plastics from the Residuals of Industrial Oil Crops. Swedish University of Agricultural Sciences. p. 1.

Newton, A.C., Cornelius, J.P., Pbaker, P., Mgillies, A.C., Mhernández, S., Ramnarine., J., Mesén, F.M. and Dwatt, A. 1996. Mahogany as a genetic resource: Is there a future for mahogany? Botanical Journal of the Linnean Society, 122(1): 61–73.

Ngobeni, N.D., Mokoena, M.L. and Funnah, S.M. 2016. Growth and yield response of fibre hemp cultivars (*Cannabis sativa* L.) under different N-levels in Eastern Cape Province of South Africa. African Journal of Agricultural Research, 11(2): 57–64.

Nguyen, A.T. and Donaldson, R.P. 2005. Metal-catalyzed oxidation induces carbonylation of peroxisomal proteins and loss of enzymatic activities. Arch. Biochem. Biophys., 439: 25–31.

Nie, G., Huang, L., Ma, X., Ji, Z., Zhang, Y., Tang, L. and Zhang, X. 2017. Enriching genomic resources and transcriptional profile analysis of Miscanthus sinensis under drought stress based on RNA sequencing. International Journal of Genomics, 2017: 12.

Nielsen, F., Hill, B. and de Jongh, J. 2011. Castor (*Ricinus communis*): Potential of Castor for Biofuel Production. 2nd edition. FACT Foundation.

Nigam, S. 2015. Groundnut at a Glance. U.S. Government's Feed the Future. pp. 1–103.

Nigam, S.N. 2014. Groundnut at a Glance. pp. 121. Available from http://oar.icrisat.org/8455/1/Groundnut%20at%20a%20Glance.pdf. Accessed 1/12/2018

Niino, T. and Arizaga, M.V. 2015. Cryopreservation for preservation of potato genetic resources. Breeding Science, 65: 41–52.

Nikkhah, A. 2012. Legumes biofarming and biopharmaceutical sciences: A review. Research Journal of Medicinal Plants, 6(7): 466–488.

Nikkhah, A. 2014a. Bio farming agrotechnologies: Inventive challenges of plant sciences. Agrotechnology, 3: 1 e110.

Nikkhah, A. 2014b. Legumes as medicine: Nature prescribes. Medicinal & Aromatic Plants, 3: 3 e153.

Nirsatmanto, A. and Gyokusen, K. 2007. Genetic transformation of *Melia azedarach* L., using *Agrobacterium* mediated transformation. Journal of Forestry Research, 4(1): 1–8.

Niu, Z., Sun, L. and Sun, T. 2009. Response of root and aerial biomass to phytoextraction of Cd and Pb by sunflower, castor bean, alfalfa and mustard. Adv. Environ. Biol., 3: 255–262.

Nkongho, R.N., Ndjogui, T.E. and Levang, P. 2015. History of partnership between agro-industries and oil palm smallholders in Cameroon: Oilseeds and fats. Crops and Lipids, OCL, pp. 55.

Noman, A., Bashir, R., Aqeel, M., Anwer, S., Iftikhar, W., Zainab, M., Zafar, S., Shahbaz Khan, S., Waqar Islam, W. and Muhammad Adnan, M. 2016. Success of transgenic cotton (*Gossypium hirsutum* L.): Fiction or reality? Cogent Food & Agriculture, 2: 1207844.

Nongdam, P. and Tikendra, L. 2014. The nutritional facts of bamboo shoots and their usage as important traditional foods of Northeast India. International Scholarly Research Notices, 2014: 17. Article ID 679073.

Norwati, A., Norlia, B., Rosli, H.M., Norwati, M. and Abdullah, R. 2011. Development of transgenic teak (*Tectona grandis*) expressing a cry1Ab gene for control of the skeletoniser. Mol. Biol. Biotechnol., 19(4): 149–156.

Nouman, W., Basra, S.M.A., Siddiqui, M.T., Yasmeen, A., Gull, T. and Alcayde, M.A.C. 2014. Potential of *Moringa oleifera* L. as livestock fodder crop: A review. Turkish Journal of Agriculture and Forestry, 38: 1–14.

Nûnez-Palenius, H.G., Gomez-Lim, M. and Ochoa-Alejo, N. 2008. Melon fruits: Genetic diversity, physiology, and biotechnology features. Critical Reviews in Biotechnology, 19: 13–55.

Nwalo, N.F. 2015. Genetic diversity of Nigerian sesame cultivars (*Sesamum indicum* L.) based on simple sequence repeat (SSR) markers and its relationship with phytochemicals. International Journal of Current Microbiology and Applied Sciences, 4(1): 898–908.

Nworu, C.S., Okoye, E.L., Ezeifeka, G.O. and Esimone, C.O. 2013. Extracts of *Moringa oleifera* Lam. showing inhibitory activity against early steps in the infectivity of HIV-1 lentiviral particles in a viral vector-based screening. African Journal of Biotechnology, 12(30): 4866–4873.

Nyadanu, D., Assuah, M.K., Adomako, B., Asiama,Y.O., Opoku, I.Y., Adu-Ampomah, Y. 2009. Efficacy of screening methods used in breeding for black pod disease resistance varieties in cocoa. African Crop Science Journal, 17(4): 175–186.

Nzelibe, H.C., Caritas, U. and Okafoagu, C.U. 2007. Optimization of ethanol production from *Garcinia kola* (bitter kola) pulp agrowaste. African Journal of Biotechnology, 6(17): 2033-2037.

Nikkhah, A. 2016. Leguminous biopharmaceutechs. Nutrition and Food Science International Journal, 1(5): 1–2.

O'Brien, R.D., Jones, L.A., King, C.C., Wakelyn, P.J. and Wan, P.J. 2005. Cottonseed oil. pp. 856, 919. *In:* Fereidoon Shahidi (ed.), Bailey's Industrial Oil and Fat Products, 6th Edition. Vol. 5. John Wiley & Sons, Inc.

Oballa, P.O., Konuche, P.K.A., Muchiri, M.N. and Kigomo, B.N. 2010. Facts on growing and use of eucalyptus in Kenya. Kenya Forestry Research Institute, pp. 1–5.

Obeng, G.Y., Amoah, D.Y., Opoku, R., Sekyere, C.K.K., Adjei, E.A. and Mensah, E. 2020. Coconut wastes as bioresource for sustainable energy: Quantifying wastes, calorific values and emissions in Ghana. Energies, 13: 2178.

Obidi, O.F., Adelowotan, A.O., Ayoola, G.A., Johnson, O.O., Hassan, M.O. and Nwachukwu, S.C.U. 2013. Antimicrobial activity of orange oil on selected pathogens. The International Journal of Biotechnology, 2(6): 113–122.

Ogbuewu, P., Odoemenam, V.U., Obikaonu, H.O., Opara, M.N., Emenalom, O.O., Uchegbu, M.C., Okoli, I.C., Esonu, B.O. and Iloeje, M.U. 2011. The growing importance of neem (*Azadirachta indica* A. Juss) in agriculture, industry, medicine and environment: A review. Research Journal of Medicinal Plant, 5: 23–24.

Ogunniyi, D.S. 2006. Castor oil: A vital industrial raw material. Bioresour. Techn., 97: 1086–1091.

Ojeda de Rodríguez, G., Ysambertt, F., Sulbarán de Ferrer, B. and Cabrera, L. 2003. Volatile fraction composition of Venezuelan sweetorange essential oil (*Citrus sinensis* (L.) Osbeck). CIENCIA, 11(1): 55–60.

Ojiako, E.N. and Okeke, C.C. 2013. Determination of antioxidant of *Moringa oleifera* seed oil and its use in the production of a body cream. Asian Journal of Plant Science and Research, 3(3): 1–4.

Ojuederie, O.B., Bullet, Igwe, D.O., Somiame, B., Okuofu, I. and Benjamin, F. 2012. Assessment of genetic diversity in some *Moringa oleifera* Lam. landraces from Western Nigeria using RAPD markers. Journal of Horticultural Science and Biotechnology, 7(1): 15–20.

Okechukwu, R.I., Iwuchukwu, A.C. and Anuforo, H.U. 2015. Production and characterization of biodiesel from *Ricinus communis* seeds. Res. J. Chem. Sci., 5(2): 1–3.

Okewale, A.O.L. and Olaitan, A. 2017. The use of rubber leaf extract as a corrosion inhibitor for mild steel in acidic solution. International Journal of Materials and Chemistry, 7(1): 5–13.

Oladoja, N.A., Aboluwoye, O.C., Oladimeji, Y.B., Ashogbon, A.O. and Otemuyiwa, I.O. 2008. Studies on castor seed shell as a sorbent in basic dye contaminated wastewater remediation. Desalination, 227: 190–203.

Olife, I.C., Beagha, O.A. and Onwualu, A.P. 2015. Citrus fruits value chain development in Nigeria. Journal of Biology, Agriculture and Healthcare, 5(4): 36–47.

Olivares, A.R., Carrillo-González, R., González-Chávez, M.D.C.A. and Hernández, R.M.S. 2013. Potential of castor bean (*Ricinus communis* L.) for phytoremediation of mine tailings and oil production. J. Environ. Manage., 114: 316–323.

Oliveira, J.S., Leite, P.M., Souza, L.B., Mello, V.M., Silva, E.C., Rubim, J.C., Meneghetti, S.M.P. and Suarez, P.A.Z. 2009. Characteristics and composition of *Jatropha gossypiifolia* and *Jatropha curcas* L. oils and application for biodiesel production. Biomass Bioenergy, 33: 449–453.

Oliveira, L.B., Araujo, M.S.M., Rosa, L.P., Barata, M. and Rovere, E.L.L. 2008. Analysis of the sustainability of using wastes in the Brazilian power industry. Renew. Sustain. Energy Rev., 12: 883–890.

Oliveiraa, S.L., Mendes, R.F., Mendes, L.M. and Freired, T.P. 2016. Particle board panels made from sugarcane bagasse: Characterization for use in the furniture industry. Materials Research, 19(4): 914–922.

Oliver, A.L. 2004. Biotechnology entrepreneurial scientists and their collaborations. Research Policy, 33: 583–597.

Ololade, Z.S. and Olawore, N.O. 2013. Chemistry and medicinal potentials of the seed essential oil of Eucalyptus toreliana F. muell grown in Nigeria. Global Journal of Science Frontier Research Chemistry, 13: 1–10.

Oloyede, G.K. 2012. Antioxidant activities of methyl ricinoleate and ricinoleic acid dominated *Ricinus communis* seeds extract using lipid peroxidation and free radical scavenging methods. Research Journal of Medicinal Plant, doi 10.3923/rjmp.

Oluwole, F.A., Aviara, N.A. and Haque, M.A. 2004. Development and performance tests of a Shea nut cracker. J. Foods Eng., 65: 117–123.

Omar, S.R., Hamsan, N.A.M.D. and Abdullah, M.N. 2020. Waste to wealth: Optimizing novel pectin acid extraction from honeydew (*Cucumis melo* L. var. inodorous) peels as a potential halal food thickener. MOJ Food Process Technology, 8(1): 13–17.

Omo-Ikerodah, E.E., Omokhafe, K.O., Akpobome, F.A. and Mokwunye, M.U. 2009. An overview of the potentials of natural rubber (*Hevea brasiliensis*) engineering for the production of valuable proteins. African Journal of Biotechnology, 8(25): 7303–7307.

Omohu, O.J. and Omale, A.C. 2017. Physicochemical properties and fatty acid composition of castor bean *Ricinus communis* L. seed oil. European Journal of Biophysics, 5(4): 62–65.

Oparaeke, A.M. 2015. Studies on insecticidal potential of extracts of *Gmelina arborea* products for control of field pests of cowpea, *Vigna unguiculata* (L.) walp: The pod borer, *Maruca vitrata* and the coreid bug, *Clavigralla tomentosicollis*. Journal of Plant Protection Research, 45(1): 1–7.

Opare-Obuobi, K. 2012. Characterisation of local and exotic accessions of Moringa (*Moringa oleifera* Lamarck). MSc thesis department of Crop Science of the Agricultural and Consumer Sciences, University of Ghana.

Opeke, L.K. 2005. Tropical Commodity Tree Crops. Ibadan, Nigeria. Spectrum Books. 503 pp.

Opuni-Frimpong, E., Tekpetey, S.L., Owusu, S.A., Obiri, B.D., Appiah Kubi, E., Samuel Opoku, Nyarko-Duah, N.Y., Essien, C., Opoku, E.M. and Storer, A.J. 2016. Managing mahogany plantations in the tropics. Field Guide for Farmers. CSIRFORIG, Kumasi, Ghana. pp. 1–2.

Orji, E.E., Falodun, A.E. and Jegede, F.I. 2018. The antioxidant properties of *Momordica charantia* extract and its protective activities against in vitro mercury intoxication. International Journal of Current Research in Biosciences and Plant Biology, 5(4): 30–35.

Ortiz, O. and Mares, V. 2017. The historical, social, and economic importance of the potato crop. pp. 1–10. *In:* Swarup Kumar Chakrabarti, Conghua Xie, Jagesh Kumar Tiwari (eds.), The Potato Genome. Springer International Publishing AG, Gewerbestrasse 11, 6330 Cham, Switzerland.

Orwa, C., Mutua, A., Kindt, R., Jamnadass, R. and Simons. A. 2009. Agroforestree Database: A tree reference and selection guide version 4.0 (http://www.worldagroforestry.org/af/treedb/)

Osava, M. 2003. Energy in a castor bean. 2003. http://www.tierramerica.net/english/2003/0526/ ianalisis.shtml (Accessed on November 18, 2016).

Oswalt, J.S., Rieff, J.M. and Severino, L.S., Auld, D.L., Bednarz, C.W. and Ritchie, G.L. 2014. Plant height and seed yield of castor (*Ricinus communis* L.) sprayed with growth retardants and harvest aid chemicals. Industrial Crops and Products, 61: 272–277.

Ovando-Medina, I., Espinosa-García, F.J., Núñez-Farfán, J.S. and Salvador-Figueroa, M. 2011. State of the art of genetic diversity research in *Jatropha curcas*. Scientific Research and Essays, 6(8): 1709–1719.

Ovenden, S.P.B., Pigott, E.J., Rochfort, S. and Bourne, D.J. 2014. Liquid chromatography–mass spectrometry and chemo metric analysis of *Ricinus communis* extracts for cultivar identification. Phytochemical Analysis, 25: 476–484.

Öztürk, Ö., Gerem, G.P., Yenici, A. and Haspolat, B. 2014. Effects of different sowing dates on oil yield of castor (*Ricinus communis* L.). International Journal of Biological, Veterinary, Agricultural and Food Engineering, 8(2): 180–184.

Padmaja, N., Bosco, S.J.D. and Rao, J.S. 2015. Physico chemical analysis of Sapota (*Manilkara zapota*) coated by edible Aloe vera gel. International Journal of Applied Sciences and Biotechnology, 3(1): 20–25.

Paguiligan, J. and Villanueva, V. 2005. Agro-industrial production of lubricating oil and castor beans. Feasibility study. Mapúa Institute of Technology School of Chemical Engineering and Chemistry.

Pal, R., Banerjee, A. and Kundu, R. 2013. Responses of castor bean (*Ricinus communis* L.) to lead stress. Proc. Nat. Acad. Sci. India Section B: Biol. Sci., 83(4): 643–650.

Palazzolo, E., Laudicina, V.A. and Germanà, M.A. 2013. Current and potential use of citrus essential oils. Current Organic Chemistry, 17: 3042–3049.

Pan, J., Chu, C., Zhao, X., Cui, Y. and Voituriez, T. 2008. Global cotton and textile product chains: Identifying challenges and opportunities for China through a global commodity chain sustainability analysis. International Institute for Sustainable Development, Manitoba, Canada. pp. 1–35.

Pandey, A., Soccol, C.R., Nigam, P. and Soccol, V.T. 2000. Biotechnological potential of agro-industrial residues. I: Sugarcane bagasse. Bioresource Technology, 74: 69–80.

Pandey, V.C. 2013. Suitability of *Ricinus communis* L. cultivation for phytoremediation of fly ash disposal sites. Ecolological Engineering, 57: 336–341.

Pandey, V.C., Bajpai, O. and Singh, N. 2016. Energy crops in sustainable phytoremediation. Renew. Sustainable Energy Review, 54: 58–73.

Pankar, S.A. and Bornare, D.T. 2018. Studies on cauliflower leaves powder and its waste utilization in traditional product. International Journal of Agricultural Engineering, 11: 95–98 (Special Issue).

Parthiban, K.T., Selvan, P., Paramathma, M., Umesh Kanna, S., Kumar, P., Subbulakshmi, V. and Vennila. 2011. Physico-chemical characterization of seed oil from *Jatropha curcas* L. genetic resources. Journal of Ecology and the Natural Environment, 3(5): 163–167.

Parthiban, K.T., Selvan, P., Paramathma, M., Umesh Kanna, S., Kumar, P., Pasricha, V., Satpathy, G. and Gupta, R.K. 2014. Phytochemical & Antioxidant activity of underutilized legume *Vicia faba* seeds and formulation of its fortified biscuits. Journal of Pharmacognosy and Phytochemistry, 3(2): 75–80.

Pate, J.S. 1976. Nutrients and metabolites of fluids recovered from xylem and phloem: Significance in relation to long distance transport. pp. 253–345. *In:* Wordlow, I.F. and Passiourn, J.B. (eds.), Transport and Transfer Process in Plants. CSIRO, Canberra, Australia.

Patel, B.P., Patel, H.S. and Patel, S.R. 2004. Modified castor oil as an epoxy resin curing agent. E-Journal of Chemistry, 1(1): 11–16.

Patel, V.R., Dumancas, G.G., Viswanath, L.C.K., Maples, R. and Subong, B.J.J. 2016. Castor oil: Properties, uses and optimization of processing parameters in commercial production. Lipid Insights, 9: 1–12.

Pathak, N., Rai, A.K., Kumari, R. and Bhat, K.V. 2014. Value addition in sesame: A perspective on bioactive components for enhancing utility and profitability. Pharmacognosy Reviews, 8(16): 147–155.

Patil, D., Roy, S., Dahake, R., Rajopadhye, S., Kothari, S., Deshmukh, R. and Chowdhary, A. 2013. Evaluation of *Jatropha curcas* Linn. leaf extracts for its cytotoxicity and potential to inhibit hemagglutinin protein of influenza virus. Indian Journal of Virology, 24(2): 220–226.

Pavingerová, D., Bříza, J., Niedermeierová, H. and Vlasák, J. 2011. Stable agrobacterium-mediated transformation of Norway spruce embryogenic tissues using somatic embryo explants. Journal of Forest Science, 57(7): 277–280.

Pavlidou, E. 2016. Alternative energy resources: *Brassica napus* for biofuel production. Advances in Plants & Agriculture Research, 4(2): 266.

Pawar, R.S., Wagh, V.M., Panaskar, D.B., Adaskar, V.A. and Pawar, P.R. 2011. A case study of soybean crop production, installed capacity and utilized capacity of oil plants in Nanded District, Maharashtra, India. Advances in Applied Science Research, 2(2): 342–350.

Perdomo, F.A., Acosto-Osorio, A.A., Herrera, G., Vasco-Leal, J.F., Mosquera-Artamonov, J.D., Millan-Malo, B. and Rodriguez-Garcia, M.E. 2013. Physicochemical characterization of seven Mexican *Ricinus communis* L. seeds & oil contents. Biomass Bioenergy, 48: 17–24.

Perea-Flores, M.J., Chanona-Pérez, J.J., Garibay-Febles, V., Calderon-Dominguez, G., Terres-Rojas, E., Mendoza-Perez, J.A. and Bucio-Herrera, R. 2011. Microscopy techniques and image analysis for evaluation of some chemical and physical properties and morphological features for seeds of the castor oil plant (*Ricinus communis*). Industrial Crops and Products, 34(1): 1057–1065.

Pereira, F.S.G., de Sobral, A.D., da Silva, A.M.R.B. and da Rocha, M.A.G. 2018. *Moringa oleifera*: A promising agricultural crop and of social inclusion for Brazil and semi-arid regions for the production of energetic biomass (biodiesel and briquettes). OCL, 25(1): D106.

Péres, E.U.X., de Souza, F.G., Silva, F.M., Chaker, J.A. and Suare, P.A.Z. 2014. Biopolyester from ricinoleic acid: Synthesis, characterization and its use as biopolymeric matrix for magnetic nanocomposites. Industrial Crops and Products, 59: 260–267.

Peuke, A.D. 2009. Correlation in concentrations, xylem and phloem flows, and partitioning of elements and ions in intact plants. A summary and statistical re-evaluation of modeling experiments in *Ricinus communis*. J. Exp. Bot., 61(3): 635–655.

Pham, T.D. 2011. Analyses of Genetic Diversity and Desirable Traits in Sesame (*Sesamum indicum* L., Pedaliaceae): Implication for Breeding and Conservation. Ph.D. thesis. Swedish University of Agricultural Sciences Alnarp.

Philp, J.C. and Pavanan, K.C. 2013. Bio-based production in a bioeconomy. Asian Biotechnology and Development Review, 15(2): 81–88.

Pichhode, M. and Nikhi, K. 2017. Teak (*Tectona grandis*) plantation towards carbon sequestration and climate change mitigation in district Balaghat, Madhya Pradesh, India. International Journal of Innovative Research in Science, Engineering and Technology, 6(9): 18673–18685.

Pidlisnyuk, V., Erickson, L.E., Wang, D., Zhao, J., Stefanovska, T. and Schlup, J.R. 2021. Miscanthus as raw materials for bio-based products. pp. 202–222. *In:* Larry E. Erickson and Valentina Pidlisnyuk (eds.), Phytotechnology with Biomass Production: Sustainable Management of Contaminated Sites. First edition. CRC Press, Boca Raton.

Pinheiro, H.A., Silva, J.V., Endres, L., Ferreira, V.M., Camara, C.A., Cabral, F.F., Oliveira, J.F., de Carvalho, L.W.T., dos Santos, J.K. and Filho, B.G.S. 2008. Leaf gas exchange, chloroplastic pigments and dry matter accumulation in castor bean (*Ricinus communis* L.) seedlings subjected to salt stress conditions. Ind. Crops Prod., 27: 385–392.

Pires-Alves, M., Grossi-de-Sá, M.F., Barcellos, G.B., Carlini, C.R. and Moraes, M.G. 2003. Characterization and expression of a novel member (JBURE-II) of the urease gene family from jackbean [*Canavalia ensiformis* (L.) DC]. Plant Cell Physiology, 44(2): 139–145.

Pîrvan, A., Jurcoane, S. and Florentina Matei, F. 2020. Life cycle assessment of *Camelina sativa* crop in a circular economy approach – A mini review. Scientific Bulletin. Series F. Biotechnologies, XXIV(2): 189–193.

Pius, C.O., Nnaemeka, S.P.O., Charles, O., Vincent, N.O. and Chinenye, A.I. 2014. Design enhancement evaluation of a castor seed shelling machine. Journal of Scientific Research & Reports, 3(7): 924–938.

Pons, E., Peris, J.E. and Peña, L. 2012. Field performance of transgenic citrus trees: Assessment of the long-term expression of uidA and nptII transgenes and its impact on relevant agronomic and phenotypic characteristics. BMC Biotechnology, 12: 41.

Poobalan, K., Lim, V., Mohamed Kamal, N.S., Yusoff, N.A., Khor, K.Z. and Abdul Samad, N. 2017. Effects of Ultrasound Assisted Sequential Extraction (UASE) of *Moringa oleifera* leaves extract on MCF 7 human breast cell line. Malaysian Journal of Medicine and Health Sciences, 14(1): 102–106.

Poovaiah , C.R., Mazarei, M., Decker, S.R., Turner, G.B., Sykes, R.W., Davis, M.F. and Stewart Jr, C.N. 2015. Transgenic switchgrass (*Panicum virgatum* L.) biomass is increased by overexpression of switchgrass sucrose synthase (PvSUS1). Biotechnology Journal, 10(4): 552–563.

Pop, I.A., Oroian, I., Lobuntiu, I. and Friss, Z. 2017. *Atemisia annua* L. culture technology in the climatic conditions in Transylvania. Procedia Engineering, 181: 433–438.

Pousga, S., Boly, H., Lindberg, J.E. and Ogle, B. 2007. Evaluation of traditional sorghum (*Sorghum bicolor*) beer residue, shea-nut (*Vitellaria paradoxa*) cake and cottonseed (*Gossypium*) cake for poultry in burkinafaso: Availability and amino acid digestibility. Int. J. Poult. Sci., 6: 666–672.

Powell, R.G. 2009. Plant seeds as sources of potential industrial chemicals, pharmaceuticals, and pest control agents. Journal of Natural Products, 72(3): 516–523.

Prakash, G., Bhojwani, S.S. and Srivastava, A.K. 2002. Production of *Azadirachtin* from plant tissue culture: State of the art and future prospects. Biotechnology and Bioprocess Engineering, 7: 185–193.

Pramote, P., Pornthip, P. and Numuen, C. 2011. Phenotypic diversity and classification of Thai bitter melon (*Momordica charantia* L.) landraces from three provinces in central region of Thailand. Journal of Agricultural Technology, 7(3): 849–856.

Prasad, K.S., Chuang, M.C. and Ho, J.A.A. 2012. Synthesis, characterization, and electrochemical applications of carbon nanoparticles derived from castor oil soot. Talanta, 88: 445–449.

Prasad, M.N.V. 2007. Sunflower (*Helinathus annuus* L.) – A potential crop for environmental industry. HELIA, 30(46): 167–174.

Prasad, M.N.V. 2011. A state-of-the-art report on bioremediation, its applications to contaminated sites in India. M/o of Environment & Forests, GOI New Delhi. pp. 90 http://moef.nic.in/downloads/public-information/BioremediationBook.pdf

Prasad, M.N.V. 2015. Phytoremediation and biofuels. Sustainable Agriculture Reviews, 17: 159–261.

Prasad, M.N.V. (Ed.). 2016. Bioremediation and Bioeconomy. Elsevier, USA. 698 pp.

Prasad, M.N.V. (2024). Bioremediation and Bioeconomy: A circular economy approach (2nd Ed). Elsevier.

Prasad, M.N.V., Freitas, H., Fraenzle, S., Wuenschmann, S. and Markert, B. 2010. Knowledge explosion in phytotechnologies for environmental solutions. Environmental Pollution, 158: 18–23.

Prasad, M.N.V. and Kiran, B.R. 2017. *Ricinus communis* L. (Castor bean), a potential multi-purpose environmental crop for improved and integrated phytoremediation. The EuroBiotech Journal, 1(2): 1–16.

Preeti, K.M. and Verma, A.B. 2014. A review on ethnopharmacological potential of *Ricinus communis* Linn. PharmaTutor, 2(3): 76–85.

Pua, T., Tan, T., Nurzatil, S.M., Jalaluddin, Othman, R.Y. and Harikrishna, J.A. 2019. Genetically engineered bananas—From laboratory to deployment. Annals of Applied Biology, 175(3): 282–301.

Pye-Smith, C. 2011. Cocoa futures an innovative programme of research and training is transforming the lives of cocoa growers in Indonesia and beyond. ICRAF Trees for Change no. 9. Nairobi: World Agroforestry Centre.

Rabadán, A., Nunes, M.A., Bessada, S.M.F., Pardo, J.E., Oliveira, M.B.P.P. and Álvarez-Ortí, M. 2020. From by-product to the food chain: Melon (*Cucumis melo* L.) seeds as potential source for oils. Foods, 9: 1341.

Radhamani, T., Ushakumari, R., Amudha, R. and Anjani, K. 2012. Response to water stress in castor (*Ricinus communis* L.) genotypes under in vitro conditions. Journal of Cereals and Oil Seeds, 3(4): 56–58.

Rahmati, H., Salehi, S., Malekpour, A. and Farhangi, F. 2015. Antimicrobial activity of castor oil plant (*Ricinus communis*) seeds extract against gram positive bacteria, gram negative bacteria and yeast. International Journal of Molecular Medicine and Advance Sciences, 11(1): 9–12.

Rahul, J., Jain, M.K., Singh, S.P., Kamal, R.K., Naz, A., Gupta, A.K. and Mrityunjay, S.K. 2015. Adansonia digitata L.(baobab): a review of traditional information and taxonomic description. Asian Pacific Journal of Tropical Biomedicine, 5(1): 79–84.

Rai, Y. 2014. Seedling behaviour and early growth status of seedlings in *Thevetia peruviana* (PERS). K. Shum. Impact: International Journal of Research in Applied, Natural and Social Sciences, 2(6): 129–134.

Raja, B.S. and Kiruba, T. 2020. De-colourisation of textile dye effluents using cost-effective *Nigella sativa* seed waste. International Journal for Modern Trends in Science and Technology, 6(9S): 78–82.

Raja, S., Balakrishna, G.B. and Tukaran, A.M. 2019. Diversity analysis of bitter gourd (*Momordica charantia*) germplasm from tribal belts of India. The Asian Journal of Plant Science and Biotechnology, 3(1): 21–25.

Rajangam, A.S., Gidda, S.K., Craddock, C., Mullen, R.T., John, M. and Eastmond, P.J. 2013. Molecular characterization of the fatty alcohol oxidation pathway for wax-ester mobilization in germinated jojoba seeds. Plant Physiology, 161: 72–80.

Rajeshkanna, A., Senthamilselvi, M.M., Prabhakaran, D., Solomon, S. and Muruganantham, N. 2017. Anti cancer activity of *Moringa oleifera* (flowers) against human liver cancer. International Journal of Innovative Science, Engineering & Technology, 4(6): 388–392.

Rajkumar, M. and Freitas, H. 2008. Influence of metal resistant plant growth-promoting bacteria on the growth of *Ricinus communis* in soil contaminated with heavy metals. Chemosphere, 71: 834–842.

Rajkumar, M., Ae, N., Prasad, M.N.V. and Helena Freitas, H. 2010. Potential of siderophore-producing bacteria for improving heavy metal phytoextraction. Trends in Biotechnology, 28(3): 142–149.

Rajkumar, M., Sandhya, S., Prasad, M.N.V. and Freitas, H. 2012. Perspectives of plant-associated microbes in heavy metal phytoremediation. Biotechnol. Adv., 30: 1562–1574.

Rajoriya, C.M., Ahmad, R., Rawat, R.S. and Jat, B.L. 2016. Studies on induction of mutation in fenugreek (*Trigonella fonum-graecum*). International Journal for Research in Applied Science & Engineering Technology, 4(X): 333–373.

Ralph, S.G., Chun, H.J.E., Kolosova, N., Cooper, D., Oddy, C., Ritland, C.E., Kirkpatrick, R., Moore, R., Barber, S., Holt, R.A., Jones, S.J.M., Marra, M.A., Douglas, C.J., Ritland, K. and Bohlmann, J. 2008. A conifer genomics resource of 200,000 spruce (Picea spp.) ESTs and 6,464 high-quality, sequence-finished full-length cDNAs for Sitka spruce Piceasitchensis. BMC Genomics, 9: 484.

Ramadan, M.F. and Morsel, J.T. 2003. Oil cactus pear (*Opuntia ficus-indica* L.). Food Chemistry, 82: 339–345.

Ramírez-Pulido, B., Bas-Bellver, C., Betoret, N., Barrera, C. and Seguí, L. 2021. Valorization of vegetable fresh-processing residues as functional powdered ingredients. A review on the potential impact of pretreatments and drying methods on bioactive compounds and their bioaccessibility. Frontiers in Sustainable Food Systems, 5: 654313.

Rampadarath, S., Puchooa, D. and Ranghoo-Sanmukhiya, V.M. 2014. A comparison of polyphenolic content, antioxidant activity and insecticidal properties of *Jatropha* species and wild *Ricinus communis* L. found in Mauritius. Asian Pacific J. Trop. Med., 7(Suppl. 1): S384–S390.

Rampadarath, S. and Puchooa, D. 2016. In vitro antimicrobial and larvicidal properties of wild *Ricinus communis* L. in Mauritius. Asian Pacific J. Trop. Biomed., 6(2): 100–107.

Ramprasad, R. and Bandopadhyay, R. 2010. Future of *Ricinus communis* after completion of the draft genome sequence. Current Science, 99(10): 1316–1318.

Rana, M., Dhamija, H., Prashar, B. and Sharma, S. 2012. *Ricinus communis* L.: A review. International Journal of Pharm Tech Research, 4: 1706–1711.

Rana, S. and Kanwar, K. 2017. Assessment of genetic diversity in *Aloe vera* L. among different provinces of H.P. Journal of Medicinal Plants Studies, 5(3): 348–354.

Rana, S., Sharma, D. and Bakshi, N. 2018. A mini review on morphological, biochemical and molecular characterization of *Aloe vera* L. International Journal of Chemical Studies, 6(4): 3109–3115.

Rana, H., Ravi Kumar, R., Chopra, A., Pundir, S. and Gautam, G.K. 2022. The various pharmacological activity of *Adansonia digitata*. Research Journal of Pharmacology and Pharmacodynamics, 14(1): 53–59.

Rani, E.A. and Arumugam, T. 2017. *Moringa oleifera* (Lam) – A nutritional powerhouse. Journal of Crop and Weed, 13(2): 238–246.

Rani, P., Kumar, A. and Arya, R.C. 2016. Phytostabilization of tannery sludge amended soil using *Ricinus communis*, *Brassica juncea* and *Nerium oleander*. J. Soils Sedim., DOI 10.1007/s11368-016-1466-6.

Rao, C.K. 2008. Biopharming: The interface of plant biotechnology, biopharmaceuticals and farming. Foundation for Biotechnology Awareness and Education, Bangalore, India. Available from http://www.plantbiotechnology.org.in/issue7.htmlAceessed 12/11/2019

Rashid, B., Kousar, S., Yousaf, M., Ali, Q.F., Parveen, S., Arshad, A., Rimsha Ahmad, N.

and Husnain, T. 2016. Biotechnology: Future tools for stable insect pest and weed control in cotton. Cotton Genomics and Genetics, 7(3): 1–14.

Regeneration International. 2021. What is Biochar? Available at https://regenerationinternational.org/2018/05/16/what-is-biochar/ Accessed 14/1/2021

Rekha, A. 2016. History, origin, domestication, and evolution. pp. 3–12. *In:* Sukhada Mohandas Kundapura V. Ravishankar (ed.), Banana: Genomics and Transgenic Approaches for Genetic Improvement. Springer Science+Business Media Singapore.

Ribeiro de Jesus, P.R. 2015. Biochemical, physiological and molecular responses of *Ricinus communis* seeds and seedlings to different temperatures: A multi-omics approach. 203 pp. PhD thesis, Wageningen University, Wageningen, NL.

Ribeiro, P.R., Fernandez, L.G., Castro, R.D., Ligterink, W. and Hilhorst, H.W.M. 2014. Physiological and biochemical responses of *Ricinus communis* seedlings to different temperatures: A metabolomics approach. BMC Plant Biol., 14: 223.

Ribeiro, P.R., Willems, L.A., Mudde, E., Fernandez, L.G., de Castro, R.D., Ligterink, W. and Hilhorst, H.W. 2015. Metabolite profiling of the oilseed crop *Ricinus communis* during early seed imbibitions reveals a specific metabolic signature in response to temperature. Industrial, Crops and Products, 67: 305–309.

Ribeiro, P.R., Zanotti, R.F., Deflers, C., Fernandez, L.G., Castro, R., Ligterink, W. and Hilhorst, H.W.M. 2015. Effect of temperature on biomass allocation in seedlings of two contrasting genotypes of the oilseed crop *Ricinus communis.* Journal of Plant Physiology, 185: 31–39.

Ribeiro, P.R., de Castro, R.D. and Fernandez, L.G. 2016. Chemical constituents of the oilseed crop *Ricinus communis* and their pharmacological activities: A review. Industrial Crops and Products, 91: 358–376.

Richner, J. 2013. Dual use switchgrass: Managing switchgrass for biomass production and summer grazing. MSc Thesis, University of Missouri-Columbia.

Rigby, N.M., McDougall, A.J., Needs, P.W. and Selvendran, R.R. 1994. Phloem translocation of a reduced oligogalacturonide in *Ricinus communis* L. Planta, 193: 536–541. https://doi.org/10.1007/BF02411559

Rind, N.A., Rafiq, M., Dahot, M.U., Faiza, H., Aksoy, O., Rind, K.H., Shar, A.H., Hidayatullah and Jatoi, A.H. 2021. Production of limonoids through callus and cell suspension cultures of chinaberry (M*elia azedarach* L.). Bangladesh Journal of Botany, 50(2): 301–309.

Rissato, S.R., Galhiane, M.S., Fernandes, J.R., Gerenutti, M., Gomes, H.M., Ribeiro, R. and de Almeida, M.V. 2015. Evaluation of *Ricinus communis* L. for the phytoremediation of polluted soil with organochlorine pesticides. BioMed Res. Int., 8. Article ID 549863.

Rivera-Madrid, R., Burnell, J., Aguilar-Espinosa, M., Rodríguez-Ávila, N.L., Lugo-Cervantes, E. and Sáenz-Carbonell, L.A. 2013. Control of carotenoid gene expression in *Bixa orellana* L. leaves treated with norflurazon. Plant Molecular Biology Rep., 1–11.

Rizvi, A.F., Rizvi, Q.F., Kumar, D., Puranik, V., Kumar, N. and Chauhan, D.K. 2018. Amelioration of therapeutic potential of *Moringa oleifera* leaves through value added products development. International Journal of Green and Herbal Chemistry, 7(3): 529–540.

Rodrigues-Corrêa, K.C.S., Lima, J.C. and Fett-Neto, A.G. 2012. Pine oleoresin: Tapping green chemicals, biofuels, food protection, and carbon sequestration from multipurpose trees. Food and Energy Security, 1(2): 81–93.

Rodrigues, C.R.F., Silva, E.N., Moura, R. and Viegas, R.A. 2014. Physiological adjustment to salt stress in *R. communis* seedlings is associated with a probable mechanism of osmotic adjustment and reduction in water lost by transpiration. Ind. Crops Prod., 54: 233–239.

Rodriguez-Bonilla, L., Cuevas, H.E., Montero-Rojas, M., Bird-Pico, F., Luciano-Rosario, D. and Dimuth Siritunga, D. 2014. Assessment of genetic diversity of sweet potato in Puerto Rico. PLoS ONE, 9(12): 1–14.

Rolim, P.M., Jucá Seabra, L.M. and de Macedo, G.R. 2019. Melon by-products: Biopotential in human health and food processing. Food Reviews International, 36(1): 1–24.

Romeiro, S., Lagôa, A.M.M.A., Furlani, P.R., de Abreu, C.A., de Abreu, M.F. and Erismann, N.M. 2006. Lead uptake and tolerance of *Ricinus communis* L. Braz. J. Plant Physiol., 18(4): 483–489.

Roshetko, J.M., Rohadi, D., Perdana, A., Sabastian, G., Nuryartono, N., Pramono, A.A., Widyani, N., Manalu, P., Fauzi, M.A. and Sumardamto, P. 2013. Teak systems' contribution to rural development in Indonesia. Paper presented at the World Teak Conference 2013. Bangkok, Thailand, 24–27 March 2013.

Roychoudhury, N. 2012. Deployment of resistance in teak to key insect pests. Indian Forester, 138(2): 123–130.

Rukhsar, Patel, M.P., Parmar, D.J. and Kumar, S. 2018. Genetic variability, character association and genetic divergence studies in castor (*Ricinus communis* L.). Annals of Agrarian Science, 16: e143-e148.

Rural Advancement Fund International. 1990. Biotechnology and castor oil. Rafi Comminique, 1–4.

Sánchez-Zapata, E., Fernández-López, J. and Pérez-Alvarez, J.A. 2012. Tiger Nut (*Cyperus esculentus*) commercialization: Health aspects, composition, properties, and food applications. Comprehensive Reviews in Food Science and Food Safety, 11: 366–377.

Saa, R.W., Fombang, E.N., Ndjantou, E.B. and Njintang, N.Y. 2019. Treatments and uses of *Moringa oleifera* seeds in human nutrition: A review. Food Science and Nutrition, 7: 1911–1919.

Saadaoui, E., Martín, J.J., Bouazizi, R., Ben Romdhane, C., Grira, M., Abdelkabir, S., Larbi Khouja, M. and Cervantes, E. 2015. Phenotypic variability and seed yield of *Jatropha curcas* L. introduced to Tunisia. Acta Botanica Mexicana, 110: 119–134.

Saadaoui, E., Martín Gómez, J.J., Ghazel, N., Romdhane, C.B., Massoudi, N. and Cervantes, E. 2015. Allelopathic effects of aqueous extracts of *Ricinus communis* L. on the germination of six cultivated species. International Journal of Plant & Soil Science, 4: 220–227.

Saba, N., Paridah, M.T., Jawaid, M., Abdan, K. and Ibrahim, N.A. 2015a. Potential utilization of kenaf biomass in different applications. pp. 1–34. *In:* K.R. Hakeem et al. (eds.), Agricultural Biomass Based Potential Materials. Springer International Publishing, Switzerland.

Saba, N., Jawaid, M., Hakeem, K.R., Paridah, M.T., Khalina, A. and Alothman, O.Y. 2015b. Potential of bioenergy production from industrial kenaf (*Hibiscus cannabinus* L.) based on Malaysian perspective. Renewable and Sustainable Energy Reviews, 42: 446–459.

Sabiha, S., Ali, H., Hasan, K., Rahman, A.S.M.S. and Islam, N. 2017. Bioactive potentials of *Melia azedarach* L. with special reference to insecticidal, larvicidal and insect repellent activities. Journal of Entomology and Zoology Studies, 5(5): 1799–1800.

Saeed, A. and Iqbal, M. 2013. Loofa (*Luffa cylindrica*) sponge: Review of development of the biomatrix as a tool for biotechnological applications. Biotechnology Progress, 29(3): 573–600.

Saeed, M.K., Shahzadi, I., Ahmad, I., Ahmad, R., Shahzad, K., Ashraf, M. and Viqar-un-Nisa. 2010. Nutritional analysis and antioxidant activity of bitter gourd (*Momordica charantia*) from Pakistan. Pharmacologyonline, 1: 252–260.

Sáenz, C. 2013. Industrial production of non-food products. pp. 89–90. *In:* Agro-industrial Utilization of Cactus Pear (ed). Food and Agriculture Organization of the United Nations (FAO).

Sáenz, C. 2013. Utilization of *Opuntia* spp. fruits in food products. pp. 31–43. *In:* Agro-industrial Utilization of Cactus Pear (ed). Food and Agriculture Organization of the United Nations (FAO).

Sáenz, C. and Sepúlveda, E. 2013. Small-scale food production for human consumption. pp. 57–88. *In:* Agro-industrial Utilization of Cactus Pear (ed). Food and Agriculture Organization of the United Nations (FAO).

Saha, P., Roy, D., Manna, S., Chowdhury, S., Banik, S., Sen, R., Jo, J., Kim, J.K. and Adhiikari, B. 2015. Biodegradation of chemically modified lignocellulosic sisal fibers: Study of the mechanism for enzymatic degradation of cellulose. e-Polymers, 15(3): 185–194.

Saha, P.D. and Sinha, K. 2012. Natural dye from bixa seeds as a potential alternative to synthetic dyes for use in textile industry. Desalination and Water Treatment, 40(1-3): 298.

Sahay, S., Yadav, U. and Srinivasamurthy, S. 2017. Potential of *Moringa oleifera* as a functional food ingredient: A review. International Journal of Food Science and Nutrition, 2(5): 31–37.

Sahu, P.K., Patel, T.S., Sahu, P., Singh, S., Tirkey, P. and Sharma, D. 2014. Molecular farming: A biotechnological approach in agriculture for production of useful metabolites. International Journal of Research in Biotechnology and Biochemistry, 4(2): 23–30.

Saingera, M., Jaiwal, A., Sainger, P.A., Chaudharya, D., Jaiwald, R. and Jaiwala, P.K. 2017. Advances in genetic improvement of *Camelina sativa* for biofuel and industrial bio-products. Renewable and Sustainable Energy Reviews, 68: 623–637.

Saini, J.K., Saini, R. and Tewar, L. 2015. Lignocellulosic agriculture wastes as biomass feedstocks for second-generation bioethanol production: Concepts and recent developments. Biotech., 5: 337–353.

Salami, S.A. and Raji, Y. 2014. Oral *Ricinus communis* oil exposure at different stages of pregnancy impaired hormonal, lipids profile and histopathology of reproductive organs in Wistar rats. Journal of Medicinal Plant Research, 8(44): 1289–1298.

Saleem, M.H., Ali, S., Rehman, M., Hasanuzzaman, M., Rizwan, M., Irshad, S., Shafiq, F., Iqbal, M., Alharbi, B.M., Alnusaire, T.S. and Qari, S.H. 2020. Jute: A potential candidate for phytoremediation of metals—A review. Plants, 9: 258.

Saleh, B. 2016. DNA changes in cotton (*Gossypium hirsutum* L.) under salt stress as revealed by RAPD marker. Advance Horticultural Science, 30(1): 13–21.

Salih, U., Özgür, A. and İbrahim, B. 2002. Selection of promising jojoba (*Simmondsia chinensis* Link Schneider) types in terms of yield and oil content. Turkish Journal of Agriculture and Forestry, 26(6): Article 2. Available at: https://journals.tubitak.gov.tr/agriculture/ vol26/iss6/2

Salihu, B.Z., Gana, A.K. and Apuyor, B.O. 2014. Castor oil plant (*Ricinus communis* L.): Botany, ecology and uses. International Journal of Science and Research, 3(5): 1333–1341.

Salimon, J., Mohd Noor, D.A., Nazrizawati, A.T., Mohd Firdaus, M.Y. and Noraishah, A. 2010. Fatty acid composition and physicochemical properties of

Malaysian castor bean *Ricinus communis* L. seed oil (komposisiasid lemak dan sifatfizikokimiaminyakbijijarak *Ricinus communis* L. Malaysia). Sains Malysiana, 39(5): 761–764.

Salimon, J., Salih, N. and Yousif, E. 2012. Biolubricant base stocks from chemically modified ricinoleic acid. Journal of King Saud University – Science, 24(1): 11–17.

Samal, S., Rout, G.R. and Lenka, P.C. 2003. Analysis of genetic relationships between population of cashew (*Anacardium occidentale* L.) by using morphological characterization and RAPD markers. Plant Soil Environment, 49(3): 176–182.

Sánchez, A.S., Almeida, M.B., Torres, E.A., Kalid, R.A., Cohim, E. and Gasparatos, A. 2018. Alternative biodiesel feedstock systems in the semi-arid region of Brazil: Implications for ecosystem services. Renewable and sustainable energy reviews, 81: 2744–2758.

Sánchez, N., Sánchez, R., Encinar, J.M., González, J.F. and Martinez, G. 2015. Complete analysis of castor oil methanolysis to obtain biodiesel. Fuel, 147: 95–99.

Santhi, T. and Manonmani, S. 2009. Removal of methylene blue from aqueous solution by bioadsorption onto *Ricinus communis* epicarp. Activated Carbon Chemical Engineering Research Bulletin, 13(1): 1–5.

Santhi, T., Manonmani, S. and Smitha, T. 2010. Removal of malachite green from aqueous solution by activated carbon prepared from the epicarp of *Ricinus communis* by adsorption. J. Hazard. Mater., 179: 178–186.

Santos, C.H., de Oliveira Garcia, A.L., Calonego, J.C., Spósito, T.H.N. and Rigolin, I.M. 2012. Pb-phytoextraction potential by castor beans in soil contaminated (Potencial de fitoextração de Pb por mamoneirasem solo contaminado). Semina Cienc. Agrar., 33(4): 1427–1433.

Sanyal, T. 2017. Jute Geotextiles and their Applications in Civil Engineering, Developments in Geotechnical Engineering. SpringerScience+Business Media Singapore. pp. 19–31.

Sapuan, S.M., Sahari, J., Ishak, M.R. and Sanyang, M.L. 2018. Kenaf Fibers and Composites. CRC Press, Taylor & Francis Group, Boca Raton, 205 pp.

Sarkar, A. and Chowdhury, R. 2016. Co-pyrolysis of paper waste and mustard press cake in a semi-batch pyrolyzer—Optimization and bio-oil characterization. International Journal of Green Energy, 13(4): 373–382.

Sarria, E. 2017. Global challenges for the future of watermelon breeding. Acta Horticulture, 1151: 5–8.

Sartova, K., Omurzak, E., Kambarova, G., Dzhumaev, I., Borkoev, B. and Abdullaeva, Z. 2018. Activated carbon obtained from the cotton processing wastes. Diamond and Related Materials, 11-13.

Satheesan, T., Sivanathawerl, T., Sivachandran, S. and Phuspakumara, D.K.N.G. 2016. Distribution, growth and aboveground biomass of teak (*Tectona grandis* L.) plantation in Mullaitivu District of Sri Lanka. International Journal of Scientific and Research Publication, 6(3): 72–76.

Sathish, S., Gowthaman, K.A. and Haaresh Augustus. 2018. Utilization of *Punica granatum* peels for the extraction of pectin. Research J. Pharm. and Tech., 11(2): 613–616. doi: 10.5958/0974-360X.2018.00113.0

Satyagopal, K., Sushil, S.N., Jeyakumar, P., Shankar, G., Sharma, O.P., Sain, S.K., Boina, D.R., Srinivasa Rao, N., Sunanda, B.S., Ram Asre, Kapoor, K.S., Sanjay Arya, Subhash Kumar, Patni, C.S., Chattopadhyay, C., Jacob, T.K., Jadav, R.G., Abhishekh Shukla, Usha Bhale, Singh, S.P., Khan, M.L., Sharma, K.C., Dohroo, N.P., Suseela Bhai, K., Eapen, Santosh J., Hanumanthaswamy, B.C., Srinivas, K.R., Thakare, A.Y.,

Halepyati, A.S., Patil, M.B. and Sreenivas, A.G. 2014. AESA based IPM package for ginger. pp. 43. National Institute of Plant Health Management, Rajendranagar, Hyderabad.

Satyanarayana, K.G. 2015. Recent developments in 'Green' composites based on plant fibers – Preparation, structure property studies. Journal of Bioprocessing & Biotechniques, 5(2): 1.

Savitha, H.N., Kale, S.M. and Prakash, M. 2014. Dynamics of fresh and dry biomass production in drumstick (*Moringa oleifera* Lam.) genotypes. Asian Journal of Bio Science, 9(1): 93–96.

Scaboo, A.M., Chen, P., David, A. Sleper, D.A. and Clark, K.M. 2010. Classical breeding and genetics of soybean. pp. 19–54. *In:* Bilyeu, K., Ratnaparkhe, M.B., Kole, C. (eds), Genetics, Genomics and Breeding in Soybean. Science Publishers, Enfield, USA.

Scarcelli, N., Cubry, P., Akakpo, R.,Thuillet, A.-C., Obidiegwu, J., Baco, M.N., Otoo, E., Scarpa, A. and Guerci, A. 1982. Various uses of the castor oil plant (*Ricinus communis* L.): A review. J. Ethnopharmacol., 5: 117–137.

Schaefer, H., Heibl, C. and Renner, S.S. 2009. Gourds afloat: A dated phylogeny reveals an Asian origin of the gourd family (Cucurbitaceae) and numerous oversea dispersal events. Proceedings of the Royal Society of Biological Sciences, 276: 843–851.

Schaefer, H. and Renner, S.S. 2011. Phylogenetic relationships in the order Cucurbitales and a new classification of the gourd family (Cucurbitaceae). Taxonomy, 60(1): 122–138.

Schieltz, D.M., McWilliams, L.G., Kuklenyik, Z., Prezioso, S.M., Carter, A.J., Williamson, Y.M., McGratha, S.C., Morse, S.A. and Barr, J.R. 2015. Quantification of ricin, RCA and comparison of enzymatic activity in 18 *Ricinus communis* cultivars by isotope dilution mass spectrometry. Toxicon, 95: 72–83.

Schneider, J., Bundschuh, J. and do Nascimento, C.W. 2016. *Arbuscular mycorrhizal* fungi-assisted phytoremediation of a lead-contaminated site. Sci. Total Environ., 572: 86–97.

Scholz, V. and da Silva, J.N. 2008. Prospects and risks of the use of castor oil as a fuel. Biomass Bioenergy, 32: 95–100.

Schultze-Kraft, R., Rao, I.M., Peters, M., Clements, R.J., Bai, C. and Liu, G. 2018. Tropical forage legumes for environmental benefits: An overview. Tropical Grasslands, 6(1): 1–14.

Schuman, K. and Siekman, K. 2005. Soap in Ullmanna's Encyclopedia of Industrial Chemistry. Wiley-VCH, Weinhei.

ScieNews. 2017. Researchers Sequence Guinea Yam Genome. http://www.sci-news.com/genetics/guinea-yam-genome-05255.html 6/12/2020

Sebii, H., Karra, S., Bchir, B., Ghribi, A.M., Danthine, S.M., Blecker, C., Attia, H. and Besbes, S. 2019. Physico-chemical, surface and thermal properties of date palm pollen as a novel nutritive ingredient. Adv. Food Technol. Nutr. Sci. Open J., 5: 84–91.

Selvakumar R. 2018. Advances in genetics and molecular breeding for nutritional quality in carrot: A review. EC Agriculture, 4(2): 133–152.

Selvam, J.N., Kumaravadivel, N., Gopikrishnan, A., Kumar, B.K., Ravikesavan, R. and Boopathi, N. 2009. Identification of a novel drought tolerance gene in *Gossypium hirsutum* L. cv KC3. Communications in Biometry and Crop Science, 4(1): 9–13.

Severino, L.S., Auld, D.L., Baldanzi, M., Candido, M.J.D., Chen, G., Crosby, W., Tan, D., He, X., Lakshmamma, P., Lavany, C., Machado, O.L.T., Mielke, T., Milani, M., Miller, T.D., Morris, J.B., Morse, S.A., Navas, A.A., Soares, D.J., Sofiatti, V., Wang,

M.L., Zanotto, M.D. and Zieler, H. 2012. A review on the challenges for increased production of castor. Agronomy Journal, 104(4): 853–880.

Severino, L.S. and Auld, D.L. 2013. Seed abortion and the individual weight of castor seed (*Ricinus communis* L.). Industrial Crops and Products, 49: 890–896.

Severino, L.S. and Auld, D.L. 2013. A framework for the study of the growth and development of castor plant. Ind. Crops Prod., 46: 25–38.

Severino, L.S. and Auld, D.L. 2014. Study on the effect of air temperature on seed development and determination of the base temperature for seed growth in castor (*Ricinus communis* L.). Australian Journal of Crops Science, 8(2): 290–295.

Severino, L.S., Mendes, B.S.S. and Lima, G.S. 2015. Seed coat specific weight and endosperm composition define the oil content of castor seed. Industrial Crops and Products, 75B: 14–19.

Shah, S.K., Patel, C.J., Rathore, B.S. and Desai, A.G. 2015. Evaluation of castor stems residue for cellulose and lignin content. International Journal of Agriculture, Environment and Biotechnology, 8(2): 331–334.

Shah, S.S.M., Luthfi, A.A.I., Low, K.O., Harun, S., Manaf, S.F.A., Illias, R., Jamaliah Md. and Jahim, J. 2018. Preparation of kenaf stem hemicellulosic hydrolysate and its fermentability in microbial production of xylitol by *Escherichia coli* BL21. Scientific Reports, 9: 4080.

Shaheen, A.M. 2002. Morphological variation within *Ricinus communis* L. in Egypt: Fruit, leaf, seed and pollen. Pakistan Journal of Biological Sciences, 5: 1202–1206.

Shahid-ul-Islam, Rather, L.J. and Mohammad, F. 2016. Phytochemistry, biological activities and potential of annatto in natural colorant production for industrial applications – A review. Journal of Advanced Research, 7: 499–514.

Shanghai Exhibition Centre. 2017. Sisal fibre: An overview. https://www.fibre2fashion.com/industry-article/2394/sisal-fibre-an-overview 31/7/2020

Sharashkin, L. and Gold, M. 2004. Pine nuts: Species, products, markets, and potential for U.S. production. *In:* Northern Nut Growers Association 95th Annual Report. Proceeding for the 95th Annual Meeting, Columbia, Missouri, August 16-19, 2004.

Sharma, P., Gaur, V.K., Sirohi, R., Larroche, C., Kim, S.H. and Pandey, A. 2020. Valorization of cashew nut processing residues for industrial applications. Industrial Crops and Products, Vol. 152.

Sharma, S.K. and Singh, H. 2013. Pharmacognostical standardisation of *Jatropha integerrima* Jacq. (Euphorbiaceae) roots. Der Pharmacia Lettre, 5(1): 155–159.

Sharry, S., Ponce, J.L.C., Estrella, L.H., Cano, R.M.R., Lede, M. and Abedini, W. 2006. An alternative pathway for plant in vitro regeneration of chinaberry-tree *Melia azedarach* L. derived from the induction of somatic embryogenesis. Electronic Journal of Biotechnology, 9(3): 188–194.

Shemfe, M.B., Whittaker, C., Gu, S. and Fidalgo, B. 2016. Comparative evaluation of GHG emissions from the use of Miscanthus for bio-hydrocarbon production via fast pyrolysis and bio-oil upgrading. Applied Energy, 176: 22–33.

Shi, G. and Cai, Q. 2009. Cadmium tolerance and accumulation in eight potential energy crops. Biotechnology Advances, 27(5): 555–561.

Shi, G., Zhu-Ge., F., Liu, Z. and Le, L. 2014. Photosynthetic responses and acclimation of two castor bean cultivars to repeated drying-wetting cycles. Journal of Plant Interactions, 9(1): 783–790.

Shi, G., Xia, S., Ye, J., Huang, Y., Liu, C. and Zhang, Z. 2015. PEG-simulated drought stress decreases cadmium accumulation in castor bean by altering root morphology. Environmental and Experimental Botany, 111: 127–134.

Shimasaki, C.D. 2009. The Business of Bioscience: What Goes into Making a Biotechnology Product? Springer. pp. 9–26.

Shirahigue, L.D. and Ceccato-Antonini, S.R. 2020. Agro-industrial wastes as sources of bioactive compounds for food and fermentation industries. Ciência Rural, Santa Maria, 50(4): 17.

Sidhu, P.O., Kumar, V. and Behl, H.M. 2003. Variability in Triterpenoids (nimbin and salanin) composition of neem among different provenances of India. Industrial Crops and Products, 19(1): 69–75.

Sidibe, M. and Williams, J.T. 2002. Baobab. *Adansonia digitata*. International Centre for Underutilised Crops, Southampton, UK. pp. 43–50.

Silitonga, A.S., Masjuki, H.H., Ong, H.C., Yusaf, T., Kusumo, F. and Mahlia, T.M. 2016. Synthesis and optimization of *Hevea brasiliensis* and *Ricinus communis* as feedstock for biodiesel production: A comparative study. Industarial Crops and Products, 85: 274–286.

Silva Rodrigues-Corrêa, K.C., de Lima, J.C. and Fett-Neto, A.G. 2013. Oleoresins from pine: Production and industrial uses. *In:* Ramawat, K. and Mérillon, J.M. (eds.), Natural Products. Springer, Berlin, Heidelberg.

Silva, B.B.R., Santana, R.M.C. and Forte, M.M.C. 2010. A solventless castor oil-based PU adhesive for wood and foam substrates. International Journal of Adhesion & Adhesives, 30: 559–565.

Silva, G.E., Ramos, F.T., de Fajra, A.P. and Franca, M.G. 2014. Seeds' physicochemical traits and mucilage protection against aluminum effect during germination and root elongation as important factors in a Biofuel seed crop (*Ricinus communis*). Environmental Science and Pollution Research International, 21(19): 11572–11579.

Silva, J.A.C., Soares, V.F., Fernandez-Lafuente, R., Habert, C. and Freire, D.M.G. 2015. Enzymatic production and characterization of potential biolubricants from castor bean biodiesel. J. Mol. Catalysis B: Enzymatic, 122: 323–329.

Silva, M., Andrade, S.A.L. and De-Campos, A.B. 2018. Phytoremediation potential of jack bean plant for multi-element contaminated soils from Ribeira Valley, Brazil. Clean – Soil, Air, Water, 11.

Singh, A., Kumar, K., Gill, M.I.S., Chhuneja, P., Arora, N.K. and Singh, K. 2013. Genotype identification and inference of genetic relatedness among different purpose grape varieties and rootstocks using microsatellite markers. African Journal of Biotechnology, 12(2): 134–141.

Singh, A.A., Afrin, S. and Karim, Z. 2017. Green composites: Versatile material for future. *In:* pp. 29-44. M. Jawaid et al. (eds.), Green Biocomposites, Green Energy and Technology. Springer International Publishing AG.

Singh, A.K., Bharati, R.C., Manibhushan, N.C. and Pedpati, A. 2013. An assessment of faba bean (*Vicia faba* L.) current status and future prospect. African Journal of Agricultural Research, 8(50): 6634–6641.

Singh, A.S., Kumari, S., Modic, A.R., Bhavesh, B., Gajera, B.B., Narayanan, S. and Kumard, N. 2015. Role of conventional and biotechnological approaches in genetic improvement of castor (*Ricinus communis* L.). Industrial Crops and Products, 74: 55–62.

Singh, B., Kaur, J. and Singh, K. 2010. Production of biodiesel from used mustard oil and its performance analysis in internal combustion engine. ASME. Journal of Energy Resources and Technology, 132(3): 031001.

Singh, D.P., Kumar, N., Bhargava, S.K. and Barman, S.C. 2010. Accumulation and translocation of heavy metals in soil and plants from fly ash contaminated area. Journal of Environmental Biology, 31: 421–430.

Singh, K., Kumar, R., Chaudhary, V., Sunil, V., Arya, A.M. and Sharma, S. 2019. Sugarcane bagasse: Foreseeable biomass of bio-products and biofuel: An overview. Journal of Pharmacognosy and Phytochemistry, 8(2): 2356–2360.

Singh, L. and Singh, J. 2019. Medicinal and nutritional values of drumstick tree (*Moringa oleifera*): – A review. International Journal of Current Microbiology and Applied Science, 8(5): 1965–1974.

Singh, R. and Geetanjali. 2015. Phytochemical and pharmacological investigations of *Ricinus communis* Linn. Algerian Journal of Natural Products, 3(1): 120–129.

Singh, S., Sharma, S. and Mohapatra, S.K. 2015. A production of biodiesel from waste cotton seed oil and testing on small capacity diesel engine. International Journal of Advance Research in Science and Engineering, 4(2): 172–178.

Singh, V.P., Arulanantham, A., Parisipogula, V., Arulanantham, S. and Biswas, A. 2018. *Moringa olifera*: Nutrient dense food source and world's most useful plant to ensure nutritional security, good health and eradication of malnutrition. European Journal of Nutrition & Food Safety, 8(4): 204–214.

Singla, N., Thind, R.K. and Mahal, A.K. 2014. Potential of eucalyptus oil as repellent against house rat, *Rattus rattus*. The Scientific World Journal, 2014: 7. Article ID 249284.

Siragi, D.B.M., Desmecht, D., Hima, H.I., Mamane, O.S. and Natatou, I. 2021. Optimization of activated carbons prepared from *Parinari macrophylla* shells. Materials Sciences and Applications, 12: 207–222.

Siritunga, D. and Sayre, R.T. 2003. Generation of cyanogen-free transgenic cassava. Planta, 217: 367–373.

Słomińska-Wojewódzka, M. and Sandvig, K. 2013 Ricin and ricin-containing immunotoxins: Insights into intracellular transport and mechanism of action in vitro. Antibodies, 2: 236–269, doi:10.3390/antib2020236.

Smith, R.E. 2014. Pomegranate: Botany, Postharvest Treatment, Biochemical Composition and Health Effects. Nova Science Publishers, Inc. New York.

Sobhani, L. and Ziarati, P. 2017. Study on potential bio-adsorption of Tangerine peel in removal of heavy metals: Pb, Cd and Ni of vegetable coriander. Journal of Science and Discovery, 1(2): 1–8.

Soetan, K.O. 2011. The role of biotechnology towards attainment of a sustainable and safe global agriculture and environment – A review. Biotechnology and Molecular Biology Review, 6(5): 109–117.

Sogan, N., Kapoor, N., Kala, S., Patanjali, P.K., Nagpal, B.N., Vikram, K. and Valecha, N. 2018. Larvicidal activity of castor oil nanoemulsion against malaria vector *Anopheles culicifacies*. International Journal of Mosquito Research, 5(3): 01–06.

Soh, A.C., Mayes, S. and Roberts, J. 2017. Oil Palm Breeding: Genetics and Genomics. CRC Press, Taylor & Francis Group, Boca Raton.

Sokoto, A.M. and Bhaskar, T. 2018. Pyrolysis of waste castor seed cake: A thermo-kinetics study. European Journal of Sustainable Development Research, 2(2): 18.

Sonké, B., Dansi, A., Djedatin, G., Mariac, C., Couderc, M., Causse, S., Alix, K., Chaïr, H., François, O. and Vigouroux. Y. 2019. Yam genomics supports West Africa as a major cradle of crop domestication. Science Advances, 5(5): 1–7.

Soto-León, S., Lopez-Camacho, E., Milan-Carrillo, J., Sanchez-Castillo, M.A., Cuevas-Rodríguez, E., Picos-Corrales, L.A. and Contreras-Andrade, I. 2014. Jatropha cinerea seed oil as a potential non-conventional feedstock for biodiesel produced by an ultrasonic process. Revista Mexicana de Ingeniería Química, 13(3): 739–747.

Souad, A.M., Jamal, P. and Olorunnisola, K.S. 2012. Effective jam preparations from watermelon waste. International Food Research Journal, 19(4): 1545–1549.

Sousa, A.F., Matos, M., Pinto, R.J.B., Freire, C.S.R. and Silvestre, A.J.D. 2014. One-pot synthesis of biofoams from castor oil and cellulose microfibers for energy absorption impact materials. Cellulose, 21: 1723–1733.

Sousa, N.L., Cabral, G.B., Vieira, P.M., Aisy, B. Baldoni and Aragão, F.J.L. 2017. Bio-detoxification of ricin in castor bean (*Ricinus communis* L.) seeds. Scientific Reports, 7: 1–9, 15385.

Souza Costa, E.T., Guilherme, L.R.G., de Melo, É.E.C., Ribeiro, B.T., dos Santos, B., Inácio, E., da Costa Severiano, E., Faquin, V. and Hale, B.A. 2012. Assessing the tolerance of castor bean to Cd and Pb for phytoremediation purposes. Biol. Trace Elem. Res., 145(1): 93–100.

Souzaa, F.V.D., Kaya, E., Vieira, L., Souzab, A., Carvalhoe, M., Emanuela Barbosa Santose, Alvesa, A.A.C. and Ellisa, D. 2017. Cryopreservation of Hamilin sweet orange (*Citrus sinensis* (L.) Osbeck) embryogenic calli using a modified aluminum cryo-plate technique. Scientia Horticulturae, 224: 302–305.

Srinivasakumar, P., Nandan, M.J., Kiran, C.U. and Rao, K.P. 2013. Sisal and its potential for creating innovative employment opportunities and economic prospects. IOSR Journal of Mechanical and Civil Engineering, 8(6): 1–8.

Srinivasarao, Ch., Shanker, A.K., Kundu, S. and Reddy, S. 2016. Chlorophyll fluorescence induction kinetics and yield responses in rainfed crops with variable potassium nutrition in K deficient semi-arid alfisols. J. Photochem. Photo Biol. B: Biol., 160: 86–95.

Srivastava, S.K. and Kumar, J. 2015. Response of castor (*Ricinus communis* L.) to sulphur under irrigated conditions of Uttar Pradesh, India. Plant Archives, 15(2): 879–881.

Steef, J., Lips, J. and Jan van Dam, E.G. 2013. Kenaf fibre crop for bioeconomic industrial development. pp. 105–143. *In:* A. Monti and E. Alexopoulou (eds.), Kenaf: A Multi-Purpose Crop for Several Industrial Applications. Green Energy and Technology, Springer-Verlag London.

Stephan, W.U., Schmidke, L. and Pich, A. 1994. Phloem translocation of Fe, Cu, Mn, and Zn in *Ricinus* seedlings in relation to the concentrations of nicotianamine, an endogenous chelator of divalent metal ions, in different seedling parts. Plant and Soil, 165: 181–188.

Steven, M., Ferguson, C.L.P., Uma, S. and Kaundinya, C.L.P. 2014. Licensing the technology: Biotechnology commercialization strategies using university and federal labs. pp. 185–206. *In:* Shimaki, G. (ed.), Biotechnology Entrepreneurship Starting, Managing, and Leading Biotech Companies. Elsevier.

Stevens, C.O., Ugese, F.D. and Baiyeri, K.P. 2015. Utilization potentials of *Moringa oleifera* in Nigeria: A preliminary assessment. Letters of Natural Sciences, 40: 30–37.

Sugandha, S. and Shashi, R. 2006. Spectroscopic determination of total phenol, flavonoid content and anti-oxidant activity in different parts of *Adansonia digitata* L.: An important medicinal tree. European Journal of Pharmaceutical and Medical Research, 4(2): 549–552.

Suh, M.C., Hyung, N. and Chung, C. 2010. Molecular biotechnology of sesame. pp. 219–243. *In:* Bedigian, D. (ed.), Medicinal and Aromatic Plants – Industrial Profiles. Sesame the Genus Sesamum. CRC Press, Taylor & Francis Group, Boca Raton, London.

Sujatha, M., Reddy, T.P. and Mahasi, M.J. 2008. Role of biotechnological interventions in the improvement of castor (*Ricinus communis* L.) and *Jatropha curcas* L. Biotechnology Advances, 26: 424–435.

Suleria, H.A.R., Butt, M.S., Khalid, N., Sultan, S., Raza, A., Aleem, M. and Abbas, M. 2015. Garlic (*Allium sativum*): Diet based therapy of 21st century – A review. Asian Pacific Journal of Tropical Disease, 5(4): 271–278.

Sultana, R.S. and Rahman, M.M. 2014. Melon crops improvement through biotechnological techniques for the changing climatic conditions of the 21st century. International Journal of Genetics and Genomics, 2(3): 30–41.

Sultana, S., Asif, H.M., Akhtar, N., Iqbal, A., Nazar, H. and Riaz Ur Rehman. 2015. *Nigella sativa*: Monograph. Journal of Pharmacognosy and Phytochemistry, 4(4): 103–106.

Sun, Y., Niu, G., Osuna, P., Ganjegunte, G., Auld, D., Zhao, L., Peralta-Videa, J.R. and Gardea-Torresdey, J.L. 2013. Seedling emergence, growth, and leaf mineral nutrition of *Ricinus communis* L. cultivars irrigated with saline solution. Ind. Crops Prod., 49: 75–80.

Sundaram, R.M., Viraktamath, B.C., Rani, N.S. and Sarla, N. 2010. Rice genetics. Current Science, 98(1): 1422–1426.

Superales, J.B. 2016. Carbon dioxide capture and storage potential of mahogany (*Swietenia macrophylla*) saplings. International Journal of Environmental Science and Development, 7(8): 611–614.

Supriyaa, R. and Priyadarshan, P.M. 2019. Genomic technologies for Hevea breeding. Advances in Genetics, 104: 1–74.

Surafel, S., Gamachu, O. and Abel, D. 2018. In vitro propagation of *Aloe vera* Linn. from shoot tip culture. GSC Biological and Pharmaceutical Sciences, 04(02): 001–006.

Suryawanshi, M.A., Mane, V.B. and Kumbhar, G.B. 2016. Methodology to extract essential oils from lemongrass leaves: Solvent extraction approach. International Research Journal of Engineering and Technology, 3(8): 1775–1780.

Swaminathan, K., Chae, W.B., Mitros, T., Varala, K., Xie, L., Barling, A., Glowacka, K., Hall, M., Jezowski, S., Ming, R., Hudson, M., Juvik, J.A., Rokhsar, D.S. and Moose, S.P. 2013. A framework genetic map for *Miscanthus sinensis* from RNA seq-based markers shows recent tetraploidy. BMC Genomics, 13: 142.

Syukur, M., Lestari, E.G., Purnamaningsih, R., Yunita, R., Aisyah, S.I. and Firdaus, R. 2011. Adaptability of mutant genotypes of artemisia (*Artemisia annua* L.) as result of gamma irradiation in three locations with different altitude. AGRIVITA, 33(3): 251–256.

Tahir, S.M., Usman, I.S., Katung, M.D. and Ishiyaku, M.F. 2013. Micro propagation of wormwood (*Artemisia annua* L.) using leaf primordia. Science World Journal, 8(1): 1–7.

Tajuddin, N.A., Rosli, N.H., Abdullah, N., Mohd Firdaus Yusoff, M.E. and Salimon, J. 2014. Estolide ester from *Ricinus communis* L. seed oil for biolubricant purpose. The Malaysian Journal of Analytical Sciences, 18(1): 85–93.

Tamiru, M., Natsume, S., Takagi, H., White, B., Yaegashi, H., Shimizu, M., Yoshida, K., Uemura, A., Oikawa, K., Abe, A. and Urasaki, N. 2017. Genome sequencing of the staple food crop white Guinea yam enables the development of a molecular marker for sex determination. BMC Biology, 15: 1–20.

Tan, Q.G., Cai, H.X., Du, Z.Z. and Luo, D.X. 2009. Three terpenoids and a tocopherol related compound from *Ricinus communis* L. Helevetica Chimica Acta, 92: 2762–2768.

Tandug, L.M. 2008. Biomass and carbon sequestration of *Gmelina arborea* Roxb. AGRIS Conferece. Available from http://agris.fao.org/agris-earch/search. do?recordID=PH2008000850 Accessed 23/4/2020

Tang, T., He, L., Feng Peng, F. and Shi, S. 2011. Habitat differentiation between estuarine and inland *Hibiscus tiliaceus* L. (Malvaceae) as revealed by retrotransposon-based SSAP marker. Australian Journal of Botany, 59: 515–522.

Tang, Z., Shi, L. and Aleid, S.M. 2014. Date and their processing byproducts as substrates for bioactive compounds production. Brazilian Archives of Biology and Technology, 57(5): 706–713.

Teixeira da Silva, J., Dobránszki, J. and Renata Rivera-Madrid, R. 2018. The biotechnology (genetic transformation and molecular biology) of *Bixa orellana* L. (achiote). Planta, 248: 287–277.

Tembe, E.T., Azeh, E. and Ekhuemelo, D.O. 2017. Effect of particle size on quality of briquettes produced from sawdust of *Daniella oliveiri* and *Gmelina arborea* in Makurdi, Benue State, Nigeria. Asian Research Journal of Agriculture, 3(4): 1–7.

Thakur, M., Kumar, R., Kanwar, R. and Mehta, D.K. 2017. Genetic studies for different seed traits in cucumber (*Cucumis sativus* L.). International Journal of Current Microbiology and Applied Sciences, 6(11): 3939–3949.

Thareja, G., Mathew, S., Mathew, L.S., Mohamoud, Y.A., Suhre, K. and Malek, J.A. 2018. Genotyping-by-sequencing identifies date palm clone preference in agronomics of the State of Qatar. PLoS ONE, 13(12): e0207299.

The Aloe vera Handbook for Product Development. Available from www.amb-wellness. com, Accessed 21/10/2019

The Max Planck Society. 2020. *Artemisia annua* to be tested against COVID-19. Available at https://www.mpg.de/14663263/artemisia-annua-to-be-tested-against-covid-19 Accessed 2/5/2020

Thimmappaiah, Shobha D., Mohana, G.S., Dinakara, J.A. and Bhat, P.G. 2016. Fingerprinting of released varieties of cashew based on DNA markers. Vegetos, 29: 4.

Thomas, R., Sah, N.K. and Sharma, P.B. 2008. Therapeutic biology of *Jatropha curcas*: A mini review. Current Pharmaceutical Biotechnology, 9(4): 315–324.

Thondaiman, V., Rajamani, K., Senthil, N., Shoba, N. and Joel, A.J. 2013. Genetic diversity in cocoa (*Theobroma cacao* L.) plus trees in Tamil Nadu by simple sequence repeat (SSR) markers. African Journal of Biotechnology, 12(30): 4747–4753.

Thwe-Thwe-Win, Hirao, T., Watanabe, A. and Goto, S. 2015. Current genetic structure of teak (*Tectona grandis*) in Myanmar based on newly developed chloroplast single nucleotide polymorphism and nuclear simple sequence repeat markers. Tropical Conservation Science, 8(1): 235–256.

Tietela, Z., Kahremany, S., Cohenb, G. and Ogen-Shternb, N. 2021. Medicinal properties of jojoba (*Simmondsia chinensis*). Israel Journal of Plant Science, 38–47.

Tiloke, C., Ananda, K., Genganb, R.M. and Chuturgoona, A.A. 2018. *Moringa oleifera* and their phytonanoparticles: Potential antiproliferative agents against cancer. Biomedicine & Pharmacotherapy, 108: 457–466.

Tiwari, S., Kumar, S. and Gontia, I. 2011. Biotechnological approaches for sesame (*Sesamum indicum* L.) and niger (*Guizotia abyssinica* L.f. Cass.). Asia Pacific Journal of Molecular Biology and Biotechnology, 19(1): 1–8.

Tofanica, B.M. 2019. Rapeseed – A valuable renewable bioresource. Cellulose Chemistry and Technology, 53(9-10): 837–849.

Tonukari, N.J. 2004. Fostering biotechnology entrepreneurship in developing countries. African Journal of Biotechnology, 3(6): 299–301.

Tounou, A.K., Mawussi, G., Amadou, S., Agboka, K., Gumedzoe, Y.M. and Dieudonné, S.K. 2011. Bio-insecticidal effects of plant extracts and oil emulsions of *Ricinus communis* L. (Malpighiales: Euphorbiaceae) on the diamondback, *Plutella xylostella* L. (Lepidoptera: Plutellidae) under laboratory and semi-field conditions. Journal of Applied Biosciences, 43: 2899–2914.

Tripathi, V., Edrisi S.A. and Abhilash, P.C. 2016. Towards the coupling of phytoremediation with bioenergy production. Renewable and Sustainable Energy Reviews, 57: 1386–1389.

Tripathi, V., Edrisi, S.A., O'Donovan, A., Gupta, V.K. and Abhilash, P.C. 2016. Bioremediation for fueling the biobased economy. Trends in Biotechnology, 34(10): 775-777.

Trivedi, M.K., Branton, A., Trivedi, D., Nayak, G., Gangwar, M. and Jana, S. 2015. Analysis of genetic diversity using simple sequence repeat (SSR) markers and growth regulator response in biofield treated cotton (*Gossypium hirsutum* L.). American Journal of Agriculture and Forestry, 3(5): 216–221.

Tropical Plants Database, Ken Fern. tropical.theferns.info. 2018-12-03. <tropical.theferns.info/viewtropical.php?id=Terminalia+mantaly>

Udgire, M.S. and Pathade, G.R. 2014. Antibacerial activity of Aloe vera against skin pathogens. American Journal of Ethnomedicine, 1(3): 147–151.

Universty of Ibadan website. 2020. COVID-19: UI scientists recommend *Euphorbia hirta* Linn. for relief. Available at https://www.ui.edu.ng/news/covid-19-ui-scientists-recommend-euphorbia-hirta-linn-relief Accessed 2/5/2020

Upma, Ashok, K., Pankaj, K. and Tarun, K. 2011. The nature's gift too mankind: Neem. Int. Res. J. Pharm., 2(10): 13–15.

Usman, L.A., Ameen, O.M., Lawal, A. and Awolola, G.V. 2009. Effect of alkaline hydrolysis on the quantity of extractable protein fractions (prolamin, albumin, globulin and glutelin) in *Jatropha curcas* seed cake. African Journal of Biotechnology, 8(22): 6374–6378.

Usman, L.A., Oluwaniyi, O.O., IbiyemI, S.A., Muhammad, N.O. and Ameen, O.M. 2009. The potential of oleander *Thevetia peruviana* in African agricultural and industrial development: A case study of Nigeria. J. Appl. Biosci., 24: 1477–1487.

Utomo, D.T. and Jumiatun, B.E. 2022. Modeling the potential of cereal crops with a smart village based GIS approach to support food security. The 4th International Conference on Food and Agriculture. IOP Conf. Series: Earth and Environmental Science, 980: 13, 01205.

Uyan, M., Alptekin, F.M., Bastabak, B., Ozgul, S., Erdogan, B., Ogut, T.C., Sezer, U. and Celiktas, M.S. 2020. Combined biofuel production from cotton stalk and seed with a biorefinery approach. Biomass Conversion and Biorefinery, 10: 393–400.

Vaithanomsat, P. and Apiwatanapiwat, W. 2009. Feasibility study on vanillin production from *Jatropha curcas* stem using steam explosion as a pretreatment. World Academy of Science, Engineering and Technology, 53: 956–959.

Varnero, M.T. and Homer, I. 2017. Biogas production. pp. 187–193. *In:* Crop Ecology, Cultivation and Uses of Cactus Pear. IX International Congress on Cactus Pear and Cochineal Cam Crops for a Hotter and Drier World Coquimbo, Chile, 26-30 March 2017.

Veeresham, C. 2012. Natural products derived from plants as a source of drugs. Journal of Advanced Pharmaceutical and Technology Research, 3(4): 200–201.

Velasco, L., Fernández-Cuesta, A., Pascual-Villalobos, M.J. and Fernández-Martínez, J.M. 2015. Variability of seed quality traits in wild and semi-wild accessions of castor collected in Spain. Industrial Crops and Products, 65: 203–209.

Vella, M., Cautela, D. and Laratta, B. 2019. Characterization of polyphenolic compounds in cantaloupe melon by-products filomena. Foods, 8: 196.

Venkatachalam, P., Jayasree, P., Sushmakumari, S., Jayashree, R., Rekha, K., Sobha, S., Priya, P., Kala, R.G. and Thulaseedharan, A. 2007. Current perspectives on application of biotechnology to assist the genetic improvement of rubber tree (*Hevea brasiliensis* Muell. Arg.): An overview. Functional Plant Science and Biotechnology, 1(1): 1–17.

Venugopalan, A., Giridhar, P. and Ravishankar, G.A. 2011. Food, ethanobotanical and diversified applications of *Bixa orellana* L.: A scope for its improvement through biotechnological mediation. Indian Journal of Fundamental and Applied Life Sciences, 1(4): 9–31.

Verghese., M., Thomson, J.C. and Willis, S. 2018. Antioxidant content and capacity of moringa leaf. Nutrition and Food Toxicology, 3.3(2018): 672–679.

Verhaegen, D., Inza, J.F., Logossa, A.Z. and Ofori, D. 2010. What is the genetic origin of teak (*Tectona grandis* L.) introduced in Africa and in Indonesia? Tree Genetics & Genomes, Springer, 6: 717–733.

Verma, P., Bijalwan, A., Shankhwar, A.K., Dobriyal, M.J.R., Jacob, V., Kumar, S. and Rathaude, S.K. 2017. Scaling up an Indigenous Tree (*Gmelina arborea*) based agroforestry systems in India. International Journal of Science and Qualitative Analysis, 3(6): 73–77.

Veronesi, F., Brummer, E.C. and Huyghe, C. 2010. Alfalfa. pp. 395–438. *In:* B. Boller et al. (eds.), Fodder Crops and Amenity Grasses, Handbook of Plant Breeding 5, Springer Science+Business Media, New York, NY 10013, U.S.A.

Verwer, C., Peña-Claros, M., Staak, D., Ohlson-Kiehn, K. and Sterck, F.J. 2008. Silviculture enhances the recovery of overexploited mahogany *Swietenia macrophylla*. Journal of Applied Ecology, 45: 1770–1779.

Viel, Q. 2013. Interface properties of bio-based composites of polylactic acid and bamboo fibers. Mechanical (and Materials) Engineering – Dissertations, Theses, and Student Research, 56. https://digitalcommons.unl.edu/mechengdiss/56

Vignesh, V., Vijayan, S. and Selvakumar, G. 2021. Nanometer-scale mechanical properties of MWCNT-mustard oil nanofluid as a potential base stoke. Journal of Chi Chemical Society, 5051–5052.

Vijayanand, C., Kamaraj, S., Sriramajayam, S. and Ramesh, D. 2016. Biochar production from arecanut waste. International Journal of Farm Sciences, 6(1): 43–48.

Vilar, D., Vilar, M.S., Accioly de Lima e Moura, T.F., Raffin, F.N., Márcia Rosa de Oliveira, Franco, C.F., Filgueiras de Athayde-Filho, P., Formiga Melo Diniz, M. and Barbosa-Filho, J. 2014. Traditional uses, chemical constituents, and biological activities of *Bixa orellana* L.: A review. The Scientific World Journal, 2014: 11.

Visser, E.M., Filho, D.O., Martins, M.A. and Steward, B.L. 2011. Bioethanol production potential from Brazillian biodiesel co-products. Biomass Bioenergy, 35: 489–494.

Vivodík, M., Balážová, Z., Gálová, Z. and Hlozáková, T.K. 2015. Evaluation of molecular diversity of new castor lines (*Ricinus communis* L.) using random amplified polymorphic DNA markers. Horticultural Biotechnology Research, 1(1): 1–4.

Vivodík, M., Balážová, Z., Gálová, Z., Chňapek, M. and Petrovicová, L. 2015. Study of DNA polymorphism of the castor new lines based on RAPD markers. J. Microbiol. Biotech. Food Sci., 2: 125–127.

Vivodík, M., Saadaoui, E., Balážová, Z., Gálová, Z. and Petrovičová, L. 2019. Genetic diversity in Tunisian castor genotypes (*Ricinus communis* L.) detected using RAPD markers. Potravinarstvo Slovak Journal of Food Sciences, 13(1): 294–300.

Vwioko, D.E. and Fashemi, D.S. 2005. Growth response of *Ricinus cummunis* L. (castor oil) in spent lubricating oil polluted soil. J. Applied Sci. Environ. Manage., 9(2): 73–79.

Vwioko, D.E., Anoliefo, G.O. and Fashemi, S.D. 2006. Metal concentration in plant tissues of *Ricinus communis* L. (Castor oil) grown in soil contaminated with spent lubricating oil. J. Appl. Sci. Environ. Manage., 10(3): 127–134.

Wadenbäck, J., von Arnold, S., Egertsdotter, U., Walter, M.H., Grima-Pettenati, J., Goffner, D., Gellerstedt, G., Gullion, T. and Clapham, D. 2008. Lignin biosynthesis in transgenic Norway spruce plants harboring an antisense construct for cinnamoyl CoA reductase (CCR). Transgenic Res., 17(3): 379–392.

Wadl, P.A., Olukolu, B.A., Branham, S.E., Jarret, R.L., Yencho, G.C. and Jackson, D.M. 2018. Genetic diversity and population structure of the USDA sweet potato (*Ipomoea batatas*) germplasm collections using GBSpoly. Frontiers in Plant Science, 9: 1166.

Wafa, G., Amadou, D., Larbi, K.M. and Héla, F.O. 2014. Larvicidal activity, phytochemical composition, and antioxidant properties of different parts of five populations of *Ricinus communis* L. Industrial Crops and Products, 56: 43–51.

Wagner, M., Kiesel, A., Hastings, A. and Lewandowski, I.Y. 2017. Novel miscanthus germplasm-based value chains: A life cycle assessment. Frontier Plant Science, 8: 990.

Wang, C., Li, G.R., Zhang, Z.Y., Peng, M., Shang, Y.S., Luo, R. and Chen, Y.S. 2013. Genetic diversity of castor bean (*Ricinus communis* L.) in Northeast China revealed by ISSR markers. Biochemical Systematics and Ecology, 51: 301–307.

Wang, K., Huang, H., Zhu, Z., Li, T., He, Z., Yang, X. and Alva, A. 2013. Phytoextraction of metals and rhizoremediation of PAHs in co-contaminated soil by co-planting of sedum alfredii with ryegrass (*Lolium perenne*) or castor (*Ricinus communis*). International Journal of Phytoremediation, 15(3): 283–298.

Wang, L., Jiang, X., Wang, L., Wang, W., Fu, C., Yan, X. and Geng, X. 2019. A survey of transcriptome complexity using PacBio single-molecule real-time analysis combined with Illumina RNA sequencing for a better understanding of ricinoleic acid biosynthesis in Ricinus communis. BMC Genomics, 20: 456.

Wang, S., Zhao, Y., Guo, J. and Zhou, L. 2016. Effects of Cd, Cu and Zn on *Ricinus communis* L. growth in single element or co-contaminated soils: Pot experiments. Ecol. Eng., 90: 347–351.

Wang, Z. and Brummer, E. 2012. Is genetic engineering ever going to take off in forage, turf and bioenergy crop breeding? Annals of Botany, 110: 1317–1325.

Wani, S.H. and Sah, S.K. 2015. Transgenic plants as expression factories for bio-pharmaceuticals. Research & Reviews: Journal of Botanical Sciences – Commentary, 1–4.

Wannapokin, A. 2016. Biogas production potential from fallen teak leaves (*Tectona grandis*). Journal of Fundamental Renewable Energy and Applictions, 6(Suppl 3) http://dx.doi.org/10.4172/2090-4541.C1.009

Warra, A.A. 2011. Sesame (*Sesamum indicum* L.) seed oil methods of extraction and its prospects in cosmetic industry: A review. Bayero Journal of Pure and Applied Sciences, 4(2): 164–168.

Warra, A.A. 2012. Extraction and saponification of gingerbread plum (*Parinari macrophylla*) seed oil. Journal of Scientific Theory and Methods, 168–188.

Warra, A.A. 2012. Production of Soap from an Indigenous *Moringa oliefera* Lam seed oil. Journal of Raw Materials Research, 7(1&2): 23–30.

Warra, A.A. 2012. The science and technology of soap production (utilizing indigenous raw materials). Raw Materials Research and Development Council, Abuja, Nigeria. pp. 76–100.

Warra, A.A. 2014. Cosmetic Potential of oil extracts from seeds and nuts commonly found in Nigeria. Vol. 1. Ahmadu Bello University Press Limited, Zaria, Nigeria. 103 pp.

Warra A.A. 2015a. Castor seed oil and its potential cosmetic and pharmaceutical applications. Achieves of Scientific Research, 1(1): 19–22.

Warra, A.A. 2015b. Physico-chemical and GC/MS analysis of wild castor (*Ricinus communis* L.) seed oil. Applied Science Reports, 9(3): 123–128.

Warra, A.A. 2015c. Physico-chemical and GC/MS analysis of castor bean (*Ricinus communis* L.) seed oil. Chemistry and Materials Research, 7(2): 56–60.

Warra, A.A., Sheshi, F., Ayurbami, H.S. and Abubakar, A. 2015. Physico-chemical, GC-MS analysis and cold saponification of canary melon (*Cucumis melo*) seed oil. Trends in Industrial Biotechnology Research, 1: 10–17.

Warra, A.A. 2016. Soap production from quality assessed gingerbread plum (*Neocarya macrophylla*) seed oil. Journal of Scientific Research in Pharmaceutical, Chemical & Biological Sciences, 1(1): 30–40.

Warra, A.A. and Prasad, M.N.V. 2016. *Jatropha curcas* L. cultivation on constrained land: Exploring the potential for economic growth and environmental protection. pp. 129–147. *In:* M.N.V. Prasad (ed.), Bioremediation and Bioeconomy. Amsterdam: Elsevier.

Warra, A.A., Babatola, L.J., Abubakar, F., Abbas, A. and Nasarawa, A.A. 2016a. Quality characteristics and cold saponification of hexane extract of two varieties of sesame seed (*Sesamun indicum* L.) oil. Scientia Agriculturae, 16(3): 83–88.

Warra, A.A., Jonathan, B.L, Ibrahim, B.D. and Adedara, A.O. 2016b. GC-MS analysis of hexane extracts of two varieties of sesame (*Sesamun indicum* L.) seed oil. International Journal of Chemistry, Pharmacy & Technology, 1(1): 1–9.

Warra, A.A. 2017. Physico-chemical, gas chromatography-mass spectrometry (GC-MS) analysis and soap production from Thervetia peruviana seed oil. Austin J. Biotechnol. Bioeng., 4(1): 1072.

Warra, A.A. 2017. Shea butter chemical composition, quality characteristics and its personal care applications. Ahmadu Bello University Press Limited, Zaria, Nigeria. pp. 16–22.

Warra, A.A., Babatola, L.J., Omodolapo, A.A. and Ibraheem, B.D. 2017. Characterization of oil extracted from two varieties of tiger nut (*Cyperus esculentus* L.) tubers. American Journal of Heterocyclic Chemistry, 3(3): 28–36.

Warra, A.A. and Prasad, M.N.V. 2018. Artisanal and small-scale gold mine waste in Nigeria – Rehabilitation with energy crops and native flora. pp. 473–491. *In:* Prasad, Favas, Maiti (eds.), BioGeotechnologies for Mine Site Rehabilitation. Elsevier, U.S.A.

Warra, A.A. and Prasad, M.N.V. 2019. Transgenic Euphorbiaceae for remediation of toxic metals and metalloids. pp. 131–154. *In:* M.N.V. Prasad, Shabir H. Wani (eds.), Transgenic Plant Technology for Phytoremediation of Toxic Metals and Metalloids. Elsevier, USA.

Warra, A.A., Hassan, L.G., Babatola, L.J., Omodolapo, A.A., Ukpanukpong, R.U. and Ayebaitari Berena, G. 2019. Characterization of *Neocarya macrophylla* seed oil using gas chromatography-mass spectrometry (GC-MS) and fourier transform infra-red (FT-IR) techniques. Chemical Science International Journal, 28(3): 1–7.

Warra, A.A., Garba, N.A. and Hassan, L.G. 2020. Castor plant: A potent crop for bioenergy medicinal and industrial applications. University Press Limited. Usmanu Danfodiyo University, Sokoto Nigeia. ISBN 978-978-980-767-3. pp. 1–63.

Warren, R., Keeling, C., Yuen, M.M., Raymond, A., Taylor, G., Vandervalk, B.P., Mohamadi, H., Paulino, D., Chiu, R., Jackman, S., Robertson, G.A., Yang, C., Boyle, B., Hoffmann, M., Weigel, D., Nelson, D., Ritland, C., Isabel, N., Jaquish, B., Yanchuk, A., Bousquet, J., Jones, S., Mackay, J., Birol, I. and Bohlmann, J. 2015. Improved white spruce (*Picea glauca*) genome assemblies and annotation of large gene families of conifer terpenoid and phenolic defense metabolism. The Plant Journal for Cell and Molecular Biology, 83(2): 189–212.

Warzecha, H. 2008. Biopharmaceuticals from plants: A multitude of options for posttranslational modifications. Biotechnology and Genetic Engineering Reviews, 25: 315–330.

Webber, C.L. III, Bhardwaj, H.L. and Bledsoe, V.K. 2002. Kenaf production: Fiber, feed, and seed. pp. 327–339. *In:* J. Janick and A. Whipkey (eds.), Trends in New Crops and New Uses. ASHS Press, Alexandria, VA.

Wei, R., Guo, Q., Wen, H., Liu, C., Yang, J., Peters, M., Hu, J., Zhu, G., Zhang, H., Tian, L., Han, X., Ma, J., Zhu, C. and Wan, Y. 2016. Fractionation of stable cadmium isotopes in the cadmium tolerant *Ricinus communis* and hyperaccumulator *Solanum nigrum*. Scientific Reports, 6. Article. No. 24309. doi:10.1038/srep24309.

Wei, X., Liu, K., Zhang, Y., Feng, Q., Wang, L., Zhao, Y., Li, D., Zhao, Q., Zhu, X., Zhu, X., Li, W., Fan, D., Gao, Y., Lu, Y., Zhang, X., Tang, X., Zhou, C., Zhu, C., Liu, L., Zhong, R., Tian, Q., Wen, Z., Weng, Q., Han, B., Huang, X. and Zhang, X. 2015. Genetic discovery for oil production and quality in sesame. Nature Communications, 1–10.

Welker, C.M., Balasubramanian, V.K., Petti, C., Rai, K.M., DeBolt, S. and Mendu, V. 2015. Engineering plant biomass lignin content and composition for biofuels and bioproducts. Energies, 8: 7654–7676.

Wendt, T. and Mullins, E. 2011. Future challenges and prospects. pp. 260–268. *In:* James, M., Bradeen, P. and Chittaranjan Kole (eds.), Genetics, Genomics and Breeding of Potato. Science Publishers, Enfield, NH 03748, USA.

Weng, Y. and Sun, Z.Y. 2011. Major cucurbits. pp. 1–15. *In:* Yi-Hong Wang, T.K. Behera, Chittaranjan Kole (eds.), Genetics, Genomics and Breeding of Cucurbits. CRC Press.

Wettasinghe, R. 2012. Development of castor (*Ricinus communis*) var. Brigham with ultra low ricin content by analyzing soluble seed proteins. PhD. Thesis. Texas Tech University. 82 pp.

Wettasinghe, R.C., Zabet-Moghaddam, M., Ritchie, G. and Auld, D.L. 2013. Relative quantitation of ricin in *Ricinus communis* seeds by image processing. Industrial Crops and Products, 50: 654–660.

Wiggins, S., Henley, G. and Keats, S. 2015. Competitive or complementary? Industrial crops and food security in sub-Saharan Africa. ODI Report. Overseas Development Institute, London. pp. 1–37.

Wiszniewska, A., Hanus-Fajerska, E., Muszynska, E. and Ciarkowska, K. 2016. Natural organic amendments for improved phytoremediation of polluted Soils: A review of recent progress. Pedosphere, 26(1): 1–12.

Wolukau, J.N., Gesimba, R.M., Nyamari, J.K., Chen, J.F. and Zhou, X.H. 2018. RAPD analysis of genetic diversity of Chinese F1 cultivated melon (*Cucumis melo* L). African Journal of Agriculture and Food Security, 6(5): 248–254.

Wong, S.K. and Chan, E.W.C. 2010. Antioxidant properties of coastal and inland populations of *Hibiscus tiliaceus*. ISME/GLOMIS Electronic Journal, 8(1): 1-2.

World Drug Report. 2016. United Nations Office on Drugs and Crime. United Nations Publication, Sales No. E.16.XI.7.

Wu, P., Zhou, C., Cheng, S., Wu, Z., Lu, W., Han, J., Chen, Y., Chen, Y., Ni, P., Wang, Y. and Xu, X. 2015. Integrated genome sequence and linkage map of physic nut (*Jatropha curcas* L.), a biodiesel plant. The Plant Journal, 81(5): 810–821.

Wu, S., Shen, C., Yang, Z., Lin, B. and Yuan, J. 2016. Tolerance of *Ricinus communis* L. to Cd and screening of high Cd accumulation varieties for remediation of Cd contaminated soils. Int. J. Phytoremed.,18(11): 1148–1154.

Wu, X.H., Zhang, H.S., Gang, L., Liu, X.C. and Qin, P. 2012. Ameliorative effect of castor bean (*Ricinus communis* L.) planting on physic-chemical and biological properties of seashore saline soil. Ecol. Eng., 38: 97–100.

Wyrwicka, A., Urbaniak, M. and Przybylski, M. 2019. The response of cucumber plants (*Cucumis sativus* L.) to the application of PCB-contaminated sewage sludge and urban sediment. Peer J., 7: e6743.

Xiaoyi, L. and Chiqu, H. 2005. Environmental behavior of zinc in *Ricinus communis* L. Env. Poll. Control, 2005: 1.

Yadav, K.S. and Singh, B.K. 2017. Role of biotechnology in floricultural crops: An overview. Trends in Biosciences, 10(17): 2992–2994.

Yakubu, F.B., Adejoh, O.P., Ogunade, J.O. and Igboanugo, A.B.I. 2014. Vegetative propagation of garcinia kola (Heckel). World Journal of Agricultural Sciences, 10(3): 85–90.

Yang, S., Ding, M., Chen, M. and Xu, Y. 2017. Proteomic analysis of latex from *Jatropha curcas* L. stems and comparison of two classic proteomic sample isolation methods: The phenolextraction and the TCA/acetone extraction. Electronic Journal of Biotechnology, 27: 14–24.

Yashim, Z.I., Agbaji, E.B., Gimba, C.E. and Idris, S.O. 2016. Phytoremediation potential of *Ricinus communis* L. (Castor oil plant) in Northern Nigeria. International Journal of Plant and Soil Science, 10(5): 1–8.

Yasur, J. and Rani, P.U. 2013. Environmental effects of nanosilver: Impact on castor seed germination, seedling growth, and plant physiology. Environ. Sci. Pollut. Res. Int., 20(12): 8636–8648.

Ye, L.W., Wood, B.A., Stroud, L.J. , Andralojc, J.P., Raab, A., McGrath, A.S., Feldmann, J. and Zhao, F.J. 2010. Arsenic speciation in phloem and xylem exudates of castor bean. Plant Physiol., 154: 1505–1513.

Yeasmin, L., Ali, N., Gantait, S. and Chakrabort, S. 2015. Bamboo: An overview on its genetic diversity and characterization. Biotech., 5: 1–11.

Yee, K.L., Rodriguez Jr, M., Thompson, O.A., Fu, C., Zeng-Yu Wang, Davison, B.H. and Mielenz, J.R. 2014. Consolidated bioprocessing of transgenic switchgrass by an engineered and evolved Clostridium thermocellum strain. Biotechnology for Biofuels, 7: 75.

Yekeen, M.O., Ajala, O.O., Adegbite, R.A. and Alarape, A.B. 2014. Physico-chemical properties and in vitro antifungal activities of *Ricinus communis* seed oil against *Lentinus sajorcaju*. Archives of Applied Science Research, 6(5): 1–6. www.scholarsresearchlibrary.com

Yessuf, A.M. 2015. Phytochemical extraction and screening of bio active compounds from black cumin (*Nigella sativa*) seeds extract. American Journal of Life Sciences, 3(5): 358–364.

Yi, X., Jiang, L., Chen, J., Liu, Q. and Yi, S. 2016. Effects of lead/zinc tailings on photosynthetic characteristics and antioxidant enzyme system of *Ricinus communis* L. Chinese J. Ecol., 35(4): 880–887.

Yi, X., Jiang, L., Liu, Q., Luo, M. and Chen, Y. 2014. Seedling emergence and growth of *Ricinus communis* L. grown in soil contaminated by lead/zinc tailing. pp. 445–452. *In:* Proc. 2014 Ann. Cong. Advanced Eng. Tech., CAET.

Yin, Z., Han, X., Li, Y., Wang, J., Wang, D., Wang, S., Fu, X. and Ye, W. 2017. Comparative analysis of cotton small RNAs and their target genes in response to salt stress. Genes, 369(8): 1–21.

Yong, J.W.H., Ge, L., Ng, Y.F. and Tan, S.N. 2009. The chemical composition and biological properties of coconut (*Cocos nucifera* L.) water. Molecules, 14: 5144–5164.

Yu, A., Li, F., Xu, W., Wang, Z., Sun, C., Han, B., Wang, Y., Wang, B., Cheng, X. and Li, A. 2019. Application of a high-resolution genetic map for chromosome-scale genome assembly and fine QTLs mapping of seed size and weight traits in castor bean. Scientific Reports, 9: 11950.

Yu, Y., Lu, X., Zhang, T., Zhao, C., Guan, S., Pu, Y. and Gao, F. 2022. Tiger nut (*Cyperus esculentus* L.): Nutrition, processing, function and applications. Foods, 11: 601.

Yuan, L. and Li, R. 2020. Metabolic engineering a model oilseed *Camelina sativa* for the sustainable production of high-value designed oils. Frontiers of Plant Science, 11: 11.

Yue, Y., Lin, Q., Muhammad, I., Chen, Q., Zhao, X. and Li, G. 2017. Characteristics and potential values of bio-products derived from switchgrass grown in a saline soil using a fixed-bed slow pyrolysis system. BioResources, 12(3): 6529–6544.

Yusuf, A.J., Abdullahi, M.I., Musa, A.M., Haruna, A.K., Mzozoyana, V. and Abubakar, H. 2019a. Bioactive (+)-Catechin-3'-O-rhamnopyranoside from *Neocarya macrophylla* (Sabine) Prance (Chrysobalanaceae). Egyptian Journal of Basic and Applied Sciences, 6(1): 124–136.

Yusuf, A.J., Abdullahi, M.I., Musa, A.M. , Haruna, A.K., Mzozoyana, V. and Abubakar, H. 2019b. Antisnake venom activity and isolation of quercetin from the leaf of *Neocarya macrophylla* (Sabine) Prance ex F. White (Malpighiales: Chrysobalanaceae). Brazilian Journal of Biological Sciences, 6(13): 381–389.

Zagórska-Dziok, M., Furman-Toczek, D., Dudra-Jastrzębska, M., Zygo, K., Stanisławek, A. and Kapka-Skrzypczak, L. 2017. Evaluation of clinical effectiveness of *Aloe vera* – A review. J. Pre-Clin. Clin. Res., 11(1): 86–93. doi: 10.26444/jpccr/74577.

Zahra'u, B., Mohammed, A.S., Ghazali, H.M. and Karim, R. 2014. Baobab tree (*Adansonia digitata* L.) parts: Nutrition, applications in food and uses in ethno-medicine – A review. Annals of Nutritional Disorders & Therapy, 1(3): 9.

Zakaria, N.A., Rahman, R.A., Zaidel, D.N., Dailin, D. and Jusoh, M. 2021. Microwave-assisted extraction of pectin from pineapple peel. Malaysian Journal of Fundamental and Applied Sciences, 17: 33–38.

Zakir, H.M., Hasan, M., Shahriar, S.M.S., Ara, T. and Hossain, M. 2016. Production of biofuel from agricultural plant wastes: Corn stover and sugarcane bagasse. Chemical Engineering and Science, 4(1): 5–11.

Zaku, S.G., Emmanual, S.A., Isa, A.H. and Kabir, A. 2012. Comparative studies on the functional properties of neem, jatropha, castor, and moringa seeds oil as potential feed stocks for biodiesel production in Nigeria. Global Journal of Science Frontier Research Chemistry, 12(7): 22–26.

Zamani, A., Marjani, A.P. and Mousavi, Z. 2019. Agricultural waste biomass-assisted nanostructures: Synthesis and application. Green Process Synth., 8: 421–429.

Zanella, K., Pinheiro, C.L., Tomaz, E. and Taranto, O.P. 2013. Study of volatiles from 'Pera' sweet orange peel using Karl Fischer Titration, GC-FID and GC/MS techniques. Chemical Engineering Transactions, 32: 409–414.

Zapata, N., Marisol Vargas, M., Reyes, J.F. and Belmar, G. 2012. Quality of biodiesel and press cake obtained from *Euphorbia lathyris, Brassica napus* and *Ricinus communis.* Industrial Crops and Products, 38: 1–5.

Zarai, Z., Chobba, I.B., Mansour, R.B., Békir, A., Gharsallah, N. and Kadri, A. 2012. Essential oil of the leaves of *Ricinus communis* L.: In vitro cytotoxicity and antimicrobial properties. Lipids in Health and Disease, 11(1): 1–7.

Zarroug, A.B., Baraket, G., Zourgui, L., Souid, S. and Hannachi, A.S. 2015. Genetic diversity and phylogenetic relationship among Tunisian cactus species (Opuntia) as revealed by random amplified microsatellite polymorphism markers. Genetics and Molecular Research, 14(1): 1423–1433.

Zayas-Viera, M.D., Vivas-Mejia, P.E. and Reyes, J. 2016. Anticancer effect of *Moringa oleifera* leaf extract in human cancer cell lines. Journal of Health Disparities Research and Practice, 9(1): 141–142.

Zhang, B., Xi, X., Wu, Z., Lei, H., Li, L. and Tian, M. 2020. An eco-friendly wood adhesive from alfalfa leaf protein. Journal of Renewable Materials, 8(11): 1429–1441.

Zhang, D., Lin, Y., Li, A. and Tarasov, V.V. 2011. Emulsification for castor biomass oil. Frontiers of Chemical Science and Engineering, 5(1): 96–101.

Zhang, H. and Mittal, H. 2010. Biodegradable protein based films from plant resources: A review. Environmental Progress & Sustainable Energy, 29(2): 203–220.

Zhang, H., Chen, X., He, C., Liang, X., Oh, K., Liu, X. and Lei, Y. 2015. Use of energy crop (*Ricinus communis* L.) for phytoextraction of heavy metals assisted with citric acid. I. J. Phytoremed, 17(7): 632–639. doi: 10.1080/15226514.2014.935287

Zhang, H., Guo, Q., Yang, J., Chen, T., Zhu, G., Peters, M., Wei, R., Tian, L., Wang, C., Tan, D., Ma, J., Wang, G. and Wan, Y. 2014. Cadmium accumulation and tolerance of two castor cultivars in relation to antioxidant systems. Journal of Environmental Science, 26(10): 2048–2055.

Zhang, H., Guo, Q., Yang, J., Shen, J., Chen, T., Zhu, G., Chen, H. and Shao, C. 2015. Subcellular cadmium distribution and antioxidant enzymatic activities in the leaves of two castor (*Ricinus communis* L.) cultivars exhibit differences in Cd accumulation. Ecotoxicol. Environ. Saf., 120: 184–192.

Zhang, H., Guo, Q., Yang, J., Ma, J., Chen, G., Chen, T., Zhu, G., Wang, J., Zhang, G., Wang, X. and Shao, C. 2016. Comparison of chelates for enhancing *Ricinus communis* L. phytoremediation of Cd and Pb contaminated soil. Ecotoxicology and Environmental Safety, 133: 57–62.

Zhang, H., Miao, H., Wang, L., Qu, L., Liu, H., Wang, Q. and Yue, M. 2013. Genome sequencing of the important oilseed crop *Sesamum indicum* L. Genome Biology, 14: 4.

Zhang, J., Lin, M., Chen, H. and Chen, X. 2017. Agrobacterium tumefaciens-mediated transformation of drumstick (*Moringa oleifera* Lam.). Biotechnology & Biotechnological Equipment, 31(6): 1126–1131.

Zhang, L., Ibrahim, A.K., Niyitanga, S., Zhang, L. and Qi, J. 2019. Jute (*Corchorus* spp.) breeding. pp. 85–113. *In:* J.M. Al-Khayri et al. (eds.), Advances in Plant Breeding Strategies: Industrial and Food Crop. Springer Nature Switzerland AG.

Zhang, L., Simmons, M.P., Kocyan, A., Susanne, S. and Renner, S.S. 2006. Phylogeny of the Cucurbitales based on DNA sequences of nineloci from three genomes: Implications for morphological and sexual system evolution. Molecular Phylogenetics and Evolution, 39: 305–332.

Zhang, L., Xu, Y., Zhang, X., Ma, X., Zhang, L., Liao, Z., Zhang, Q., Wan, X., Cheng, Y., Zhang, J., Li, D., Zhang, L., Xu, J., Tao, A., Lin, L., Fang, P., Chen, S., Qi, R., Xiuming Xu, X., Qi, J. and Ming, R. 2020. The genome of kenaf (*Hibiscus cannabinus* L.) provides insights into bast fibre and leaf shape biogenesis. Plant Biotechnology Journal, 18: 1796–1809.

Zhang, Y., Zhang, X., Che, Z., Wang, L., Wei, W. and Li, D. 2012. Genetic diversity assessment of sesame core collection in China by phenotype and molecular markers and extraction of a mini-core collection. BMC Genetics, 13(102): 1–14.

Zhao, J., Wang, D., Pidlisnyuk, V. and Erickson, L.E. 2021. Miscanthus biomass for alternative energy production. pp. 177–199. *In:* Larry E. Erickson and Valentina Pidlisnyuk (eds.), Phytotechnology with Biomass Production: Sustainable Management of Contaminated Sites. First edition. CRC Press, Boca Raton.

Zhao, S., Li, X., Cho, D.H., Arasu, M.V., Al-Dhabi, N.A. and Park, S.U. 2014. Accumulation of kaempferitrin and expression of phenyl-propanoid biosynthetic genes in kenaf (*Hibiscus cannabinus*). Molecules, 19: 16987–16997.

Zhao, X., Chen, J. and Du, F. 2012. Potential use of peanut by-products in food processing: A review. Journal of Food Science and Technology, 49(5): 521–529.

Zhi-Xin, N., Sun, L.N., Sun, T.H., Li, Y.S. and Wang, H. 2007. Evaluation of phytoextracting cadmium and lead by sunflower, *Ricinus*, alfalfa and mustard in hydroponic culture. J. Environ. Sci. (China), 19: 961–967.

Zhou, J.Z.M., Zhu, J.W.L., Chen, Y. and Ming, R. 2014. Sugarcane genetics and genomics. pp. 623–643. *In:* Paul H. Moore and Frederik C. Botha (eds.), Sugarcane: Physiology, Biochemistry, and Functional Biology. First Edition. John Wiley & Sons, Inc.

Zhou, S.M., Cheng, L., Guo, S.J., Wang, Y., Czajkowsky, D.M., Gao, H., Hu, X.F. and Tao, S.C. 2015. Lectin RCA-I specifically binds to metastasis-associated cell surface glycans in triple negative breast cancer. Breast Cancer Research, 17(1): 1–4.

Biodiversity Index

Subject Index

For Product Safety Concerns and Information please contact our EU
representative GPSR@taylorandfrancis.com
Taylor & Francis Verlag GmbH, Kaufingerstraße 24, 80331 München, Germany